W0049922

Cell Biology Monographs

Volume 12

Springer-Verlag Wien New York

The Nucleolus and Ribosome Biogenesis

A. A. Hadjiolov

Springer-Verlag Wien New York

Prof. Dr. Asen A. Hadjiolov
Department of Molecular Genetics,
Institute of Molecular Biology,
Bulgarian Academy of Sciences, Sofia, Bulgaria

Softcover reprint of the hardcover 1st edition 1985

With 46 Figures

Library of Congress Cataloging in Publication Data. Hadjiolov, A. A. The nucleolus and ribosome biogenesis. (Cell biology monographs; vol. 12.) Bibliography: p. Includes index. 1. Nucleolus. 2. Ribosomes. I. Title. II. Series: Cell biology monographs; v. 12. QH596.H33. 1985. 574.87'32. 84-25356

ISSN 0172-4665

ISBN-13:978-3-7091-8744-9 e-ISBN-13:978-3-7091-8742-5
DOI: 10.1007/978-3-7091-8742-5

To the memory of my son Asen,
to my wife Krassimira
and my daughters Helen and Maria

Preface

The nucleolus had consistently attracted the attention of investigators in the fields of cell biology and pathology. Because of its ubiquitous presence in the nucleus of eukaryotic cells, its rapid changes during their life cycle, and its rapid response to noxious agents, this organelle has been the subject of a large number of studies. Yet, the exact function and the very reason for the existence of the nucleolus (the only large cellular structure not delimited by a membrane) remain largely unknown.

The ribosomes were discovered relatively late in the study of cells, but due to their crucial involvement in the protein synthesis machinery of all living organisms, the elucidation of their structure and function quickly became one of the major goals of molecular biology. The relatively simple structure of the ribosome strengthens the hope that a full understanding of the structure and function of this organelle in molecular terms is within the reach of contemporary research. Since each of the rRNA and protein molecules embodied in the ribosome is the product of a distinct gene, studies on the biogenesis of ribosomes expanded rapidly to become a core topic in molecular genetics.

For a long time the molecular and cellular aspects of ribosome biogenesis were pursued along independent lines. It is only in the mid-1960's that it became evident that *the nucleolus is the main site of ribosome biogenesis* in the cell. This generalization opened the way for concerted efforts to understand the molecular essence and the biological role of the nucleolus. Moreover, it became gradually clear that the nucleolus is in fact the visible correlate of the activity of ribosomal genes.

The purpose of this book is to provide an integrated view of the present day knowledge about the nucleolus as the morphological visualization of ribosome biogenesis. It aims to present under a common molecular denominator the wealth of experimental data collected by the efforts of the numerous investigators using primarily the approach of molecular genetics, biochemistry, immunology, and ultrastructural cytology. This task was facilitated by the existence of the classic monograph by BUSCH and SMETANA—"The Nucleolus"—in which most of the information available till 1970 is expertly presented. Still, the rapid expansion during the last decade and a half of studies on the nucleolus and ribosome biogenesis created unanticipated difficulties for the single author trying to encompass and evaluate the present status of the problem. Many expert reviews, some of them gathered by JORDAN and CULLIS in a recent book "The Nucleolus" (1982) and by BIELKA in "The Eukaryote Ribosome" (1982), cover various aspects of the field. These reviews may help the reader of the present book to gain a deeper insight into selected specific topics. Here, I have attempted to give primarily the basic information available by late 1983 and to outline the numerous important and still unsolved questions and problems. Some tentative answers and models are also proposed which may hopefully serve as working hypotheses in future studies within this central field of modern molecular and cell biology. For it is my firm belief that understanding ribosome biogenesis and its control will play a leading role in the elucidation of such basic biological phenomena as cell growth and differentiation, senescence, and cancer.

In my lifetime in science I was lucky to have enlightened and patient teachers to whom I am greatly indebted for criticism and support. First of all I am thankful to Dr. FRITZ LIPMANN, under whose expert guidance I entered the field of ribosome biogenesis. During the last two decades I was also lucky to have many skilled and enthusiastic coworkers and students with whom many questions were asked and a few answers obtained. To all of them, past and present, goes my profound gratitude. The competent and meticulous editorial work of Professor LESTER GOLDSTEIN helped me to avoid many imperfections. Finally, I am greatly obliged to Mrs. S. NIKOLOVA and Mrs. R. FILIPOVA for the technical help in the preparation of the present manuscript. The book is now in the hands of the reader. While many people made possible the appearance of this book, I am the only one to blame for involuntary errors and omissions. If within the next pages the reader finds the needed information and some helpful ideas, I shall be greatly rewarded.

Sofia, Bulgaria, October 1984 A. A. HADJIOLOV

Contents

I. Introduction

The nucleolus is an intranuclear organelle that has attracted the attention and imagination of scientists for a little over two centuries. Its first description in 1781 is attributed to FONTANA (see MILLER 1981) and since then many cytologists have contributed findings to establish that the nucleolus is ubiquitous in eukaryotic cells. However, the early descriptive work did not go far beyond establishing the existence of variations in the number, size, and shape of nucleoli in various plant and animal cells. It is only 50 years ago that the next step was made when HEITZ (1931, 1933) and MCCLINTOCK (1934), followed by many cytologists and cytogeneticists, firmly established that the nucleolus is related to "nucleolus organizers" in some chromosomes and is therefore the likely structural expression of chromosomal activity. The next step was due largely to the cytochemical work of CASPERSSON and BRACHET, who showed in the 1940's that the nucleolus contains RNA, related in some hypothetical way to cytoplasmic RNA, particularly abundant in cells actively engaged in protein synthesis (see CASPERSSON 1950, BRACHET 1957). These findings placed the nucleolus in a prominent position for those trying to understand the then intriguing relationship between DNA, RNA, and protein. This was the reason why many cytochemists and electron microscopists of the 1950's, perfecting continuously the ultrastructural analysis of cellular organelles, devoted their skills to the structure of the nucleolus in a broad variety of eukaryotic cells studied under different physiological and pathological conditions (see BERNHARD 1966, BERNHARD and GRANBOULAN 1968, BUSCH and SMETANA 1970).

It was also in the 1950's that the cytoplasmic "basophilia" of old cytologists was related to the presence of ribonucleoprotein granules, either free or membrane-bound in the cytoplasm, which were christened *ribosomes* in 1958 (see PALADE 1958). Subsequent studies of ribosomes from bacteria and animal cells clearly established their direct involvement in protein synthesis. After the discovery of messenger RNA, its association with ribosomes to form *polyribosomes* as the basic cytoplasmic structure engaged in protein synthesis was demonstrated.

Two principal discoveries in the 1960's permitted the identification of the nucleolus as the main cellular center of ribosome biogenesis in eukaryotes:

1. Conclusive evidence was obtained that the "nucleolus organizer" is the site of rRNA genes (BROWN and GURDON 1964, RITOSSA *et al.* 1966, BIRNSTIEL *et al.* 1966) and

2. Detailed studies showed that the nucleolus is the source of cytoplasmic RNA (PERRY *et al.* 1961, EDSTRÖM *et al.* 1961, PERRY 1962) now known to be ribosomal RNA.

With the establishment of the nucleolus as the organelle responsible for ribosome biogenesis, the stage of contemporary investigations that constitute the subject of this monograph began. The history of the subsequent steps in our deeper understanding of the nucleolus and ribosome biogenesis is outlined in previous reviews (PERRY 1967, BUSCH and SMETANA 1970, MILLER 1981) and will be considered later in this book. Here, I would like to outline some of the broader biological problems related to the studies on the nucleolus and ribosome biogenesis.

The ribosome is a complex molecular structure constituted, in fact, by two distinct components—the large and the small particle. These two components are normally made and present in the cell in equimolar amounts. Moreover, they function in the polyribosome $\leftrightarrows$ ribosome cycle in a strict 1 : 1 ratio. We know now that the precise stoichiometry in the production of the two particles is determined by the presence of the two genes coding for constituent rRNAs within a common transcription unit. However, we still have to learn how the cell succeeds in preserving the equimolar amounts of the two particles throughout their continuous recycling. Clearly, this phenomenon involves both a complementarity in their function in protein synthesis and selective interactions when in the pool of free particles. Elucidation of these interactions will certainly provide an example of the functional and metabolic synchrony of cellular structures.

The two ribosomal particles are complex molecular structures built up of 4 distinct rRNAs and about 80 proteins. It is evident that L-rRNA and S-rRNA molecules serve as the backbones for the construction of these complex molecular structures. In all likelihood, the ribosome will be the first cellular organelle to be fully understood at the molecular level. Yet, understanding the detailed RNA : protein and protein : protein recognition phenomena will certainly be enlightened by a more detailed knowledge of the subsequent events of ribosome assembly taking place during ribosome biogenesis.

Ribosomal RNA genes exist as multigene families in the eukaryotic genome. Their sequences display a complex pattern of evolutionary conservation. An understanding of these genes is likely to provide valuable clues to the detection of functionally important domains in rRNA structure. Moreover, it will certainly help to understand better the complex factors participating in defining the comportment of genes in the course of evolution. Further, the spacer sequences in the multiple rDNA repeating units display a remarkable similarity within the species, but are widely divergent in different, even closely related, species. Clearly, important new facets of genome evolution will be uncovered in studies on the structure of rDNA.

Ribosomal RNA genes occupy an unique position in the eukaryotic genome. They are organized to be in a constantly working state. Also, sequences homologous to some spacer segments in rDNA are found at often unexpected positions within the genome, i.e., in centromeric chromatin. Thus, it may be anticipated that understanding the structure and interactions of rRNA genes is likely to clarify many important problems regarding the organization and the function of the eukaryotic genome.

Ribosome biogenesis requires the synchronous and interdependent expression of three sets of genes, those coding for rRNA, for 5 S rRNA and for about 80 r-proteins. In this case we have an example of intracellular synchronization of great complexity. Yet, we now possess the experimental tools to understand the specific devices used by the cell in achieving the harmonious work of the ribosome producing machinery. Moreover, ribosome biogenesis is finely tuned to the protein synthesis requirements of the cell during its entire life cycle. How the whole process is adapted to the needs of the cell is a crucial question in cell biology. Let us add that the genes involved in ribosome biogenesis belong to the class of "household genes", whose structure and expression is the least known at present.

The continuous formation of ribosomes is an intriguing aspect of cell life. This was noted by the early cytologists who stated that a cell devoid of its nucleolus is either dead or dying. But why do even non-growing cells (for example, a neuron) continuously make ribosomes? That ribosomes turnover in non-cycling cells is now firmly established. Still, we do not understand the factors determining the life cycle of a ribosome. What determines its stability or triggers its degradation? There is little doubt that the solution of these questions, touching fundamental features of the life phenomenon, is likely to come from our understanding of the molecular mechanisms of nucleolar activity.

The continuous process of ribosome biogenesis is clearly related to the growth state of the cell. We know that rapidly growing cells are characterized by a fully active nucleolus and an intensive ribosome biogenesis. We also know that cell differentiation correlates with profound changes in ribosome biogenesis and reorganization of nucleolar structure. But we do not know the molecular phenomena that trigger and direct these changes. Whether the nucleolus is directly involved in defining growth rates of cycling cells remains to be determined. We also have to learn the role of the nucleolus during growth arrest and formation of non-cycling, differentiated cells. Still, there is little doubt that without an adequate tuning of ribosome biogenesis no changes in the growth or differentiation status of the cell could be attained.

Finally, numerous cytological studies have firmly established that many important pathological phenomena, including viral infections, toxic lesions, aging and cancer, have a profound repercussion on nucleolar structure. We know that in molecular terms these alterations reflect changes in ribosome biogenesis. But we do not know yet how to relate an alteration in nucleolar structure to the complex and multistep process of ribosome production. Even less is known about the likely alterations in the regulatory factors switching on the changes characteristic of a pathological process. Hopefully, the answer to many of these questions will come from our understanding of the relationship between nucleolar structure and ribosome biogenesis.

The above enumeration of some solved and unsolved problems in molecular and cell biology outlines the scope of this book. It is the author's conviction that the understanding of nucleolar structure and function and its relation to the molecular mechanisms of ribosome biogenesis will be rewarded by the solution of many important aspects of the biology and pathology of the whole spectrum of eukaryotic cells and organisms.

II. Ribosomal Genes

II.1. Definitions

The genes involved in the biogenesis of eukaryotic ribosomes include:

1. The genes of the four rRNAs: S-rRNA, L-rRNA, 5.8 S rRNA, and 5 S rRNA,

2. The genes for the structural ribosomal proteins (r-proteins),

3. The genes for the specific enzymes and other proteins specialized in transcription, modification and processing mechanisms resulting in the formation of mature ribosomes.

The genes coding for rRNA species are among the best studied eukaryotic genes. With the advent of modern molecular cloning techniques important new information on these genes is rapidly accumulating. It is now firmly established that in all eukaryotes the genes for S-rRNA, L-rRNA, and 5.8 S rRNA are adjacent and transcribed as a single pre-rRNA molecule, *i.e.*, these three genes constitute one *transcription unit*. Accordingly, some authors use the term "pre-rRNA gene" to designate this set of genes. To avoid confusion, I shall designate these genes as the *rRNA genes*, while considering separately the *5 S rRNA genes*. Both types of rRNA genes are highly repeated, each being present in 10^2 to 10^3 copies per haploid genome. The repeated genes for rRNA species are normally grouped in one or several clusters located at specific sites in one or a few chromosomes. Thus, rRNA genes may be considered as typical examples of *clustered repeated genes* in the genome of eukaryotic cells. Several competent reviews cover different aspects of earlier and recent knowledge about rRNA genes (BIRNSTIEL *et al.* 1971, REEDER 1974, TARTOF 1975, LONG and DAWID 1980).

The study of the genes coding for r-proteins in eukaryotes is still at an early stage. Extensive studies with *E. coli* provided conclusive evidence that practically all r-proteins have different primary structures (see BRIMACOMBE *et al.* 1978) and are encoded by distinct genes present in single (or a few) copies that are clustered in the genome (see NOMURA *et al.* 1977). At present, about 85 individual r-proteins have been identified in eukaryotic ribosomes (MCCONKEY *et al.* 1979, WOOL 1980). If the *E. coli* pattern is perpetuated in eukaryotes, we may expect that numerous distinct r-protein genes will be present in unique (or a few) copies per genome. Unlike the unique eukaryotic genes, related to the function of differentiated cells, the r-protein genes are expected to operate throughout the whole life cycle of almost all cell types. In this respect, their organization, expression, and replication may be considered as representative of the sparsely studied unique genes responsible for the background metabolism of the cell ("household genes").

The third group of ribosomal genes is the least known. In fact, we do not know yet to what extent the eukaryotic cell has elaborated enzymes and proteins strictly specialized in ribosome biogenesis. That such specialization has been achieved is evidenced by the example of RNA polymerase I and, partly, of RNA polymerase III. The apparently strict confinement of Ag-staining protein(s) to r-chromatin in the nucleolus organizer seems to provide another example of specialization. On the other hand, the substrate specificity of the enzymes involved in the modification and processing of rRNA and r-proteins (*i.e.* methylases, nucleases, protein kinases etc.) is still unclear.

II.2. Ribosomal RNA Genes

The rRNA genes are present in one to several hundred copies per haploid genome and are clustered in one or a few chromosomal loci. During mitosis the clusters of rRNA genes are seen in most cases as "secondary constrictions" in the respective chromosomes. The rRNA genes and their respective chromosomal sites (the "nucleolus organizers") possess the capacity to form nucleoli at telophase. Thus, rRNA genes are the only eukaryotic genes whose activity in the interphase nucleus results in the formation of the easily identifiable nucleolar structures. It is considered that rRNA genes are covalently linked to the DNA of a given "nucleolar" chromosome. Yet, in several organisms and cell types multiple extra-chromosomal rRNA genes have been identified.

II.2.1. Multiplicity

The number of rRNA genes has been determined in many eukaryotic species and the available data collected (BIRNSTIEL *et al.* 1971, LONG and DAWID 1980). These data are listed in Table 1. In general terms the number of rRNA genes correlates with the genome size (BIRNSTIEL *et al.* 1971). It is plausible that a higher redundancy of rRNA genes confers definite evolutionary advantages to a given species, but the mechanisms controlling rRNA gene number are still obscure. It is remarkable that considerable differences in the number of rRNA genes exist even among closely related species. Moreover, it has been shown that variations in the number of rRNA genes exist among individuals of a species and possibly among cell types in an individual (see LONG and DAWID 1980, FLAVELL and MARTINI 1982).

Table 1. *Multiplicity of ribosomal RNA genes in eukaryotes [from* LONG and DAWID *(1980)* [a]*; reproduced with permission from the authors and Annual Reviews of Biochemistry*

Species	N^b
Protists	
Dictyostelium discoideum	200[c]
Leischmania donovani	170
Physarum polycephalum	80/280[c, d]
Algae	
Acetabularia mediterranea	1,900
Chlamydomonas reinhardii	150
Polytoma obtusum	380
Mycophyta (fungi)	
Achlya bisexulais	430
Allomyces arbuscula	270
Neurospora crassa	100/180[e]
Saccharomyces cerevisiae	140/100[f]
S. carlsbergensis	140
Embryophyta [g]	
Gymnospermae (cone-bearing plants)	
Juniperus chinensis pyramidalis	2,050
Larix decidua (larch)	13,400
Picea abies (Norway spruce)	9,650
P. albertiana conica	6,950
P. sitchensis	12,350
Pinus sylvestris (Scots pine)	5,350
Pseudotsuga douglasii	3,600
Taxus baccata (yew)	1,250
Angiospermae (flowering plants)	
Allium cepa (onion)	6,950
Aquilegia alpina	2,300
Bellevalia dubia	4,900
B. romana	1,600
Beta vulgaris (swiss chard)	1,150
Brassica rapa (turnip)	4,300
Citrus sinensis (orange)	630
Cucumis melo (melon)	1,000
C. sativum (cucumber)	4,400
Cucurbita pepo (pumpkin)	4,900
Helianthus annuus (sunflower)	3,350
H. tuberosus (artichoke)	790
Hordeum vulgare (barley)	4,200
H. bulbosum	2,870
H. parodii	4,830

Table 1 (continued)

Species	N^b
H. procerum	3,560
Hyacinthus orientalis	8,400
Lagenaria vulgaris (gourd)	1,050
Linum usitatissimum (flax)	990
Luffa cylindrica	1,800
Momordica charantia	2,750
Nicotiania glutinosa	1,600
N. otophora	3,400
N. paniculata	6,150
N. rustica	1,100/5,650
N. sylvestris	2,450/3,150
N. tabacum (tobacco)	1,100/1,000/1,560
N. tomentosiformis	2,320
Oenothera fructicosa	1,700
Passiflora antioquiensis (passion flower)	900
Phaseolus coccineus (runner bean)	2,000
Pisum sativum (pea)	3,900
Secale cereale (rye)	2,850
Thalictrum aquilegiifolium	700
Tradescantia paludosa	2,400
T. virginiana	4,300
Trillium grandiflorum	3,150
Triticum aestivum (wheat)	6,350
Tulbaghia violaceae	2,330
Vicia benghalensis	950
V. faba (broad bean)	4,750
V. narbonensis	3,130
V. sativa	1,875
V. villosa	1,250
Zea mays (maize)	8,500/3,100/5,900–9,450[h]
Protozoa	
Euglena gracilis[i]	1,000/800
Tetrahymena pyriformis[j]	200/290
T. thermophila[j]	600
Metazoa	
Nematoda	
Ascaris lumbricoides	300
Caenorhabditis elegans	55
Panagrellus silusiae	280

Table 1 (continued)

Species	N^b
Mollusca	
Ilyanassa obsoleta	800
Mulinia lateralis (clam)	120
Mytilus edulis	220
Spitula solidissima (clam)	200
Echiuroidea	
Urechis caupo	290
Arthropoda	
Insecta	
Acheta domesticus (cricket)	170
Bombyx mori (silk moth)	240
Chironomus tentans	100/40
Drosophila erecta[k]	80
D. hydei	280
D. mauritiana[k]	250
D. melanogaster	120/180/100/230/240
D. simulans[k]	200/230
D. teissieri	150
D. yakuba	130
Dytiscus marginalis (water beetle)	220
Rhynosciara angelae	100
Sciara coprophila	45
Echinodermata	
Lytechinus variegatus (sea urchin)	260
Chordata	
Chondrichtyes (elasmobranchs)	
Squalus acanthias (shark)	960
Osteichtyes (bony fish)	
Abramis brama	125
Barbus barbus	250
Carassius auratus	240/280
Clupea harengus	300
Coregonus fera	520
Cyprinus carpio	290
Leuciscus cephalus	140
Neoceratodus forsteri (lungfish)	4,800
Osmerus esperlanus	330
Pseudopleuronectes americanus	280
Rutilus rutilus	150
Salmo gairdneri (trout)	730
S. irideus	570/740
S. salar	710

Table 1 (continued)

Species	N^b
S. truta	1,080
Salvelinus fontinalis	1,190
Sprattus sprattus	340
Thymallus thymallus	610
Tinca tinca	120
Amphibia	
Urodeles	
Ambystoma mexicanum	4,100/4,500
A. opacum	4,500/2,500
A. talpoideum	3,950
A. tigrinum[1]	4,870
Amphiuma means[1]	19,300
Desmognathus fuscus	4,300
Necturus maculosus	2,700/5,400/4,000
Notophtalmus (Triturus) viridescens	5,800/5,100
Oedipina uniformis	3,720
Plethodon cinereus	2,060
P. dunni	2,440
P. elongatus	2,400
P. glutinosus	2,000
P. jordani melaventris	4,300
P. vehiculum	2,630
Pseudotriton ruber schenki	3,900
Rhycotriton olympicus	2,430
Triturus carnifex	5,460
T. cristatus	4,100
T. taeniatus	3,900
Anura	
Bombina variegata	2,050
Bufo americanus	940/950
B. bufo[1]	1,050/830
B. marinus[1]	380–850[h]
B. viridis	500
B. woodhoussei fowleri	1,100
Hyla cynerea	500
H. versicolor	950
Rana agilis	600
R. catesbeiana	600
R. clamitans	890/600
R. esculenta	1,100/1,230
R. pipiens	1,290/950
Xenopus borealis	500
Xenopus laevis	600/500/600/760

Table 1 (continued)

Species	N^b
Aves	
Gallus domesticus (chicken)	190/200
Mammalia	
Rodentia	
Cricetulus griseus (Chinese hamster)	250
Mus musculus (mouse)	100
Rattus norvegicus (rat)	150/170
Oryctolagus cuniculus (rabbit)	250
Primates	
Homo sapiens	200/160/190/50

[a] The references to the original publications are given in LONG and DAWID (1980). Additional entries are indicated below. When more than one figure is given, the order is according to year of publication.

[b] N = number of genes per haploid genome.

[c] rDNA in these species is extrachromosomal. Given is the number of nuclear genes divided by the ploidy of the nucleus.

[d] Order according to year of publication.

[e] Data from SELKER *et al.* (1981 a).

[f] Data from PHILIPPSEN *et al.* (1978).

[g] The DNA present in germ cells is considered as haploid genome although many plants are polyploid in an evolutionary sense.

[h] Numbers for different lines or individuals. Extremes are given.

[i] The ploidy of this organism is not known. The numbers of genes per nucleus are listed.

[j] The haploid number for the micronucleus is 1. The numbers of extrachromosomal genes in the macronucleus divided by the ploidy of the macronucleus are listed.

[k] The gene numbers in these species are calculated with the assumption that their genome size is identical to that of *Drosophila melanogaster*.

[l] Only the hybridization saturation value is determined. The number of genes is calculated on the basis of the respective genome size.

Such variations may be an inevitable consequence of the tandem arrangement of multiple rRNA genes. It has been proposed that tandemly arrayed identical sequences should result in unequal crossing-over events during mitosis or meiosis (SMITH 1973, 1976, TARTOF 1973, 1975). Recently, the occurrence of unequal crossing-over in the array of rRNA genes in sister chromatids was shown to take place during mitosis (SZOSTAK and WU 1980) and meiosis (PETES 1980) in *Saccharomyces cerevisiae*. These experiments provide support to the concept that unequal crossing-over at the clusters of rRNA genes may be a major factor in determining the number of rRNA genes in species and individuals. It is also noteworthy that variations in the cellular rDNA content were observed even in different parts of a plant (flax) following growth in particular environments (CULLIS 1979, CULLIS and CHARLTON 1981).

II.2.2. Chromosomal Location

Clustering of rRNA genes in the chromosome is typical of prokaryotes (NOMURA *et al.* 1977) and this pattern is preserved in eukaryotes although the number of gene sets per cell is at least one order of magnitude higher in the latter. In lower eukaryotes considerable variations in the organization of the genome are observed which makes generalizations precarious. Nevertheless, in all thoroughly studied cases evidence for the clustering of the multiple rRNA genes was obtained. Thus, primitive eukaryotes from the *Saccharomyces* group contain 100–120 rRNA genes per haploid genome (RETEL and PLANTA 1968, SCHWEIZER *et al.* 1969, PHILIPPSEN *et al.* 1978). Experiments with monosomic (2 n—1) strains of *Saccharomyces cerevisiae* demonstrated that about 70% of the rRNA genes are confined to chromosome I, about 4.0—4.5 × 10^8 daltons of DNA (FINKELSTEIN *et al.* 1972, ØYEN 1973, KABACK *et al.* 1973). Estimations based on a molecular mass of 6 × 10^6 per rRNA gene set (PHILIPPSEN *et al.* 1978) indicate that almost the entire chromosome is constituted by clustered rRNA genes. An independent genetic analysis approach showed that about 90% of the rRNA genes are located in one chromosome (PETES and BOTSTEIN 1977), identified, however, as chromosome XII (PETES 1979 a, b). The mapping of the rRNA gene cluster to chromosome XII is: centromere—*asp 5*—*gal 2*—rRNA genes—*ura 4* (PETES 1979 a, b), but the precise physical linkage of the rRNA genes within the cluster remains to be clarified. Physico-chemical analyses (KABACK *et al.* 1973, CRAMER *et al.* 1976) and electron microscopy of R-looped rDNA (KABACK and DAVIDSON 1980) provided evidence that at least 15–20 rRNA gene sets are arranged in tandem. Further, the genetic evidence points to the possibility that all 100 rRNA gene sets constitute a single tandem array (ZAMB and PETES 1982).

The chromosomal location of rRNA gene clusters in a broad variety of higher eukaryotes has been established. Since the pioneer studies of HEITZ (1931, 1933) and MCCLINTOCK (1934) it has been known that "secondary constrictions", observed during mitosis in one or a few chromosomes, are the sites of nucleolus formation at telophase. Experiments with *Zea mays* showed that the "nucleolar genes" are located at the secondary constriction of chromosome 6. Deletion of the secondary constriction resulted in homozygous anucleolate (O_{nu}) mutants, which fail to form a nucleolus (MCCLINTOCK 1934). Accordingly, the secondary constriction site was designated as "nucleolus organizer" (NO) or "nucleolus organizer region"

(NOR). That the nucleolus organizer is the chromosomal site of clustered rRNA has been proven in studies with *Drosophila melanogaster* (RITOSSA and SPIEGELMAN 1965, RITOSSA *et al.* 1966) and *Xenopus laevis* (BIRNSTIEL *et al.* 1966) mutants carrying a varying number of NOR complements. Thus, it was shown that *X. laevis* O_{nu} mutants do not synthesize rRNA (BROWN and GURDON 1964) and lack rRNA genes (BIRNSTIEL *et al.* 1966). Direct proof for the location of rRNA genes in the NOR site of chromosomes was obtained by *in situ* rRNA : DNA hybridization (GALL and PARDUE 1969, JOHN *et al.* 1969). These early studies provided conclusive evidence that the multiple rRNA genes are clustered in the nucleolus organizer. The clustering of rRNA genes has since been observed in all eukaryotic cells and organisms studied till now. However, recent, more powerful techniques permitted the identification of single rRNA genes dispersed throughout the genome of *Saccharomyces cerevisiae* and *Drosophila melanogaster* (CHILDS *et al.* 1981). The presence of such solitary rRNA genes called "orphons" was not detected in other strains of *Saccharomyces* (ZAMB and PETES 1982) and therefore their ubiquitous existence in eukaryotes remains to be ascertained.

The chromosomal location of rRNA gene clusters is now known for a broad variety of plant and animal cells and organisms (LIMA-DE-FARIA 1976, 1980, LONG and DAWID 1980). The *in situ* hybridization studies provided direct evidence that the rRNA gene clusters are found in the nucleolus organizer of only one or a few chromosomes. These studies also showed that all true nucleoli in the interphase nucleus contain rDNA. However, in some cases clusters of rRNA genes were detected at chromosome loci that do not organize a nucleolus (PARDUE *et al.* 1970, BATISTONI *et al.* 1978), thus suggesting the existence of regulatory mechanisms switching on the rRNA genes in a given cluster.

Generally, one to six chromosomes carry the nucleolus organizer (BUSCH and SMETANA 1970, LIMA-DE-FARIA 1976, 1980). Comparative studies led to the proposal that eukaryotes with a higher number of "nucleolar" chromosomes are a later event in evolution (HSU *et al.* 1975). Many exceptions to this "rule" are now observed making uncertain its validity. For example, the class of *Primates* show large variations in the number of "nucleolar" chromosomes. Thus, five human chromosomes (13, 14, 15, 21, 22) carry rRNA gene clusters (HENDERSON *et al.* 1972, EVANS *et al.* 1974, GOSDEN *et al.* 1978) and the same number is typical for two chimpanzee species (*Pan paniscus* and *Pan troglodytes*) (HENDERSON *et al.* 1974, 1976). The orangutan (*Pongo pygmaeus*) has 9 "nucleolar" chromosomes (GOSDEN *et al.* 1978), the largest number uncovered till now. On the other hand, two gorilla species possess only two "nucleolar" chromosomes (HENDERSON *et al.* 1976, GOSDEN *et al.* 1978) and a single NOR characterizes other primates (WARBURTON *et al.* 1975, HENDERSON *et al.* 1974, 1976).

In many cases the rRNA gene clusters are located in autosomes and there is not an apparent regularity in the selection of a given type of chromosome. Location of the NOR in sex chromosomes is observed in many insects, thoroughly studied cases being *Drosophila* (COOPER 1959, RITOSSA *et al.* 1966, cf. RITOSSA 1976), *Sciara coprophila* (PARDUE *et al.* 1970, CROUSE *et al.* 1977), *Blatella germanica* (CAVE 1976) and others. Sex chromosome linkage of the rRNA gene cluster is found also in some mammalian species: *Potorous tridactylis, Carollia castanea,* (HSU *et al.* 1975) and *Muntiacus mintjak* (PARDUE and HSU 1975). Whether the location of the NOR in the sex chromosomes confers some phenotypic advantage or disadvantage is not known at present.

As a general rule, a single rRNA gene cluster is observed in a given "nucleolar" chromosome, although some exceptions have been reported. For example, all 150 rRNA genes of *Drosophila melanogaster* are clustered in a single locus of the X and Y chromosome (see RITOSSA 1976, SPEAR 1974), but two separate NOR sites were identified in the Y chromosome of *D. hydei* (MEYER and HENNIG 1974, SCHÄFER and KUNZ 1975).

Analysis of the location of nucleolus organizers within the chromosomes of large number of plant and animal species strongly suggests that they do not occur at random. In the vast majority of cases the cluster of rRNA genes is located in the short arm of the "nucleolar" chromosome near the telomere (LIMA-DE-FARIA 1973, 1976, 1980). This characteristic location of the nucleolus organizer is obviously unrelated to the arm length of the chromosome (Fig. 1). It has been argued that rRNA genes tend to occupy a defined site in the centromere-telomere "field", which is considered as optimal for their expression and regulation (LIMA-DE-FARIA 1976, 1980).

The factors leading to the spreading of rRNA gene clusters on different chromosomes are totally unknown. However, restriction analysis of multichromosomal rRNA gene clusters in *Homo sapiens* and other primates (ARNHEIM *et al.* 1980 a), as well as in human-mouse hybrids (KRYSTAL *et al.* 1981), has revealed that the entire rRNA gene family of a species evolves in a concerted fashion. This suggests that exchanges of rRNA gene clusters on non-homologous chromosomes may take place. Unequal crossing-over involving rRNA genes on homologous and non-homologous chromosomes might explain both the concerted evolution of rRNA gene sequences (cf. TARTOF 1975, SMITH 1976) and the variations in the number of rRNA genes in a given "nucleolar" chromosome cluster. It is noteworthy that exchanges of rRNA genes in non-homologous chromosomes in mice are apparently much more rare. The difference may be attributed to the chromosomal location of rRNA gene clusters: near the telomere in human, and close to the centromere in mouse "nucleolar" chromosomes (HENDERSON *et al.* 1974, 1976 a, ELSEVIER and RUDDLE 1975). Translocation of material from telomere loci in chromosomes is conceivably less likely to be phenotypically deleterious. If this interpretation is confirmed, it may provide a plausible explanation for the preferred location of rRNA gene clusters near the telomere of "nucleolar" chromosomes (LIMA-DE-FARIA 1976, 1980).

The studies on the chromosomal location of rRNA gene clusters clearly demonstrate that this is not a random event. However, the factors underlying the selection of a given chromosome as the site of the nucleolus organizer are obscure. It is also not clear whether multichromosomal

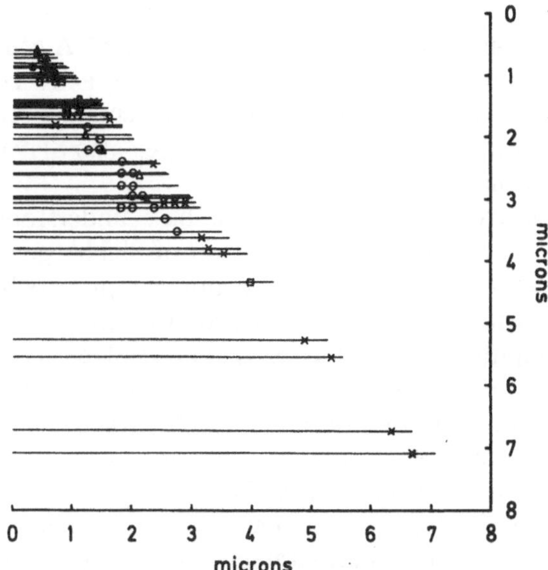

Fig. 1. Location of the rRNA genes in the chromosomes of mammals (△), worms (□), *Compositae* (○), and *Liliaceae* (×). Only the chromosome arms carrying rRNA genes are represented. The rRNA genes (NORs) are at a constant distance from the telomeres, irrespective of: species, phylum, and chromosome arm length. *Abscisse*—Distance of NOR from centromere; *Ordinate*—Length of chromosome arm. [From LIMA-DE-FARIA (1979). Reproduced with the permission of the author]

locations of rRNA gene clusters confers any phenotypical advantage. The presence of multiple tandem copies of identical rRNA genes creates the possibility for unequal crossing-over events during mitosis (sister chromatid exchanges) and meiosis (sister and non-sister chromatid exchanges). The multichromosomal presence of rRNA genes also makes possible crossing-over events among non-homologous chromosomes. These phenomena confer considerable plasticity to the number of rRNA genes in nucleolus organizers among species, individuals and even cell types of a given individual. Also, converging evidence points to the role of unequal crossing-over as a major factor determining the sequence homogeneity of the multiple sets of rRNA genes (see below).

II.2.3. Extrachromosomal rRNA Genes

The rRNA genes possess the capacity of differential replication. As a result, in some cases the cell has more or less rRNA genes than correspond to its ploidy. Generally, the differential replication of rRNA genes takes place in order to meet unusually high demands for ribosome production raised by a given cell. Also, differential replication of rRNA genes seems to take place in some cells and organisms as a correction mechanism, aimed at keeping the number of rRNA genes above a critical level. This mechanism may be switched on in cases when the number of rRNA genes is reduced by unequal crossing-over or mutational events. The extra copies of rRNA genes may be integrated in the chromosome or remain extrachromosomal. Here, I shall consider mainly the latter case (see also SPEAR 1974, TOBLER 1975, MACGREGOR 1972, 1982).

The first discovered and one of the best understood examples of differential replication is the amplification of rRNA genes (BROWN and DAWID 1968, GALL 1968). The amplification is a process by which rRNA genes multiply within the nucleus to produce many extrachromosomal copies. Gene amplification is a process that takes place in the germ cells of a large number of species, amphibia and insects in particular (GALL 1969, BIRD 1980). The presence of large amounts of extrachromosomal rDNA in the oocytes of many species is now amply documented (BIRNSTIEL *et al.* 1971, STRELKOV and KAFFIANI, 1978, BIRD 1980). Amplification of rRNA genes begins at early stages of the maturation of oogonia and involves a broad variety of developmental strategies, with a given species having a characteristics strategy (MACGREGOR 1982). In fact, at the early stages of maturation of gametes in some amphibia, a small amount of amplified rDNA can be detected in both oogonia and spermatogonia (PARDUE and GALL 1969). When the gametes enter mitosis most of the amplified rDNA is lost. Subsequently, during meiosis in oocytes (but not in spermatocytes) a second, and markedly more efficient, round of amplification takes place resulting in the formation of a large amount of extrachromosomal rDNA. Thus, the amplification of rRNA genes seems to be an intriguing feature in the process of male-female differentiation of gametes.

In many cases, the amounts of amplified, extrachromosomal rDNA in oocytes are exceedingly large. For example, the amount of amplified rDNA in *Xenopus laevis* is about 3,000-fold higher than the amount of haploid rDNA and corresponds to 1.5×10^6 rRNA genes forming up to 1,500–2,000 additional nucleoli (BIRNSTIEL *et al.* 1971). This extent of rRNA gene amplification seems to be fairly typical for oocytes of most *Amphibia,* but it is encountered also in other species, like the Reptilian *Testudo hermanii* (MACGREGOR 1982). Extensive amplification of rRNA genes is observed also in the oocytes of insects. Thus, about $1.0 \times 2.5 \times 10^5$ rRNA genes are forming during amplification in *Achaeta domesticus* (TRENDELENBURG *et al.*

1976, 1977) and a record of more than 3.0×10^6 extrachromosomal copies is established in *Dytiscus marginalis* (GALL and ROCHAIX 1974, TRENDELENBURG *et al.* 1974, 1976, 1977).

The biological role of rRNA gene amplification seems to reside in the necessity to supply an enormous reserve of ribosomes during maturation of the oocyte. This in turn provides for a fast and independent growth upon fertilization and early embryogenesis. It is noteworthy, that in some species the large pool of ribosomes in the germinal vesicle of oocytes is ensured by the activity of rRNA genes in the usually polyploid ovarian nurse cells, rather than by amplification of these genes in the oocyte itself (MACGREGOR 1982). Thus, it seems likely that signals for rRNA gene amplification should be provided by the evolutionarily fixed balance of oocytes and nurse cells in the ovary. The molecular character of such putative intercellular signals is unknown.

The bulk of amplified rDNA in *Xenopus laevis* is in the form of linear molecules, but up to about 10% is in the form of closed circles, containing 1–10 rRNA gene sets (HOURCADE *et al.* 1973, ROCHAIX *et al.* 1974, BOUNGIORNO-NARDELLI *et al.* 1976). Different replicative forms including "tailed" circles are observed, leading to the generally accepted view that rRNA gene amplification is an extrachromosomal process occuring by a "rolling circle" mechanism (see TOBLER 1975, STRELKOV and KAFFIANI, 1978, BIRD 1980). The rather widespread occurrence of a "rolling circle" mechanism of rDNA amplification is evidenced by the presence of circular rDNA molecules in the oocytes of many other amphibia, like *Triturus alpestris* (SCHEER *et al.* 1976), *Rana pipiens* (TRENDELENBURG and MCKINNELL 1979) and various insects like *Achaeta domesticus* (TRENDELENBURG *et al.* 1976), *Dytiscus marginalis* (TRENDELENBURG *et al.* 1974, GALL and ROCHAIX 1974), *Colymbetes fuscus* (GALL and ROCHAIX 1974) and others. That only some rDNA sequences are selected for extrachromosomal replication is strongly supported by the observation that in the oocytes of *X. laevis* × *X. borealis* hybrids the amplified rDNA is always of the *X. laevis* type which obviously supplied the original extrachromosomal rRNA gene set(s) (BROWN *et al.* 1972).

Drosophila melanogaster and other *Drosophila* species offer interesting cases of differential replication of rRNA genes, the genetics of which has been studied in considerable detail (see RITOSSA 1976, SPEAR 1974, TARTOF 1975). The rRNA genes of these species are in the X and Y chromosomes and coincide with the site of the bobbed (*bb*) gene. The bobbed gene is pleiotropic and its phenotype includes slower development, reduced fertility and shortened bristles. Evidence was presented that in bobbed mutants partial deletion of rRNA genes in NOR has taken place (RITOSSA *et al.* 1966). However, the bobbed phenotype in both *D. melanogaster* (SHERMOEN and KIEFER 1975) and *D. hydei* (SCHÄFER and KUNZ 1976) is related to rRNA synthesis rather than to rRNA gene redundancy. It was shown that in bobbed *Drosophila melanogaster* males an increase in the number of rRNA genes takes place in the germ cells of subsequent generations (RITOSSA *et al.* 1971). This process is designated as *magnification* of rRNA genes and may result in the stable restoration of wild-type amounts of rDNA in the *bb* locus. Magnification is reversible (to the original *bb* mutation) and stabilization of the increased amount of rDNA requires several generations and favorable conditions (BONCINELLI and FURIA 1979). It was proposed that magnification involves the formation and replication of extrachromosomal rDNA copies which may be reintegrated subsequently into the chromosome (RITOSSA 1972). Indeed, circular rDNA molecules have been observed in the germ cells of male flies, thereby indicating that the formation of extrachromosomal rDNA may be a stage in the magnification process (GRAZIANI *et al.* 1977). The possibility that a higher amount of rRNA genes in bobbed *Drosophila* mutants is attained by unequal sister chromatid exchange (TARTOF 1974) cannot be ruled out at present.

Elucidation of the genetic and other control mechanisms leading to the formation of extrachromosomal rDNA in *Drosophila* are not yet clarified. It is likely that the chromosomal rDNA cluster may respond to the extent of its transcription (SCHERMOEN and KIEFER 1975, SCHÄFER and KUNZ 1976, FRANZ and KUNZ 1981), but the relevant molecular signals are totally unknown. It should be pointed out that the formation of extrachromosomal rRNA genes may involve more complicated mechanisms. Thus, a high percentage of extrachromosomal rDNA is found in *Drosophila melanogaster* females heterozygous for a X chromosome inversion transposing the rRNA gene cluster to the telomere region of the X chromosome (ZUCHOWSKI and HARFORD 1977, HARFORD and ZUCHOWSKI 1977).

Differential replication of rRNA genes of germ cells is now firmly established and the process studied in considerable detail. On the other hand, differential replication in somatic cells in the whole organism or tissue culture is still controversial. It was reported (KOCH and CRUCEANU 1971) that growth of a culture of Cheng human cells for 48 h in the presence of 1.6×10^{-6}M 3,3′,5-triodothyronine resulted in a 1.8-fold higher content of rRNA genes than in controls. A moderate increase of the number of rRNA genes was reported also for human diploid somatic cells (MILLER *et al.* 1978). Also, a human hepatoma cell line was described carrying a 10-fold higher amount of rRNA genes (MILLER *et al.* 1979, TENTRAVAHI *et al.* 1981). These examples indicate that under some conditions amplification of rRNA genes may take place also in somatic cells. In none of these cases has evidence for the stable or transient existence of extrachromosomal rDNA been obtained.

Differential replication of rRNA genes is observed also in various species of lower eukaryotes. A large number of extrachromosomal rRNA genes is found in *Tetrahymena* (GALL 1974, ENGBERG *et al.* 1974), *Physarum polycephalum* (BROWN and EVANS 1969, VOGT and BRAUN 1976), *Stylonychia* (PRESCOTT *et al.* 1973, LIPPS and STEINBRUCK 1978) and *Dictyostelium* (COCKBURN *et al.* 1978). The case of *Tetrahymena* has been studied in considerable detail. Analysis of the location of rRNA genes in *T. pyriformis* showed that only a single gene is integrated in the micronuclear haploid genome (YAO and GALL 1977). The remaining rRNA genes, about 200 per haploid equivalent (ENGBERG and PEARLMAN 1972), are extrachromosomal and located in the macronucleus. The single micronuclear rRNA gene is amplified during conjugation to yield the multiple, extrachromosomal rRNA genes of the macronucleus (ENGBERG *et al.* 1976, KARRER and GALL 1976, YAO *et al.* 1978). These characteristics of rRNA gene amplification in *Tetrahymena* raise the question: is the first extrachromosomal rDNA formed by selective excision or by selective replication? By studying the structure of rDNA and its flanking sequences in *T. thermophila* micro- and macro-nuclei evidence was obtained that gene amplification is associated with chromosome breakage near the junction between rDNA and its neighboring sequences (YAO 1981). However, the general occurrence of such a mechanism is unlikely since it would disrupt the genetic continuity of the organisms capable of amplifying their rRNA genes. Therefore, selective replication of rRNA genes has to be envisaged as taking place in other protists and higher eukaryotes.

The existence of extrachromosomal rRNA genes was found also in *Saccharomyces*. Free 3 μm circles of DNA are found in these organisms and each contains a single set of rRNA and 5 S rRNA genes (MEYERINK *et al.* 1979, CLARK-WALKER and AZAD 1980, LARIONOV *et al.* 1980). Up to 10% of the rRNA genes are found as extrachromosomal 3 μm circles in some yeast strains, each circle containing a replication origin (LARIONOV and SHOUBOCHKINA 1982). Analysis of the generation of extrachromosomal rDNA molecules in *Saccharomyces* is likely to provide further insight into the mechanisms of differential replication of rRNA genes.

The widespread occurrence of extrachromosomal rRNA genes in eukaryotes strongly suggests that differential replication of these genes may play an important role in maintaining the genetic balance of rRNA genes. The generation of linear or circular extrachromosomal rDNA is obviously a basic mechanism in ensuring the adequate production of ribosomes in unicellular organisms, metazoan oocytes and, possibly, some somatic cells. Reintegration of some of these rRNA genes in the chromosomal rDNA cluster provides an attractive, yet still unproven, possibility.

II.2.4. Organization and Structure

Several powerful techniques introduced in the last decade led to a booming expansion of our knowledge on the organization and structure of eukaryotic rRNA genes. Isolation, cloning and mapping of rRNA genes from an ever-increasing number of organisms has been achieved. Combined with electron microscopic techniques for the analysis of RNA-DNA hybrids and visualization of active rRNA genes, these techniques now permit us to outline in considerable detail the basic features in the organization of rRNA genes. Also, knowledge of the sequence of important rDNA segments from different eukaryotes has led to conclusions about the evolutionary, genetic, and functional features of rRNA genes. Here, I shall summarize the presently available information, emphasizing the best studied and presumably most typical cases.

The rRNA genes are organized into transcription units. A *transcription unit* is a DNA segment containing a set of genes transcribed as one RNA molecule. The rRNA transcription unit contains two elements: (1) the sequences corresponding to the mature rRNA species and (2) *transcribed spacer* (tS) sequences that are present in the pre-rRNA molecule, but not in the mature rRNAs. The multiple sets of rRNA transcription units are arranged in tandem along the DNA separated by *nontranscribed spacer* (ntS) sequences. One transcription unit and the adjacent ntS constitute the rDNA *repeating unit*. The bulk of rRNA transcription units contain the genes for S-rRNA, 5.8 S rRNA and L-rRNA. This observation shows that the rRNA transcription units (and possibly also the repeating units) tend to behave as an entity in recombinational events during phylogenesis and ontogenesis. However, large deletions of 18 S rRNA genes were observed in the tandemly arrayed rDNA repeating units of *Drosophila melanogaster* (DAWID *et al.* 1978) and *Caenorhabditis elegans* (FILES and HIRSH 1981), indicating that the formation of *pseudogenes* may affect also the family of eukaryotic rRNA genes.

The tandem arrangement of the multiple repeating units constitute a basic principle in the organization of rRNA genes in the nucleolar chromosomes and also in most extrachromosomal gene sets.

II.2.4.1. Saccharomyces cerevisiae

Several yeast rDNA fragments were cloned and mapped leading to the elucidation of the structure of the yeast rDNA repeating unit (Fig. 2). The results obtained with *S. cerevisiae* by different groups (NATH and BOLLON 1977, 1978, CRAMER *et al.* 1977, BELL *et al.* 1977, PHILIPPSEN *et al.* 1978, PETES *et al.* 1978) are in agreement with each other and may be summarized as follows. The yeast rDNA repeating unit contains the sequences coding for the primary pre-rRNA (including 18 S, 5.8 S, and 25 S rRNA segments) and for 5 S rRNA. The length of the repeating unit is 9.1 knp. There is not gross length heterogeneity among individual repeating units, a fact suggesting that spacer sequences (ntS in particular) are identical in size. Although included in the same repeating unit, the coding sequences of the rRNA and 5 S rRNA genes are in different DNA strands and are transcribed in opposite directions (AARSTAD and ØYEN 1975, KRAMER *et al.* 1978, ZAMB and PETES 1982).

Analysis of the junction between chromosomal DNA and the cluster of rRNA genes reveals that it is located in the ntS region (ZAMB and PETES 1982). This finding may be correlated with the location of a replication origin in the ntS region of the extrachromosomal 3 μm (about 9 knp) rDNA circles (LARIONOV and SHOUBOCHKINA 1982). Accordingly, differential replication starting at the ntS region may be conceived as the possible origin of extrachromosomal rDNA in *Saccharomyces*.

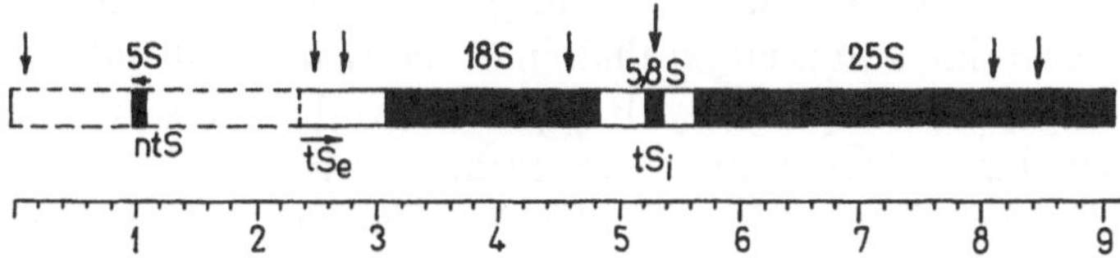

Fig. 2. Organization of the rDNA + 5 S rDNA repeating unit of *Saccharomyces cerevisiae*. Vertical arrows designate *EcoR I* sites. Horizontal arrows show the start and the direction of transcription. The scale is in knp. *ntS* non-transcribed spacer; tS_e external transcribed spacer: tS_i internal transcribed spacer. The complete sequence of the *Saccharomyces* rDNA repeating unit is now known (see the text)

The complete sequence of the *S. cerevisiae* rDNA repeating unit is now known. Its determination involved sequencing studies on the 5 S rRNA gene and its surroundings (MAXAM *et al.* 1977, VALENZUELA *et al.* 1977), the internal transcribed spacer (tS_i) and the 5.8 S rRNA gene (BELL *et al.* 1977, SKRYABIN *et al.* 1979), as well as the mapping (BAYEV *et al.* 1981) and determination of the complete sequences of the 18 S (RUBTSOV *et al.* 1980) and 25 S (GEORGIEV *et al.* 1981) rRNA genes. The identification of the primary transcript in *S. cerevisiae* (HADJIOLOV *et al.* 1978, NIKOLAEV *et al.* 1979) made possible the mapping of the rRNA transcription unit boundaries (BAYEV *et al.* 1980, 1981, KLEMENZ and GEIDUSCHEK 1980). The size of the transcription unit in *S. cerevisiae* is 6.63 knp, while that of the ntS is 2.5 knp. Several interesting sequence features are noteworthy, including the presence of numerous rather long T-stretches (20–30 np) in the ntS segment. Comparison of known sequences from different rDNA clones shows the almost complete identity in the rRNA gene region, but considerable differences in the ntS sequence (VALENZUELA *et al.* 1977, BAYEV *et al.* 1980). These findings indicate that the rRNA gene sequences are subject to stronger genetic pressure than the respective ntS and tS sequences. The observed deletion of an *EcoR I* site in the ntS of some *S. cerevisiae* strains (PETES *et al.* 1978) further supports this contention.

The presence of both rRNA and 5 S rRNA genes in one repeating unit seems to be restricted to only a few lower eukaryotes. The organization of the rDNA repeating unit in *Saccharomyces carlsbergensis* (MEYERINK and RETEL 1977, MEYERINK *et al.* 1978, PLANTA *et al.* 1978) is the same as that of *S. cerevisiae*. Considerable parts of the *S. carlsbergensis* rDNA have been sequenced including portions from the ntS (VELDMAN *et al.* 1980 b), the tS_i (VELDMAN *et al.* 1981 a) and the whole 25 S rRNA gene (VELDMAN *et al.* 1981 b).

The *Saccharomyces* type of organization of rRNA and 5 S rRNA genes within a common repeating unit is found also in *Torulopsis utilis* (TABATA 1980), the phycomycete *Mucor racemosus* (CIHLAR and SYPHERD 1980) and the protist *Dictyostelium discoideum* (MAIZELS 1976). In all three cases the size of the repeating unit is larger than in *S. cerevisiae*, with values of about 12.6, 10, and 38 knp, respectively. It appears that the two sets of ribosomal RNA genes have drifted apart at rather early stages of evolution. Remarkably, the rRNA and 5 S rRNA genes are already in different repeating units in the yeast *Schizosaccharomyces pombe* (TABATA 1981). There is not an apparent relation between the size of the ntS and the presence or absence of a 5 S rRNA gene within its sequence since the size of the ntS in *Schizosaccharomyces pombe* and *Torulopsis utilis* is identical (TABATA 1981). The absence of linkage between 5 S rRNA and rRNA genes is established also in other primitive eukaryotes like *Neurospora crassa* (FREE *et al.* 1979, COX and PEDEN 1979) and *Acanthamoeba castellanii* (D'ALESSIO *et al.* 1981). The separation of rRNA and 5 S rRNA genes in most eukaryotes is an intriguing phenomenon. Whether it confers some advantage in evolution remains to be elucidated.

II.2.4.2. Tetrahymena

The extrachromosomal rRNA genes in the *Tetrahymena* macronucleus have attracted considerable interest because they offer a rather peculiar organization. About 90% of these genes constitute a homogeneous population of linear rDNA molecules of about 20 knp. Most of the remaining 10% are DNA circles of similar size (GALL 1974, ENGBERG *et al.* 1974). Analysis of the structure of the linear rDNA molecules revealed that they are built of two identical halves (about 10 knp each) constituting a giant palindrome (KARRER and GALL 1976, ENGBERG 1976). Detailed study of this rDNA and its transcription map showed that each half contains ntS sequences plus one transcription unit (ENGBERG *et al.* 1980) and thus follows the general pattern of rDNA repeating unit structure. The palindromic rDNA is generated by amplification of a single rRNA gene integrated into the chromosome of the micronucleus (YAO and GALL 1977). Numerous copies of free, single, and linear rDNA repeating units are found during the formation of the macronucleus (PAN and BLACKBURN 1981); these units may be intermediates in the formation of palindromic rDNA (Pan *et al.* 1982). It is known that the palindromic rDNA molecules may replicate autonomously, showing bidirectional replication starting at about 0.6 knp on either side from the center of the molecule (Truett and Gall 1977, Cech and Brehm 1981). Transcription is also divergent starting from centrally located initiation points (Grainger and Ogle 1978, Engberg *et al.* 1980) and using opposite coding strands (Fig. 3). The rRNA genes follow the order S-

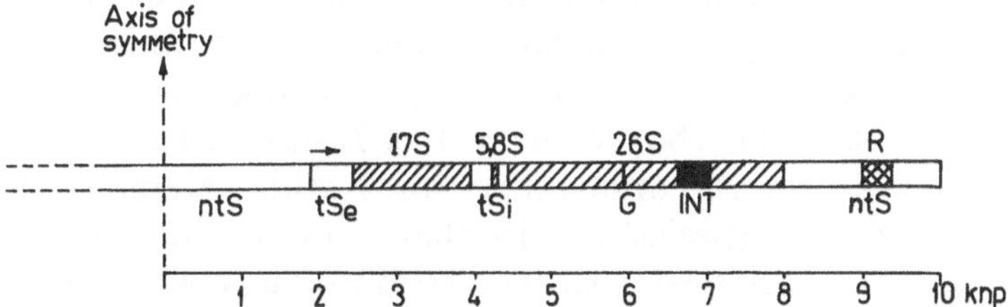

Fig. 3. Organization of one of the arms of the palindromic extrachromosomal rDNA of *Tetrahymena*. The hatched boxes represent 17 S, 5.8 S, and 26 S rRNA genes. *R* repeated sequences in the non-transcribed spacer (*ntS*); *G* location of the sequence gap in mature 26 S rRNA; *INT* position of the 0.4 knp intron in some *Tetrahymena* strains

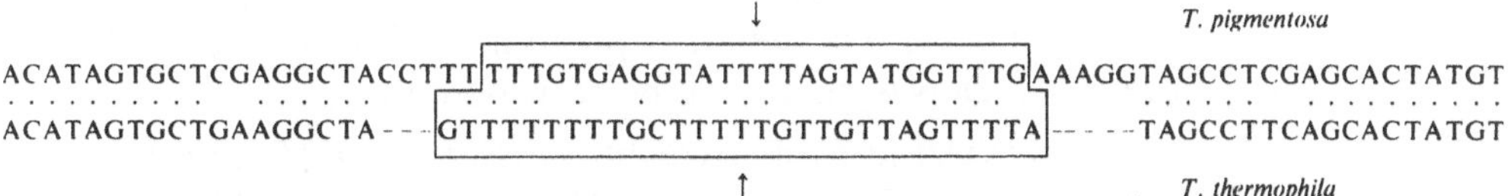

Fig. 4. Sequence of rDNA in the center of the palindromic extrachromosomal rDNA repeating units of two *Tetrahymena* species. The non-palindromic sequences around the axis of symmetry (vertical arrows) are boxed. Sequence data from Kiss and Pearlman (1981) and Kan and Gall (1981)

rRNA—5,8 S rRNA—L-rRNA, identical to that of all other eukaryotes, the 5,8 S rRNA gene being located within a tS_i sequence of only about 0.45 knp in *T. thermophila* and probably in other *Tetrahymena* species (Engberg *et al.* 1980). The tS_e sequence in *T. thermophila* and *T. pyriformis* is about 0.5 knp (Engberg *et al.* 1980, Higashinakagawa *et al.* 1981), while the 3′-end of the rRNA transcription unit in both *T. pyriformis* (Niles *et al.* 1981 b) and *T. thermophila* (Din *et al.* 1982) coincides with or extends with 15 nucleotides the 3′-end of the L-rRNA gene. The ntS located at the proximal part of the transcription unit is about 2.1 knp in length. The sequence near the center of the palindromic rDNA from different *Tetrahymena* species has been determined (Kiss and Pearlman 1981, Kan and Gall 1981, Engberg 1983) (Fig. 4). It is remarkable that in all species, the

T-rich nucleotide stretch (24 or 26 np) located at the very center is not palindromic and may form a single stranded loop at the end of the snapped-back rDNA. This structure may play an important role in the excision and/or replication of the single rRNA gene integrated in the chromosome. It seems that this central non-palindromic sequence is a vestige from a longer one (about 0.3 knp) observed in the initial gene excision product (PAN and BLACKBURN 1981, PAN *et al.* 1982). The distal ntS sequences of about 2.1 knp also reveal an unusual structure: approximately 1 knp from the 3′-end of the transcription unit in *T. thermophila* a tandemly repeated sequence 5′-$(CCCCAA)_n$-3′ is found ("coding" strand), where n is between 20 and 70 (BLACKBURN and GALL 1978). The role of this sequence is unknown, but it is likely to be present in other *Tetrahymena* species too (DIN and ENGBERG 1979). A comparative study with several strains from different *Tetrahymena* species revealed that all extrachromosomal rDNAs of a given strain are identical in size (DIN and ENGBERG 1979). The absence of ntS length heterogeneity is apparently due to the origin of all extrachromosomal rDNAs from a single integrated gene in the micronucleus as shown for *T. thermophila* (YAO and GALL 1977).

An important feature of rRNA gene structure in many, but not all, *Tetrahymena* species and strains is the presence of an about 0.4 knp intron within the L-rRNA gene (WILD and GALL 1979, DIN and ENGBERG 1979). All extrachromosomal rDNA molecules of the intron$^+$ *Tetrahymena* strains possess the intron sequence. The detailed position and structure of the introns in *Tetrahymena* rRNA genes will be considered below.

The palindromic organization of extrachromosomal rDNA repeating units is found also in *Physarum polycephalum* (VOGT and BRAUN 1976, 1977, MOLGAARD *et al.* 1976) and *Dictyostelium discoideum* (COCKBURN *et al.* 1978). In the case of *Physarum* the repeating unit is markedly larger (28–30 knp) and a large ntS segment (about 28 knp per dimer rDNA) is located at the center of the rDNA molecule (VOGT and BRAUN 1976, HALL and BROWN 1977, STEER *et al.* 1978). A more detailed R-loop study showed that the central ntS is about 18 knp per repeating unit (total = 36 knp). The tS_i (containing the 5.8 S rRNA gene) is of 1.7 knp, followed by the L-rRNA gene (CAMPBELL *et al.* 1979). Bidirectional replication (VOGT and BRAUN 1977) and transcription (GRAINGER and OGLE 1978) starts from specific symmetrically located points within the ntS. Terminal spacers with a variable length (average 5.4 knp) are located distally from the rRNA transcription unit. This spacer contains multiple inverted repeats of about 100 np near the termini of the molecule (JOHNSON 1980). The L-rRNA gene is also interrupted, but in the case of *Physarum* there are two introns (0.68 and 1.21 knp) within this gene, a unique situation encountered for eukaryotic rRNA gene introns (see below) (CAMPBELL *et al.* 1979). The studies on the organization and structure of rRNA genes in *Physarum* reveal many similarities to the pattern typical of *Tetrahymena*. On the other hand, the organization of the extrachromosomal rRNA genes in the macronucleus of another ciliated protozoan *Paramecium tetraurelia* is markedly different. In this case the rDNA is constituted by tandem repeating units with an average size of about 6.2 knp (FINDLY and GALL 1978, 1980). The tandem arrangement of rDNA repeating units in extrachromosomal rDNA is observed also in many lower eukaryotes (see FRANKE *et al.* 1979) so that the presence of extrachromosomal palindromic rDNA seems to be a comparatively rare device used in evolution.

II.2.4.3. Drosophila

Drosophila melanogaster has about 150 rRNA genes clustered in the X chromosome and a somewhat smaller number in the Y chromosome (see

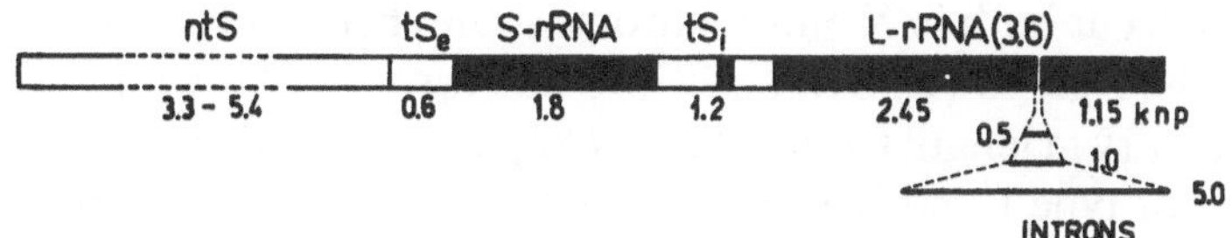

Fig. 5. Organization of the rDNA repeating unit of *Drosophila melanogaster*. The rDNA segments coding for the mature rRNA species are shown in black. The location and the common sizes of type I introns in L-rRNA are represented. The white dot in the L-rRNA gene indicates the position of the sequence gap in mature L-rRNA. The broken line in ntS indicates variations in size of this rDNA segment observed in different rDNA repeating units

TARTOF 1975). Cloning of these genes (GLOVER *et al.* 1975) resulted in several important observations on their organization. The "normal" repeating unit is moderately heterogeneous showing length variations in the range of 10.5 to 12.5 knp (GLOVER and HOGNESS 1977, WELLAUER and DAWID 1977, DAWID *et al.* 1978). The transcription units in *Drosophila melanogaster* are organized in tandem repeats (MCKNIGHT and MILLER 1976) as in all other eukaryotes. The primary transcript has been identified as a 38 S pre-rRNA (LEVIS and PENMAN 1978) and the transcription initiation point precisely mapped within a completely sequenced rDNA fragment of about 500 np (LONG *et al.* 1981 a). It has been shown also that the 3′-end of L-rRNA and of the transcription unit coincide exactly (MANDAL and DAWID 1981). The tS_e in *D. melanogaster* is among the shortest known (about 0.8 knp) and is markedly shorter than tS_i. The whole "normal" transcription unit is about 7.8 knp (Fig. 5). The variable length of the ntS is due, at least in part, to the presence of a 240 np repeat of varying redundancy (LONG and DAWID 1979 a, SIMEONE *et al.* 1982). The region of these repeats starts at about 150 np upstream from the transcription initiation site (LONG and DAWID 1981 a). Remarkably, a 42 np sequence surrounding the transcription initiation site is repeated within the ntS (SIMEONE *et al.* 1982), a situation studied in more detail in the *Xenopus* rDNA repeating unit (see below).

The most striking feature of *D. melanogaster* rRNA genes is the presence of introns splitting the L-rRNA gene into two unequal parts (GLOVER and HOGNESS 1977, WHITE and HOGNESS 1977, WELLAUER and DAWID 1977, PELLEGRINI *et al.* 1977). The insertions (designated also as IS or inserted sequences) have been classified in two types (I and II) depending on the absence or presence of nuclease *EcoRI* restriction sites (WELLAUER *et al.* 1978, DAWID *et al.* 1978, WELLAUER and DAWID 1978). The type I insertions appear in three sizes of 0.5, 1.0, and 5.0 knp respectively. They are found almost exclusively in the X chromosome and occur in about 50% of the rRNA genes (TARTOF and DAWID 1976, WELLAUER *et al.* 1978). The type II insertions occur both in X and Y chromosomes being present in about 15% of the rRNA genes (GLOVER 1977, WELLAUER *et al.* 1978). The type II insertions vary in length from 2.7 to 3.4 knp (ROIHA and GLOVER 1980). The insertions of type I and II share no homology and occur at different sites within the L-rRNA gene separated by 51 np (ROIHA *et al.* 1981). It is of particular interest that the non-ribosomal type I insertions belong to a class of repeated DNA sequences encountered outside the rRNA gene cluster with an estimated total amount in the genome equivalent of about 400 knp (0.2%) of its DNA (DAWID and BOTCHAN 1977). The type I intron sequences located outside the rDNA are found mainly in the centromere-associated heterochromatin (KIDD and GLOVER 1980, PEACOCK *et al.* 1981) where they are flanked by short sequences (20 np) from the 28 S rRNA gene (ROIHA *et al.* 1981). These observations strongly suggest that the type I insertion can undergo transposition from the rDNA to the centromere heterochromatin. They also suggest that the rRNA genes carrying type I introns are not transcribed, a view corroborated by the extremely low amount of transcripts hybridizing to intron sequences (LEVIS and PENMAN 1978, LONG and DAWID 1979 b, JOLLY and THOMAS 1980). The rRNA genes carrying type II introns appear also to be silent or transcribed at a very low, although barely detectable, rate (KIDD and GLOVER 1981). The mechanism for the blocked (type I) or limited (type II) transcription of intron$^+$ rRNA genes remains unknown. Whether an insertion mediated shut-off of rRNA genes may play some regulatory role during possible developmental rearrangements of rRNA genes in *Drosophila* remains unknown. It is however noteworthy that the capacity for differential replication is typical for rRNA genes in the X chromosome (WILLIAMSON and PROCUNIER 1975) as is the exclusive presence of type I introns. Whether these two features are somehow related remains to be clarified.

The presence of interrupted rRNA genes, as established in *Drosophila melanogaster*, seems to be a rare event in metazoa. Until now it has been found only in different insect species. Such intron$^+$ rRNA genes were reported in several species of *Drosophila* (BARNETT and RAE 1979, KUNZ and GLÄTZER 1979), *Calliphora erythrocephala* (BECKINGHAM and WHITE 1980, BECKINGHAM 1981), *Sciara coprophila* (RENKAWITZ *et al.* 1979) and *Sarcophaga bullata* (FRENCH *et al.* 1979). It is interesting to note that the intron$^+$ rRNA genes in *Calliphora* are clustered within the nucleolus organizer (BECKINGHAM 1981), an observation indicating that unequal crossing-over occurs preferentially at short distances within the tandemly arrayed rRNA genes, as shown in *S. cerevisiae* (SZOSTAK and WU 1980). In many insect species considerable length heterogeneity of the ntS was found (see, *e.g.*, ISRAELEWSKI and SCHMIDT 1982). An interesting case is *Calliphora erythrocephala*, in which it was shown that the ntS length heterogeneity is due to a 350 np repeat, located however much further upstream from the 5'-end of the transcription unit than in the case of *D. melanogaster* (SCHÄFER *et al.* 1981).

The presence of intron$^+$ rRNA genes does not seem to be ubiquitous to insects. Studies with *Bombyx mori* not only did not uncover splitting of rRNA genes, but also established that the 10.5 knp repeating units are remarkably homogeneous in length (MANNING *et al.* 1978).

II.2.4.4. Xenopus laevis

Studies with *Xenopus laevis* have contributed greatly to the pioneering elucidation of the basic principles in rRNA gene organization and structure. Owing to peculiarities in the structure of rRNA genes (e.g. the markedly higher GC content than the rest of the genome) and the availability of anucleolate (O_{nu}) mutants, the 500–600 (haploid) gene cluster in *X. laevis* supplied the starting material for the first unequivocal isolation of rRNA genes (BIRNSTIEL *et al.* 1966). It is also with *X. laevis* that the first rDNA repeating unit was characterized in this being a DNA segment of about 8.7×10^6 Da or 13 knp (WENSINK and BROWN 1971). Recent studies have led to a deeper understanding of the organization and structure of *X. laevis* repeating units, which are likely to be good models for rDNA repeating units in higher eukaryotes (Fig. 6).

The organization of the transcription unit (tS_e—S-rRNA—tS_i—L-rRNA) in *X. laevis* follows the typical eukaryotic pattern. Its size (approx. 7.5 knp) is constant within cells and individuals (WELLAUER *et al.* 1974 b, BOTCHAN *et al.* 1977). The position of the 5.8 S rRNA gene between S-rRNA and L-rRNA has been conclusively proven (SPEIRS and BIRNSTIEL 1974, WALKER and PACE 1977) and is typical of all rRNA transcription units studied till now. It splits the tS_i sequence into two unequal segments designated as tS_i1 and tS_i2. The complete sequence of the 1825 np 18 S rRNA gene is now established (SALIM and MADEN 1980, 1981) as is that of the tS_i1 (557 np) and tS_i2 (262 np) (BOSELEY *et al.* 1978, HALL and MADEN 1980). In contrast to the *S. cerevisiae* tS_i, the corresponding segment in *X. laevis* is extremely GC-rich, a feature, shared by most tS_i segments of vertebrates. This finding shows that tS_i sequences have rapidly evolved in evolution. The sequences of considerable parts of the 28 S rRNA gene are also determined (SOLLNER-WEBB and REEDER 1979, GOURSE and GERBI 1980b, HALL and MADEN 1980). The boundaries of the *X. laevis* rRNA transcription unit are precisely mapped. The transcription initiation start is about 710 np upstream from the 18 S rRNA locus (SOLLNER-WEBB and REEDER 1979, BOSELEY *et al.* 1979, MOSS and BIRNSTIEL1979, MADEN *et al.* 1982b), while transcription termination coincides with the 3′-end of the L-rRNA gene (SOLLNER-WEBB and REEDER 1979).

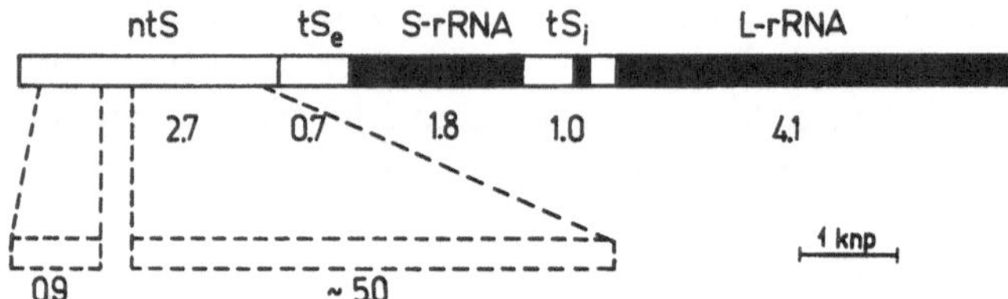

Fig. 6. Organization of the rDNA repeating unit of *Xenopus laevis*. The variable regions in ntS and the maximal length observed in cloned rDNA are indicated by broken lines. The numbers denote the size (in knp) of the respective rDNA segment

The non-transcribed spacer in *X. laevis* is of particular interest because it is at present the best studied ntS of higher eukaryotes. The ntS of *X. laevis* shows considerable length heterogeneity (WELLAUER *et al.* 1974b, WELLAUER *et al.* 1976a). The length heterogeneity is observed also im amplified rDNA presumably originating from a single nucleolus organizer (WELLAUER *et al.* 1976a, b). Detailed heteroduplex mapping analysis permitted the identification of at least two zones of internally repetitive sequences within the *X. laevis* ntS (WELLAUER *et al.* 1976a). Mapping with restriction enzymes of cloned rDNA, containing ntS segments of different

---

Fig. 7. Sequence of the non-transcribed spacer from one *Xenopus laevis* rDNA clone. The units of the repetitive segments in ntS are aligned beneath each other. Transcribed rDNA segments are indicated by shading. Homologous sequences between "Bam islands" and the ntS segment preceding the transcription start site are underlined with wavy lines (for details see the text) [From Moss *et al.* (1980). Reproduced with permission by Dr. M. L. BIRNSTIEL and the Editors *Nucleic Acids Res.*]

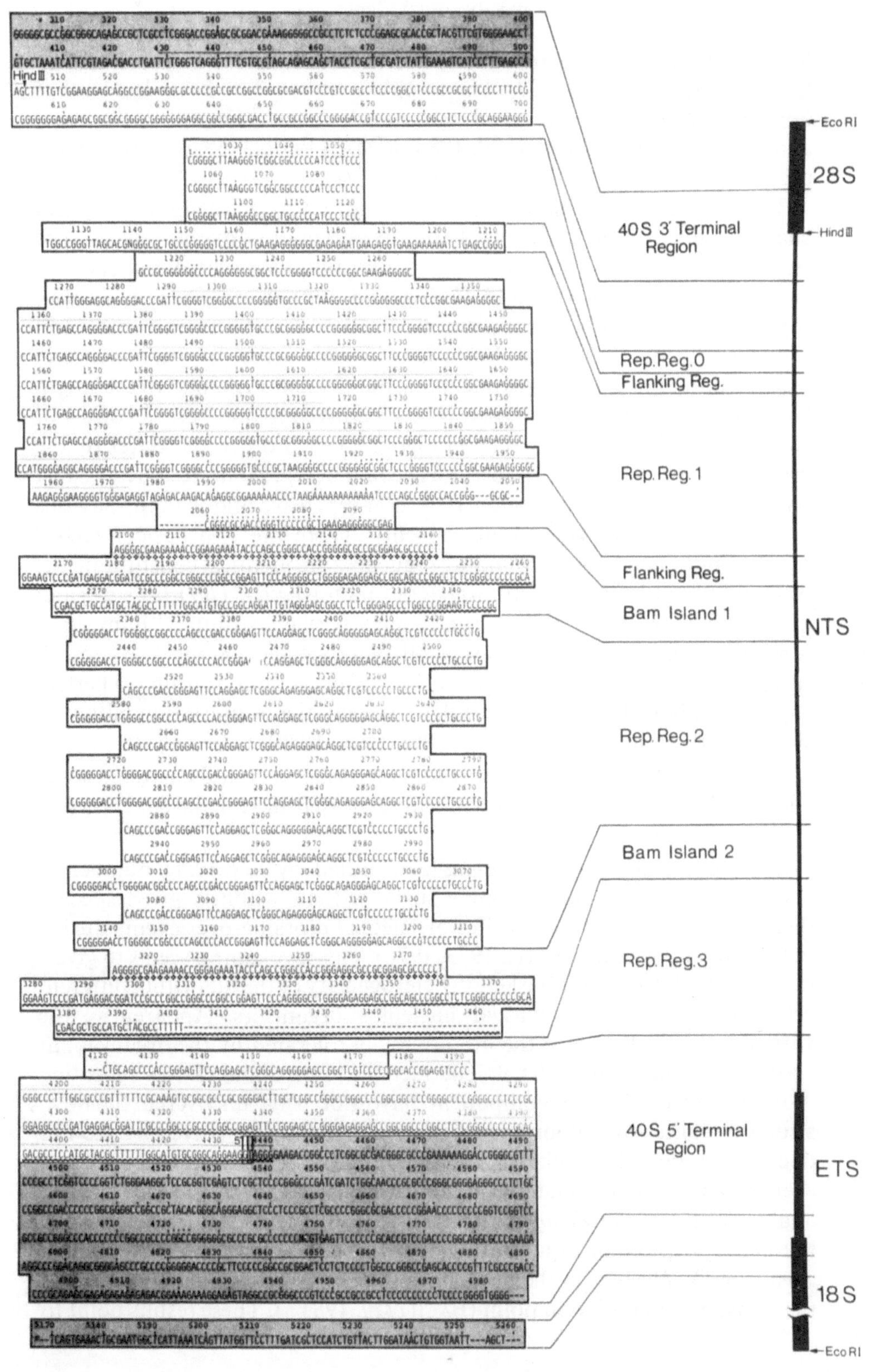
Eco RI
28S
Hind III
40S 3' Terminal Region
Rep. Reg. 0
Flanking Reg.
Rep. Reg. 1
Flanking Reg.
Bam Island 1
NTS
Rep. Reg. 2
Bam Island 2
Rep. Reg. 3
40S 5' Terminal Region
ETS
18S
Eco RI

length, revealed that repetitive and unique elements are interspersed within the ntS (BOTCHAN *et al.* 1977). The complete sequence of the *X. laevis* ntS is now known (BOSELEY *et al.* 1979, SOLLNER-WEBB and REEDER 1979, MOSS and BIRNSTIEL 1979, MOSS *et al.* 1980, MADEN *et al.* 1982b), permitting a deeper understanding of its structure (Fig. 7).

The transcription termination site is followed by a 500 np ntS segment characterized by high GC content and the absence of obvious repeats. The remainder of the ntS contains a high proportion of repeated sequences of different length and structure (MOSS and BIRNSTIEL 1982). Repetitive regions 0 and 1 (see Fig. 7) consist of repeated 35 and 100 np units, each unit showing a distinct sequence. Variations in the number of these units can occur in different rDNA repeating units, but their contribution to the length heterogeneity of the ntS is limited. The repetitive regions 2 and 3 consist of 60 and 81 np units which are homologous to each other and differ by the presence of an additional 21 np sequence in the larger unit. Relatively short regions flanking repeated region 1 are non-repetitive and without obvious homology to the repeated unit elements. There are two basic *BamH I* restriction sites and the surrounding sequences are identical constituting the so-called "Bam-islands". The number of "Bam-islands" (*BamH I* sites) and the flanking repetitive regions 2 and 3 may vary constituting "Bam super repeats" and generating the major length heterogeneity among *X. laevis* ntS. It is remarkable that the "Bam-islands" sequences contain an almost perfect copy of an about 150 np sequence that precedes the transcription initiation site (BOSLEY *et al.* 1979, MOSS and BIRNSTIEL 1979).

The elucidation of the structure of the *Xenopus laevis* ntS answered some and raised other questions concerning its genetic behavior and function (see FEDOROFF 1979).

The identification of the ntS as constituted by a varying number of dissimilar repetitive units strongly suggests that, within a given species, these sequences may serve as sites of unequal crossing-over events during mitosis and meiosis (see SMITH 1973, 1976). This mechanism seems to give a plausible explanation for the length heterogeneity of the ntS within different cell types and even individual nucleolus organizers. Apparently, the variations in the length of the ntS are easily tolerated by the organism, at least above some critical level. The fact that functional rDNA repeating units devoid of the ntS component have not been uncovered till now, strongly suggests its importance. It seems that the role played by the ntS does not reside in the bulk of its sequence, since rapid divergence is found even among closely related species. For example, restriction enzyme analysis of the ntS in *Xenopus clivii* and *X. borealis* revealed that, although the basic principle of organization is preserved, considerable differences exist from *X. laevis* with the exception of a conserved segment preceeding the transcription initiation point (BACH *et al.* 1981a). These results point to the possibility that the function of the ntS is either species-specific or related to some more conserved sequences within the variable framework.

The observation that the putative transcription control sequences are duplicated within the ntS strongly suggests that the designation "non-transcribed" may not be absolute. Evidence for transcription within the ntS has been obtained by different approaches (FRANKE *et al.* 1976, TRENDELENBURG 1981, 1983, MOSS 1982) and will be considered later.

An interesting and very important function seems to be the presence of replication start signals within the ntS. Recently, electron microscopic studies provided evidence for the occurrence of replication origins (visualized as "microbubbles") within the ntS of *Drosophila melanogaster* (McKNIGHT *et al.* 1978) and *Xenopus laevis* (BOZZONI *et al.* 1981). Synchronization of the replication initiation by 5-fluoro-2-deoxyuridine and the slow recovery of replication in *X. laevis* revealed also a preferential incorporation of [^{3}H]deoxyadenosine into replication origins located within the rDNA ntS (BOZZONI *et al.* 1981). Elucidation of the role of ntS structures as replication origins may help in understanding the basic fact that the rDNA repeating unit behaves as an entity in the organization of rRNA genes within the chromosome and in the generation of extrachromosomal rDNA copies. Possible relationship between sequences serving as transcription and replication signals appears as a challenging field for future investigations. Recently, the presence of a short (320 np) DNA fragment in a vector plasmid was found to enhance 15-fold the replication of the plasmid injected into *X. laevis* oocytes (CHAMBERS *et al.* 1982). This fragment has been sequenced and comparison with the structures of known *X. laevis* ntS sequences may help to unravel some characteristics of replication origin signals.

To what extent the basic structural features of *Xenopus* rDNA repeating units are preserved in other lower vertebrate species remains unknown. However analysis of the organization of rRNA genes in the loach (*Misgurnus fossilis*) revealed that the size heterogeneity of rDNA repeating units results mainly from variations in ntS length (KUPRIJANOVA *et al.* 1982). Thus, the *X. laevis* model seems to be followed throughout the realm of vertebrates.

II.2.4.5. Higher Plants

Higher plants are characterized by an about 5–10-fold higher number of rRNA genes than found in most eukaryotes (see Table 1). Unfortunately, the amount of information available about their organization and structure is relatively small. Recently, the rRNA genes of *Hordeum vulgare* (barley), *Triticum aestivum* (wheat) (GERLACH and BEDBROOK 1979), *Linum usitatissimum* (flax) (GOLDSBROUGH and CULLIS 1981), *Citrus lemon* (FODOR and BERIDZE 1980), *Glycine max* (soybean) (FRIEDRICH *et al.* 1979, VARSANYI-

BREINER *et al.* 1979) and *Cucurbita pepo* (pumpkin) (SIEGEL and KOLASZ 1983) have been cloned and their organization determined by analysis with restriction nucleases. The basic organization of eukaryotic rDNA repeating units is found also in plants. The size of the transcription units is not known, but is estimated to be about 6 knp (GERLACH and BEDBROOK 1979). It is remarkable that the size of the repeating unit is relatively small, varying from 8.6 to 9.9 knp. Thus, it appears that the ntS in higher plants will be in the modest range of 2.5–3.0 knp. The ntS sequences of wheat were found only in rDNA, but not elsewhere in the genome (GERLACH and BEDBROOK 1979). The rather short ntS in plants correlates with the absence (as in flax and wheat) of, or a limited (as in pumpkin), length heterogeneity of rDNA repeating units. Two major types of repeating units (9.0 and 9.9 knp) were observed in barley (GERLACH and BEDBROOK 1979) and lemon (9.8 and 9.4 knp) (FODOR and BERIDZE 1980). Given the large number of rRNA genes in plants, the small size and the relative homogeneity of their repeating units is somewhat unexpected. Whether these characteristics are of more general occurrence remains to be established. Whatever the case, the study of rDNA repeating units in higher plants now appears to be unduly neglected. There is little doubt that elucidation of their structure, correlated with the broad information on the genetics of many plant rRNA genes, will soon be forthcoming and will help us to understand better the general features in the organization and structure of rRNA genes.

II.2.4.6. Mammalia

The initial restriction enzyme studies revealed exceedingly long and moderately heterogeneous ntS in mouse and man (ARNHEIM and SOUTHERN 1977, CORY and ADAMS 1977, KRYSTAL and ARNHEIM 1978). The presence in mammals of 45 S pre-rRNA (about 12.5×10^3 nucleotides), the largest known, putative primary transcript (see HADJIOLOV and NIKOLAEV 1976), combined with a 2–3 fold longer ntS, puts the size of rDNA repeating units in the range of 35–45 knp. Such a size creates considerable difficulties in the cloning of the entire rDNA, although several authors have reported the cloning of rDNA fragments from the gene region. Investigations by electron microscopy (RNA : DNA heteroduplex mapping) and restriction enzyme analyses have confirmed the unexpectedly large size of mammalian rDNA repeating units. Thus, the repeating units in mouse (CORY and ADAMS 1977, KOMINAMI *et al.* 1981) and man (WELLAUER and DAWID 1979) are both of an average size of 44 knp, while those of the rat are about 37 knp (STUMPH *et al.* 1979). The transcription initiation site has been mapped at about 4.0–4.4 knp upstream from the 18 S rRNA gene in mice (MISHIMA *et al.* 1980, BACH *et al.* 1981 b), rats (FINANCSEK *et al.* 1982 a, ROTHBLUM *et al.* 1982) and humans (FINANCSEK *et al.* 1982 b, MIESFELD and ARNHEIM 1982). It is also reported that in the mouse (and possibly in other mammalian species) the 3′-ends of the transcription unit and of the 28 S rRNA gene are separated by 30 np (KOMINAMI *et al.* 1982). These studies fix the boundaries of the rRNA transcription unit in mammals and confirm that the order of rRNA genes is common to all eukaryotes. Partial sequencing of the rat and mouse 18 S and 28 S rRNA genes (SUBRAHMANYAM *et al.* 1982, KOMINAMI *et al.* 1982, MICHOT *et al.* 1982) reveals the existence of moderate homology with other eukaryotes at least at the termini of 18 S and 28 S rRNA genes. In contrast, the tS_i sequences show considerable divergence from other eukaryotes and are characterized by very high GC contents (SUBRAHMANYAM *et al.* 1982, GOLDMAN *et al.* 1983, MICHOT *et al.* 1983).

The studies of mammalian ntS are still at an early stage. R-loop analysis of rat rDNA repeating units revealed only a limited length heterogeneity (STUMPH *et al.* 1979). A more extensive study on human repeating units also showed that most are rather uniform in length, although some heterogeneity in the ntS was noted (WELLAUER and DAWID 1979). Variations in the length of the region adjacent to the 3′-end of 28 S rDNA in the ntS of humans were also reported (KRYSTAL and ARNHEIM 1978). Restriction enzyme analysis revealed the presence of some internally repetitive sequences in the ntS's of mouse and rat (GRUMMT and GROSS 1980, ARNHEIM 1979, BRAGA *et al.* 1982), but further studies are needed before a structural comparison with the ntS's of other higher eukaryotes becomes feasible. The existence in mammalian species of an exceedingly large ntS raises the possibility that it has functions beyond providing signal sequences for transcription initiation and termination. In this respect, it is interesting to note that mouse ntS sequences were found scattered throughout the genome flanking C_H (μ, α and γ 2 b) immunoglobulin genes (ARNHEIM *et al.* 1980 b).

The presence of a large ntS does not seem to be restricted to mammalian species. Thus, a recent study on rDNA repeating units in the chicken (*Gallus domesticus*) showed that there two size classes of 37 and 25 knp exist, the heterogeneity being due to differences in the ntS's (MATTAJ *et al.* 1982). These studies also confirmed previous less direct evidence (SCHIBLER *et al.* 1975) that the size of the tS_i in chicken is 4.4. knp, the longest of all eukaryotes studied till now.

II.2.5. General Features

The investigations carried out in the last few years resulted in a remarkable expansion of our knowledge on the organization and structure of rRNA genes in eukaryotes. This in turn led to a deeper understanding of the behavior of rRNA genes in phylogenesis and ontogenesis. The accepted dynamic picture of rRNA genes contributes also to our understanding of the eukaryotic genome. Here, I shall outline some general features of eukaryotic rRNA genes that have emerged.

a) rRNA and 5 S rRNA Genes Drift Apart and a New rRNA Gene Is Formed

All three rRNA genes in *E. coli* (for 16 S, 23 S, and 5 S rRNA) are organized as a single transcription unit and their arrangement in the respective repeating unit is: ntS—tS_e—16 S rRNA—tS_i1—23 S rRNA—tS_i2—5 S rRNA (see HADJIOLOV and NIKOLAEV 1976). This basic organization is preserved in eukaryotes, but three major changes evolved: (a) the S-rRNA and L-rRNA genes are larger; (b) the 5 S rRNA gene drifted apart from close linkage to rRNA genes and (c) a new, 5.8 S rRNA gene

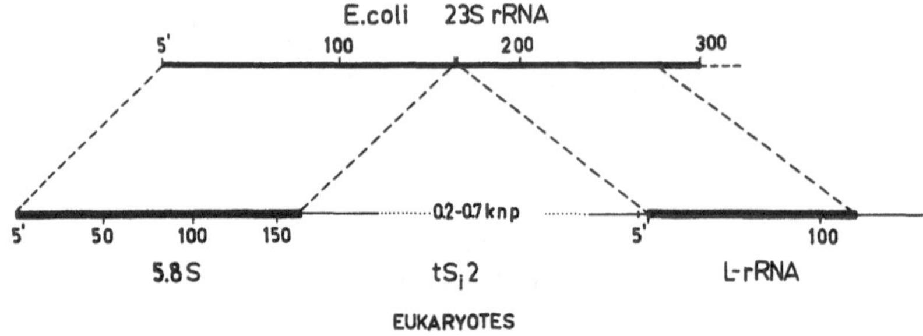

Fig. 8. Possible evolutionary origin of the 5.8 S rRNA gene in eukaryotes. The scheme illustrates the existence of sequence homology between eukaryotic 5.8 S rDNA plus L-rRNA and the respective segments in *Escherichia coli* 23 S rRNA indicated by the broken lines. There are no sequences homologous to eukaryotic tS_i 1 or tS_i 2 within the rDNA repeating unit of *E. coli*. The numbers denote nucleotides in the respective mature rRNA species

located in the tS_i formed. These changes are already present in primitive eukaryotes and characteristic of all species studied till now. It seems that initially (*Saccharomyces*, *Mucor racemosus, Dictyostelium*) the 5 S rRNA gene remained in the same repeating unit, although it was transcribed separately. However, this pattern was broken early in evolution: in many simple eukaryotes (i.e. *Schizosaccharomyces pombe, Neurospora crassa*) the 5 S rRNA gene is no longer linked to rRNA genes. The advantage conferred by the independence of the 5 S rRNA gene, if any, is still unknown. But as a result of separate evolution, in many cases the multiplicity of 5 S rRNA genes now exceeds that of rRNA genes.

The emergence of the new 5.8 S rRNA gene is typical of all eukaryotes. Moreover, the extensive sequence homology between the 5′-terminal segment of *E. coli* 23 S rRNA and eukaryotic 5.8 S rRNA + 5′-terminal segment of L-rRNA (Fig. 8) clearly demonstrates the origin of the 5.8 S rRNA gene (NAZAR 1980, WALKER 1981, COX and KELLY 1981, JACQ 1981,

CLARK and GERBI 1982, OLSEN and SOGIN 1982). It is remarkable, that the tS_i2 sequence in eukaryotes has not an *E. coli* counterpart, suggesting an independent (?) origin for the spacer sequence.

b) Semper Repeating Units

The rRNA genes are always organized in *repeating units* usually arranged in tandem and located at the specific chromosome loci of nucleolus organizers. The rDNA repeating units contain *nontranscribed spacer* (ntS) sequences and a *transcription unit* containing *transcribed spacer* (tS) and rRNA gene sequences. The organization in repeating units is also typical of extrachromosomal rRNA genes, even in rDNA molecules carrying a single set of rRNA genes as for instance in the *Saccharomyces* 3 μm rDNA circles (MEYERINK *et al.* 1979, CLARK-WALKER and AZAD 1980, LARIONOV *et al.* 1980) or in the linear single copy rDNA intermediates observed during the formation of the macronucleus in *Tetrahymena* (PAN and BLACKBURN 1981, PAN *et al.* 1982). These observations strongly suggest the existence of genetic mechanisms ensuring the comportment of rDNA repeating units as an entity. It is also noteworthy, that while the rDNA repeating units display considerable variations in size, partial or complete deletions of individual rRNA genes are seldom found (DAWID *et al.* 1978, FILES and HIRSCH 1981). The rarity of "pseudogene" type rDNA repeating units, whose presence is unlikely to be phenotypically sensed above the background of the multiple rRNA genes, indicates that some correction mechanisms eliminate aberrant repeating units from the genome.

c) rRNA Genes Are Clustered in the Nucleolus Organizer

Since the pioneer studies of MCCLINTOCK (1934) the nucleolus organizer is defined as the chromosomal site of nucleolos formation. This notion has been substantiated by a variety of cytogenetic and cytochemical approaches. In particular, studies with NO mutants of *Drosophila* and *Xenopus* (BROWN and GURDON 1964, RITOSSA and SPIEGELMAN 1965, RITOSSA *et al.* 1966, BIRNSTIEL *et al.* 1966) and introduction of *in situ* hybridization techniques (GALL and PARDUE 1969) provided conclusive evidence that rRNA genes are a major constituent of the nucleolus organizer (cf. HADJIOLOV 1980, STAHL 1982). Accordingly, the nucleolus may now be envisaged as the product of the activity of rRNA genes or more precisely of rDNA repeating units. This is true also for the nucleoli formed by amplified extrachromosomal rDNA, as illustrated beautifully by the detailed gene spreading studies with amphibian oocytes (MILLER and BEATTY 1969 b, TRENDELENBURG and MCKINNEL 1979). The formation of the nucleolus around a cluster of tandemly arrayed rRNA genes may confer additional advantages in nucleologenesis, as for example higher local concentrations of RNA polymerase I, ribosomal proteins, processing enzymes, and other factors needed in ribosome biogenesis. From this point of view rDNA orphons, even if shown to be ubiquitous and active in eukaryotes, are likely to operate under very unfavorable conditions. Whether the presence of multiple nucleolus organizers in some species is advantageous is not known at present. It is also unknown whether there are differences in the structure and behavior of autosomal and sex-linked nucleolus organizers. The role of the preferential location of nucleolus organizers at the telomere region of chromosomes (LIMA-DE-FARIA 1976, 1980) is also unknown, but seems to allow more flexibility in intra- and interchromosomal recombination events.

d) Transcription Units Follow a Common Pattern of Organization

The transcription unit of all rDNA repeating units studied until now displays a common pattern of organization. Constituent sequences follow the order:

(promoter)—tS_e—S-rRNA—tS_i1—5.8 S rRNA—tS_i2—L-rRNA.

Although the tS_e may be as short as about the 100 np observed in *Sciara coprophila* (RENKAWITZ *et al.* 1979), no eukaryote has been found to lack this obviously important sequence. On the other hand, the distal end of the transcription unit in different organisms coincides with the 3′-end of the L-rRNA gene (SOLLNER-WEBB and REEDER 1979, BAYEV *et al.* 1981, MANDAL and DAWID 1981). Although 10 to 30 extra nucleotides may be present in some primary transcripts (KLOOTWIJK *et al.* 1979, BAYEV *et al.* 1981, KOMINAMI *et al.* 1982) it now seems that sizeable tS_e sequences are not present at the distal end of eukaryotic transcription units.

The 5.8 rRNA gene is invariably located within the tS_i and is constant in size, while the length of the tS_i sequences, vary from 0.65 knp in *Sciara coprophila* (RENKAWITZ *et al.* 1979) and 0.6 knp in *Saccharomyces cerevisiae* (SKRYABIN *et al.* 1979) up to 4.4 knp in *Gallus domesticus* (MATTAJ *et al.* 1982). Usually tS_i2 is shorter than tS_i1, but the general occurrence of this regularity remains to be ascertained.

The sizes of the genes for S-rRNA, 5.8 S rRNA, and L-rRNA are markedly more conserved in evolution than those of the tS_e and tS_i sequences (Fig. 9). There is an evolutionary increase in size of rRNA genes, in particular for the L-rRNA gene, but this increase is small compared to variations of the transcribed spacer units.

e) rRNA Gene Sequences Are Conserved Following a Mosaic Pattern

The observed great variations in the size of tS_e and tS_i suggest that these sequences change considerably in evolution. This has been confirmed by direct sequencing data on tS_e and tS_i in eukaryotes. The complete sequence of tS_e is known only for *S. cerevisiae* and *X. laevis*, but many authors have determined tS_e sequences near the transcription initiation point. It is remarkable that even among closely related species tS_e sequences reveal little homology, as illustrated by a comparison of rat, mouse, and human (FINANCSEK *et al.* 1982b) or even of *X. laevis* and *X. borealis* (BACH *et al.* 1981a). A similar situation is found for tS_i sequences. For example, no apparent homology can be detected among the tS_i sequences of *S. cerevisiae* (SKRYABIN *et al.* 1979), *X. laevis* (HALL and MADEN 1980) and the rat (SUBRAHMANYAM *et al.* 1982).

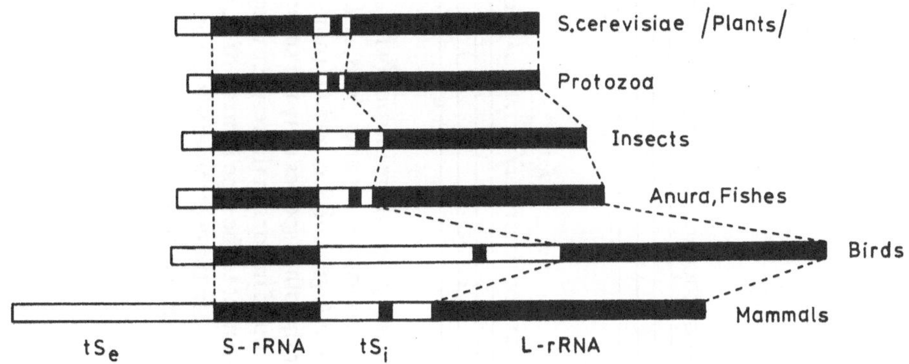

Fig. 9. Evolution of rRNA transcription units in eukaryotes. The S-rRNA, 5.8 S rRNA, and L-rRNA genes are indicated by black boxes. tS_e—external transcribed spacer; tS_i—internal transcribed spacer split into two parts (tS_i 1 and tS_i 2) by the 5.8 S rRNA gene. The species considered as representative for a given class are: *Tetrahymena* for *Protozoa*; *Drosophila* for Insects; *Xenopus laevis* for *Anura*; *Gallus domesticus* for Birds and *Rattus norvegicus* for Mammals

In contrast to transcribed spacer sequences, considerable interspecies homology is found within the rRNA gene boundaries. Comparison of the 5.8 S rRNA genes from a broad variety of eukaryotic species demonstrates that, aside from many single base substitutions, a clear-cut homology is preserved throughout the eukaryotic realm, in particular for the first 110 nucleotides of the gene (OLSEN and SOGIN 1982, CLARK and GERBI 1982). Considerable sequence homology is found between the 200 np 3′-end segments of the S-rRNA genes in *S. cerevisiae* (RUBTSOV *et al.* 1980), *D. melanogaster* (JORDAN *et al.* 1980a), *Bombix mori* (SAMOLS *et al.* 1979), *Xenopus laevis* (SALIM and MADEN 1981), *Mus musculus* (MICHOT *et al.* 1982c, GOLDMAN *et al.* 1983) and *Rattus norvegicus* (SUBRAHMANYAM *et al.* 1982). It is of particular interest that in the case of the S-rRNA gene two highly conserved stretches in this 3′-terminal domain are separated by a

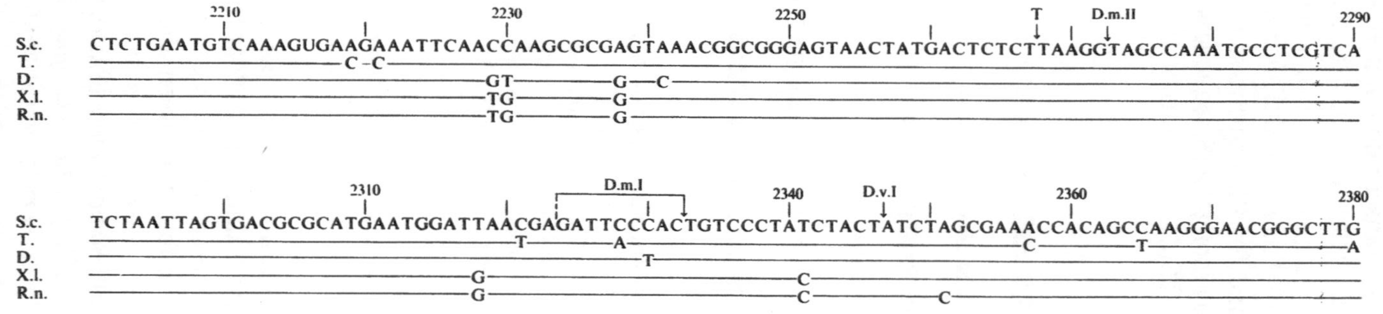

Fig. 10. Strongly conserved sequence in the L-rRNA gene of eukaryotes where introns (when present) are usually inserted. *S.c. Saccharomyces cerevisiae*; *T Tetrahymena*; *D Drosophila melanogaster* and *virilis*; *X.l. Xenopus laevis*; *R.n. Rattus norvegicus*. The sequence of the respective segment of the L-rRNA gene of *S. cerevisiae* (non-coding strand) is represented with the numbering according to Georgiev *et al.* (1981). For the other species only the substitutions in the L-rRNA gene sequence are indicated. The vertical arrows indicate the location of introns in the respective species: *T. Tetrahymena*; *D.m. Drosophila melanogaster*; *D.v. Drosophila virilis*. The sequence next to the type I intron of *D. melanogaster* (dotted vertical line) is found translocated together with the intron sequence within the centromere-associated heterochromatin (Roiha *et al.* 1981). For sources of sequence data see the text

widely divergent sequence of about 60 np (cf. Michot *et al.* 1982c, Goldman *et al.* 1983). This observation sheds some light on the possible sequence background in the evolution of rRNA genes. In this respect the L-rRNA gene may be more informative, since it displays much broader size changes. The limited available information shows considerable conservation at the 5'- and 3'-end of the gene (Michot *et al.* 1982b, Kominami *et al.* 1982). This pattern is altered after the 150th nucleotide, where GC-rich segments appear in vertebrates without detectable homology in *Saccharomyces* (Georgiev *et al.* 1981, Veldman *et al.* 1981, Michot *et al.* 1982a, Hadjiolov *et al.* 1984). The alternation of conserved and non-conserved segments in rRNA genes and spacers creates a mosaic pattern of rDNA sequences conserved throughout evolution but whose biological role remains to be deciphered.

f) IS Elements Where Present, Prefer a Highly Conserved Sequence Within the L-rRNA Gene

Till now IS (inserted sequences) elements are found only in some species of the unicellular eukaryotes *Tetrahymena* and *Physarum* and in *Drosophila* and some other insect species. The IS's in *Tetrahymena* are relatively short (0.4 knp). When present, they occur in all rDNA repeating units, are transcribed, but excised during the processing that yields the mature L-rRNA. It is interesting that the exon/intron junctions in *Tetrahymena* do not share the consensus sequences found at these junctions in protein-coding split genes (Wild and Sommer 1980). The IS elements in *Drosophila* and other insects are much larger in size (usually varying in the range of 1 to 10 knp) than those of the protozoa and are not transcribed at an appreciable rate (Long and Dawid 1979, Jolly and Thomas 1980, Kidd and Glover 1980). The sequences inserted in the L-rRNA gene of *D. melanogaster* and *D. virilis* are found also outside the NOR in the centrometric heterochromatin (Dawid and Botchan 1977, Barnett and Rae 1979, Kidd and Glover 1980, Dawid *et al.* 1981) and the presence of direct repeats at their ends strongly suggests that the rDNA IS may behave as transposable elements (Roiha *et al.* 1981).

An interesting observation is that the IS elements in *Tetrahymena* and *Drosophila* are located within an about 200 np segment of L-rRNA which is remarkably conserved (Fig. 10) in all eukaryotes thus far studied (Gourse and Gerbi 1980b). However, the two *Physarum polycephalum* introns provide a partial exception to this rule (Campbell *et al.* 1979, Nomiyama *et al.* 1981). Why transposition of IS elements should display such an affinity for a remarkably stable L-rRNA sequence remains obscure, but there is little doubt that further analysis of this phenomenon is likely to provide valuable information regarding the general comportment of movable elements in the eukaryotic genome.

g) Non-Transcribed Spacers Possess Intriguing Structural and Functional Properties

Broad species variations in size and sequence of ntS in eukaryotic rDNA repeating units are found (Fig. 11). As with tS, ntS of even closely related species show little homology, although there might be exceptions as the reportedly conserved ntS in six *D. melanogaster* sibling species (TARTOF, 1979). The available data summarized in Fig. 11 illustrate the variations in ntS size among species. It is noteworthy that the ntS is rather short (2–3 knp)

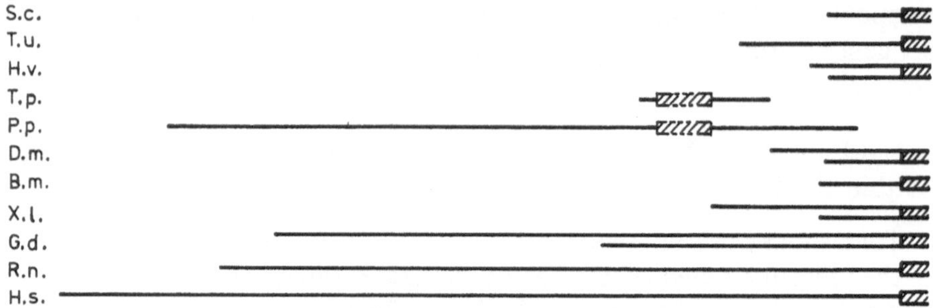

Fig. 11. Comparison of non-transcribed spacer size in some eukaryotes. Considerable variations in size of ntS is found within some species. The size of maximal and minimal ntS is represented by two horizontal lines. The hatched boxes represent tS_e segments or the whole transcription unit when surrounded by ntS segments. *S.c. Saccharomyces cerevisiae*; *T.u. Torulopsis utilis*; *H.v. Hordeum vulgare*; *T.p. Tetrahymena pyriformis*; *P.p. Physarum polycephalum*; *B.m. Bombyx mori*; *D.m. Drosophila melanogaster*; *X.l. Xenopus laevis*; *G.d. Gallus domesticus*; *R.n. Rattus norvegicus*; *H.s. Homo sapiens*

in several primitive eukaryotes (*i.e. Saccharomyces*), some insects and most higher plants. On the other hand the ntS's of vertebrates vary in the range of 10–30 knp. Considerable size heterogeneity is observed in many cases, an extreme one being the ntS of *Xenopus laevis*, where length differences among repeating units of several knp are found (WELLAUER and DAWID 1974, BOTCHAN *et al.* 1977, BOSELEY *et al.* 1979, MOSS *et al.* 1980). The length heterogeneity of ntS in other vertebrates is less pronounced, but restriction enzyme and partial sequence analyses indicate also the presence of repeating sequences. On the other hand, the ntS's of some eukaryotic species display a remarkable homogeneity in size and probably in sequence. Thus, at present it is very difficult to generalize about the sequence organization of rDNA in eukaryotic ntS's and even more about their origins and significance.

In most species (e.g., *Drosophila*, *X. laevis*) the ntS sequences are confined to rDNA repeating units. However, mouse sequences homologous to ntS were found scattered throughout the genome and in particular flanking the immunoglobulin C_H (μ, α and $\gamma 2b$) genes (ARNHEIM *et al.* 1980). More recently, ntS elements were shown to be also associated with the centromeric chromatin of *Chironomus thummi* chromosomes (ISRAELEWSKI and SCHMIDT 1982). These findings raise several interesting questions on the origin and comportment of ntS sequences (IS elements?) in some species and there is little doubt that relevant information about these characteristics will be soon forthcoming.

The function of ntS's is still unknown. Obviously, signal sequences for transcription initiation and termination may flank the transcription unit boundaries or be present deeper in the ntS (see Chapter III). It should be stressed that the designation "non-transcribed" is a conditional one. Electron microscopy of spread chromosomes from amphibian oocytes demonstrates the occurrence of transcription within the ntS (FRANKE *et al.* 1978, TRENDELENBURG 1981). The identification of duplications of the putative promoter sequences in the *X. laevis* (BOSELEY *et al.* 1979, MOSS and BIRNSTIEL 1979) and *D. melanogaster* (SIMEONE *et al.* 1982) ntS indicates also that they may serve as additional transcription initiation sites or act as an RNA polymerase I trapping structure (MOSS and BIRNSTIEL 1982). The possibility that the genes for some RNAs (or proteins) are located in the ntS sequences is supported by the recent finding that some snRNA's are encoded in the ntS of the mouse repeating unit (REICHEL *et al.* 1982). The identity and role of putative coding sequences in the ntS remains to be elucidated, but it is already clear that the "non-transcribed" spacer may not be a completely silent structure. This possibility is supported also by recent findings showing that replication origins may be in the ntS sequences of *Saccharomyces* (LARIONOV and SHOUBOCHKINA 1982), *D. melanogaster* (MCKNIGHT *et al.* 1978) and *X. laevis* (BOZZONI *et al.* 1981). These observations require further study in order to determine whether replication origins are ubiquitous in eukaryotic ntS and to elucidate their structure. Whatever the case, the presence of variable and relatively short repetitive sequences makes it likely that the coding or control sequences in the ntS are rather short or are species-specific.

h) rDNA Repeating Units Possess the Capacity for Differential Replication

rRNA genes of eukaryotes possess the capacity for differential replication. Amplification of rDNA repeating units in the oocytes of many amphibians, fishes, and insects results in the formation of up to several thousand extrachromosomal rRNA gene sets. A broad variety of complex cellular mechanisms leading to amplification of rDNA repeating units operates during oogenesis in different species (see MACGREGOR 1982). Differential replication of rDNA repeating units is observed also in somatic cells, as for instance during polytenization in *Drosophila*. The capacity for differential replication seems to be related to the selective excision or replication of rDNA repeating units, but the exact molecular mechanisms by which this is accomplished remain obscure. The identification of replication origins in the ntS of different species could be related to differential replication mechanisms, but much remains to be learned about the triggering and control factors involved. It seems that in most cases a "rolling circle" mechanism of rDNA replication could account for the

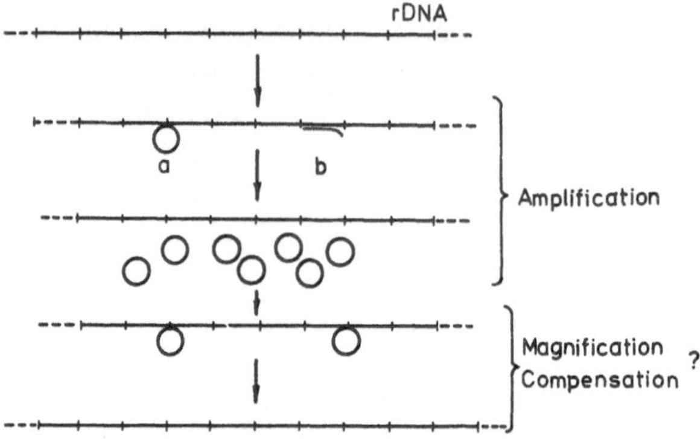

Fig. 12. Scheme of differential replication of rRNA genes. The amplification of a given rRNA gene may start by excision (*a*) or selective replication (*b*). Amplified rDNA is represented in circular form (length of circles × 2), but linear molecules may be present too. Reinsertion of rDNA circles (bottom part of figure) as a mechanism for magnification or compensation phenomena is hypothetical

generation of a large number of rDNA repeating units during amplification (see Tobler 1975, Bird 1980). The production of numerous extra-chromosomal rDNA repeating units organized as giant palindromes is another form of differential replication exploited in some lower eukaryotes (*e.g. Tetrahymena, Physarum, Stylonychia*). The fate of the extrachromosomal rDNA remains unknown. The possibility that under some conditions extrachromosomal rDNA repeating units may be reintegrated in the chromosomes of germ and somatic cells is an attractive field for further exploration.

A general view of the process of generation of extrachromosomal rDNA may be outlined (Fig. 12) and its further elucidation is likely to provide a better understanding of the dynamics of rRNA genes in eukaryotes.

i) Rectification Mechanisms Maintain the Sequence Homogeneity of rDNA Repeating Units Within Species

Ever since the discovery of the multiplicity of rRNA genes, the mechanisms maintaining the homogeneity of their sequences have posed an

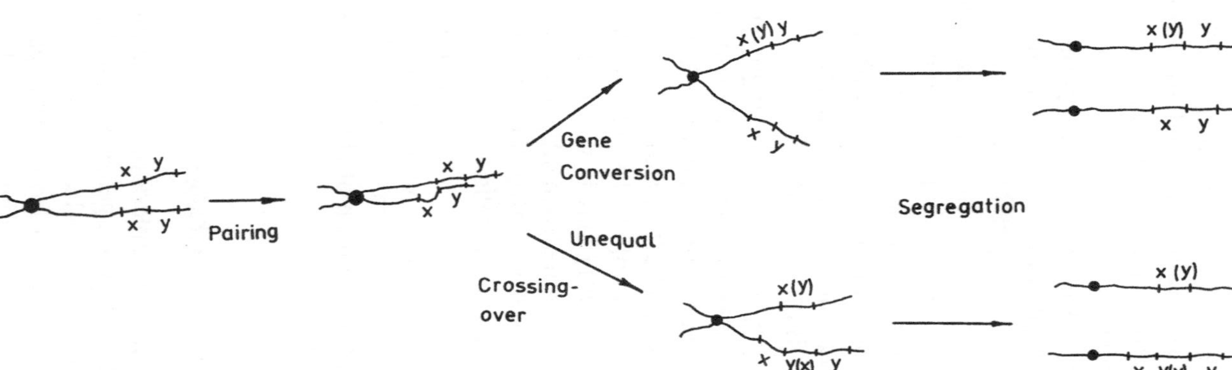

Fig. 13. Possible role of gene conversion and unequal crossing-over in the intraspecies homogenization of rDNA sequences. The scheme illustrates possible interactions between two rDNA repeating units (*x* and *y*) located in the long arm of sister chromatids. In gene conversion, repeating unit *x* interacts with *y* in such a way that part or all of its sequence becomes identical to that of gene *y*. The sequence alteration is nonreciprocal and involves only one of the partner rDNA repeating units. [According to BALTIMORE (1981)]

evolutionary puzzle. Although considerable change in size and sequence of rRNA genes has taken place during evolution of eukaryotes, the multiple copies of these genes within a species are remarkably similar. The interspecies divergence of spacer sequences is even more marked. The terms *horizontal* (WELLAUER and REEDER 1975, BROWN *et al.* 1972) or *concerted* (ARNHEIM *et al.* 1980a) *evolution* were proposed to account for these phenomena. The detailed analysis of multigene families have provided strong evidence on the mechanisms of rDNA rectification processes. The sequence homology of rDNA repeating units should permit recombination events between: (a) sister chromatids during mitosis; (b) sister and non-sister chromatids during meiosis and (c) rDNA units in non-homologous chromosomes either within or outside NOR loci. All these events are known to take place in eukaryotes. It seems that maintenance of sequence homology is achieved by two basic mechanisms: *unequal crossing-over* and *gene conversion* (Fig. 13). Both mechanisms could account for the relatively rapid horizontal spread of rRNA gene and spacer sequences within a species (see TARTOF 1975, SMITH 1976, BALTIMORE 1981).

Direct proof for the occurrence of unequal crossing-over involving the tandemly arrayed rRNA genes was obtained in elegant studies with *S. cerevisiae* that showed that the phenomenon takes place both during mitosis (WU and SZOSTAK 1980) and meiosis (PETES 1980). More recently, evidence for the occurrence of gene conversion of rRNA genes in *S. cerevisiae* was also obtained (KLEIN and PETES 1981) and it was emphasized that this rectification process may take place more frequently than unequal crossing-over. Gene conversion may be better tolerated since it can occur without altering the rRNA gene number in progeny cells. At present, the relative role played by unequal crossing-over and gene conversion as sequence rectification mechanisms in different species remains unknown. Nevertheless, both mechanisms could be conceived to operate at a sufficient rate in order to account for the rapid horizontal evolution of rDNA repeating units and maintenance of their homogeneity.

II.3. 5 S rRNA Genes

The 5 S rRNA genes are present in multiple copies and in most eukaryotes their number is different from that of rRNA genes. Detailed studies on their location within the eukaryotic genome revealed a remarkable evolutionary tendency of 5 S rRNA genes to drift apart from rRNA genes (see above). Even in the few cases (*Saccharomyces, Torulopsis utilis, Mucor racemosus, Dictyostelium discoideum*) where 5 S rRNA and rRNA genes are within the same repeating unit, they are transcribed by two distinct RNA polymerases. The separation of 5 S rRNA genes in most eukaryotes seems to have important repercussions on their number, organization, and structure. Here, I shall consider the available information on some eukaryotic 5 S rRNA genes.

Table 2. *Multiplicity of 5 S rRNA genes in some eukaryotes*[a]

Species	Number of genes per haploid genome
Physarum polycephalum	690
Neurospora crassa	100[b]
Saccharomyces cerevisiae	150
Saccharomyces carlsbergensis	200
Tetrahymena pyriformis	330/780
Drosophila hydei	320
Drosophila melanogaster	170/200/180/100/160
Ambystoma mexicanum	61,000[c]
Notophtalmus (Triturus) viridescens	300,000
Triturus cristatus carnifex	32,000[c]
Xenopus borealis	9,000
Xenopus laevis	24,000/9,000
Rattus norvegicus	830
Homo sapiens	2,000

[a] Unless indicated otherwise the data are those compiled by LONG and DAWID (1980); the original references can be found in their paper. Reproduced with permission of the authors and Annual Reviews of Biochemistry.

[b] Data from SELKER *et al.* (1981 a).

[c] Data from HILDER *et al.* (1983).

II.3.1. Number and Chromosomal Location

The number of 5 S rRNA genes is usually higher than that of rRNA genes (Table 2). Thus, the extrachromosomal 5 S rRNA genes in *Tetrahymena pyriformis* (TONNESEN *et al.* 1976, KIMMEL and GOROVSKY 1978) and *Physarum polycephalum* (HALL and BRAUN 1977) are about 3-fold more abundant than the rRNA genes. A large excess of chromosomal 5 S rRNA genes is found also in many cases. Thus, a 5–10-fold higher number of 5 S rRNA genes is found in the rat (QUINCEY and WILSON 1969) or human (HATLEN and ATTARDI 1971) genome. Even higher numbers of chromosomal 5 S rRNA genes are found in some cells of animal or plant species. Thoroughly documented examples are *Xenopus laevis* (BROWN *et al.* 1971,

Birnstiel *et al.* 1972) and *Triturus viridescens* (Pukkila 1975) and the plant *Linum usitatissimum* (Goldsbrough *et al.* 1981) where a 40- to 70-fold excess of chromosomal 5 S rRNA genes were reported. The large excess of 5 S rRNA over rRNA genes does not seem to be a general rule. Thus, the estimated number of 5 S rRNA genes in *Drosophila* varies in the range of 100 to 200 per haploid genome (Tartof and Perry 1970, Quincey 1971, Procunier and Tartof 1975, Weber and Berger 1976, Procunier and Dunn 1978) which is about equal to rRNA genes. Moreover, in the case of *Neurospora crassa* the number of 5 S rRNA genes seems to be only half that of rRNA genes (Selker *et al.* 1981 a). At present, it is difficult to judge what factors control the number of copies within the 5 S rRNA multigene family. In any case, the notion about a tendency for 5 S rRNA genes to outgrow rRNA genes seems to be an oversimplification. Since the two multigene families operate synchronously, we are intrigued by what genetic mechanisms the balance is maintained. Alternatively, it may be that the operation of two distinct RNA polymerases has made superfluous the coordinate control of gene numbers.

In almost all eukaryotes studied till now the 5 S rRNA genes are clustered within the genome and usually located on other than "nucleolar" chromosomes. The 5 S rRNA genes cluster may be on a single chromosome, as in *Drosophila melanogaster*, where it is in the band 56 F of chromosome 2 (Wimber and Steffensen 1970, Quincey 1971, Procunier and Tartof 1975), *Drosophila hydei* (Alonso and Berendes 1975), *Chironomus tentans* (Wieslander *et al.* 1975), *C. thummi* (Bäumlein and Wobus 1976) and others. However, even within the *Drosophila* genus, two or more loci are found in other species (Cohen 1976, Wimber and Wimber 1977). Also, unlike rRNA genes, the 5 S rRNA gene cluster does not display a preferential location along the chromosome. In humans, the bulk of 5 S rRNA genes is packed in chromosome 1, while smaller clusters may exist in chromosome 9 and 16 (Steffensen *et al.* 1974, Johnson *et al.* 1974). An analogous clustering of 5 S rRNA genes is observed also in some *Primates* (Henderson *et al.* 1976 b). An extreme case seems to be *Xenopus laevis*, where the 5 S rRNA gene clusters are located at the telomere regions of most, if not all, of its chromosomes (Pardue *et al.* 1973). Again, such a wide distribution of 5 S rRNA genes is not found in some other species of *Amphibia* (Pilone *et al.* 1974, Leon 1976).

Recently, the 5 S rRNA genes in *Neurospora crassa* (Free *et al.* 1979, Selker *et al.* 1981 a) and *Schizosaccharomyces pombe* (Tabata 1981) were reported to be dispersed (as single copies) throughout the genome. This situation poses several important problems concerning the evolution and function of 5 S rRNA genes, but shows also that clustering of 5 S rRNA genes (and perhaps of other multigene families) is not an absolute requirement for survival.

II.3.2. Organization and Structure

The complete sequences of fifty 5 S rRNA molecules from different eukaryotes are now known (see ERDMANN *et al.* 1983). The presence of 5 S rRNA in all living organisms permits the construction of rather precise phylogenetic trees and the elucidation of important problems of molecular evolution (*e.g.* KIMURA and OHTA 1973, SANKOFF *et al.* 1976, HORI and OSAWA 1979, KUNTZEL *et al.* 1981). An important outcome of these studies is the conclusion that the characteristic secondary structure of the 5 S rRNA molecule seems to be more stable in evolution than its primary structure (SELKER *et al.* 1981 a, DELIHAS and ANDERSEN 1982). This implies that organisms can tolerate single base substitutions (with or without G·U intermediates) until an appropriate compensatory mutation restores the base-pair involved in secondary structure maintenance. How the changed 5 S rRNA gene structure is horizontally spread remains to be determined. The possibility that compensatory mutations preserve vital secondary structure in rRNA genes as well deserves study.

The 5 S rRNA genes of *Xenopus laevis* were the first isolated (BROWN *et al.* 1971) and have been the best studied. Their organization and structure may be a useful model for eukaryotic 5 S rRNA genes in general.

The clusters of 5 S rRNA genes of *Xenopus laevis* are constituted by tandemly arrayed repeating units. Sequence analysis of 5 S rRNAs revealed that somatic cells synthesize a molecule that differs by several nucleotides from that made in oocytes (WEGNEZ *et al.* 1972, FORD and SOUTHERN 1973). Therefore, two types of differentially expressed 5 S rRNA genes should exist in *X. laevis*. These genes are designated as somatic (Xls) and oocyte (Xlo) 5 S rDNA. In addition, a minor class of oocyte 5 S rDNA (Xlt) that differs from Xlo 5 S rDNA by the smaller size of its repeating unit has been identified (BROWN *et al.* 1977). Thus, a haploid complement of *X. laevis* chromosomes contains about 24,000 5 S rRNA genes organized in three different multigene families (PETERSEON *et al.* 1980). The Xlo (22,000 copies) and Xlt (2,000 copies) 5 S rRNA genes are expressed only in oocytes, while Xls (about 400 copies) 5 S rRNA genes function only in somatic cells. Since it is known that 5 S rRNA genes are not amplified (BROWN and DAWID 1968), the large number of 5 S rRNA genes expressed only in oocytes are sufficient to match the production of rRNA's by the amplified rRNA genes. The oocyte 5 S rRNA genes are switched off during embryogenesis and remain silent in somatic cells. The Xlt 5 S rDNA repeating unit is markedly shorter and its ntS sequences do not cross-hybridize with Xlo 5 S rDNA (BROWN *et al.* 1977).

The 5 S rRNA transcription unit in *X. laevis* coincides almost exactly with the 5 S rRNA gene. Transcription starts at the 5′-terminal nucleotide of 5 S rRNA (KORN *et al.* 1979) and terminates at 2 or 3 nucleotides beyond the 3′-

end of the gene (BOGENHAGEN and BROWN 1981). The identity of the transcription unit with the 5S rRNA gene seems to be valid for all eukaryotes, since in many cases 5′-terminal triphosphates were identified in 5S rRNA (see ERDMANN *et al.* 1983) and 5S pre-rRNA molecules with only a few extra U-residues at the 3′-end were identified both *in vivo* (JACQ *et al.* 1977, RINKE and STEITZ 1982) and *in vitro* (YAMAMOTO and SEIFERT 1977, 1978). These and other results on 5S rDNA transcription will be considered in detail elsewhere (see Chapter III).

The organization of the *X. laevis* families of 5S rDNA repeating units is now known (Fig. 14) and individual repeats have been sequenced. The 5S

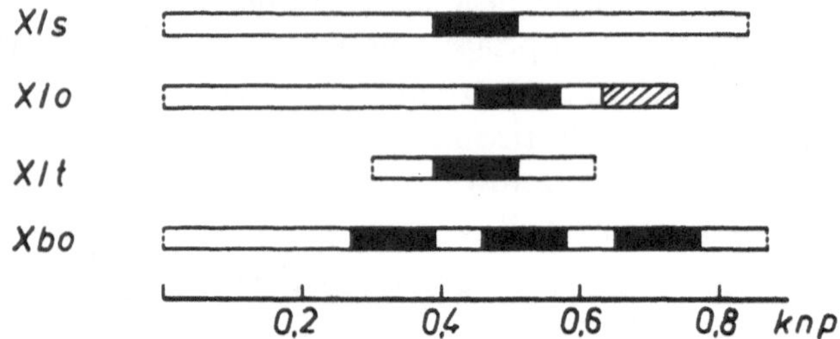

Fig. 14. Arrangement of 5 S rRNA genes (solid bar) of *Xenopus laevis* (Xl) and *Xenopus borealis* (Xb) within 5 S rDNA repeating units. *Xls* Somatic *X. laevis* 5 S rDNA; *Xlo* major oocyte *X. laevis* 5S rDNA; *Xlt* minor oocyte *X. laevis* 5 S rDNA; *Xbo X. borealis* oocyte 5 S rDNA. The hatched box represents the position of the 5 S rRNA pseudogene encountered in *X. laevis* oocyte 5 S rDNA. [According to BROWN (1982)]

rDNA repeating units are about 700 np long and the complete sequence of several individual repeats is known (FEDOROFF and BROWN 1978, MILLER *et al.* 1978, MILLER and BROWNLEE 1978). The repeating unit has two distinct regions which differ in nucleotide composition. The ntS (about 360 np) varies in length and is AT-rich, whereas the remaining part (containing the 5S rRNA gene of 121 np) has a high GC content. Much of the ntS heterogeneity is accounted for by the several fold repeated sequence: 5′-CAAAGTTTGAGTTTT-3′ (non-coding strand) encountered at the 3′-ends of the gene and the "pseudogene" (see below) and spreading to about 50 np before the beginning of the gene.

A remarkable feature of all Xlo 5 S rDNA repeating units is the presence of a "pseudogene" which has a sequence 85% homologous to that of the 5 S rRNA gene, but lacks the last 20 np (JACQ *et al.* 1977, MILLER and BROWNLEE 1978, MILLER *et al.* 1978). Transcripts of the "pseudogene" are not found *in vivo* (JACQ *et al.* 1977), but accurate transcription was observed after microinjection into oocyte nuclei (MILLER and MELTON 1981). Thus, the origin and the role of the Xlo 5 S rRNA "pseudogene" remains conjectural, but its presence, as well as that of many other "pseudogenes" is a challenge to the contemporary concepts of molecular genetics. The Xlt rDNA repeating unit is much shorter (about 350 np). It is homogeneous in size and does not show either internal repeats or "pseudogene" structures. The Xls 5 S rDNA repeating unit is about 900 np in length and sequencing of individual repeats did not reveal the presence of internal repeat or "pseudogene" structures (PETERSON *et al.* 1980). It is interesting to note that the *Xenopus borealis* repeating unit is similar in structure to that of *X. laevis*. On the other hand, the oocyte 5 S rDNA in *X. borealis* contains one or a few genes separated only by short DNA segments. This 5 S rRNA gene cluster is separated from the next by a variable in length ntS. The sequences flanking the different types of *Xenopus laevis* 5 S rRNA genes are of particular interest but it should be mentioned here that although strongly conserved within the species (MILLER and BROWNLEE 1978) they do not show any apparent similarities to the sequences flanking 5 S rRNA genes in *S. cerevisiae* (MAXAM *et al.* 1977, VALENZUELA *et al.* 1977, KRAMER *et al.* 1978) or *D. melanogaster* (TSCHUDI and PIRROTTA 1980).

The organization and structure of 5 S rDNA repeating units in other eukaryotes is less explored, but they seem to follow the pattern of *Xenopus* 5 S rDNA, being closer to the one typical of somatic cells. Thus, the 160 5 S rRNA genes of *Drosophila melanogaster* are grouped in one or two clusters containing rather homogenous 5 S rDNA repeating units of about 370 np (PROCUNIER and TARTOF 1976, HERSHEY *et al.* 1977, ARTVANIS-TSAKONAS *et al.* 1977). The ntS in *D. melanogaster* is also characterized by its high AT content (HERSHEY *et al.* 1977). The sequence of several *D. melanogaster* 5 S rDNA repeating units has been determined (TSCHUDI and PIRROTTA 1980). Unlike Xlo 5 S rDNA, the *Drosophila* repeating units do not contain "pseudogene" structures. However, a heptanucleotide (GCTGCCT, non-coding strand), located next to the putative termination site, is repeated several fold, thus creating the observed heterogeneity among the different 5 S rDNA repeating units. In this respect the 5 S rDNA repeating units in *Drosophila* and *Xenopus* are similar. The basic structure of the 5 S rDNA repeating unit seems to be followed also in plants, where the sequences of wheat (GERLACH and DYER 1980) and flax (GOLDSBROUGH *et al.* 1982) 5 S rDNA have been determined. Some length heterogeneity in flax 5 S rDNA is again due to the presence of a repeated 21 np ntS sequence.

An important observation reveals the presence in *Drosophila melanogaster* of six tRNA genes, including three $tRNA^{Glu}$ genes, adjacent to the 3'-end of the 5 S rRNA gene cluster. It is noteworthy that these genes are transcribed in the same direction as the 5 S rRNA genes (INDIK and TARTOF 1982).

A novel type of 5 S rRNA gene organization was uncovered recently in *Neurospora crassa* (SELKER *et al.* 1981 a) and *Schizosaccharomyces pombe*

(TABATA 1981, MAO *et al.* 1982). In these organisms the 5 S rRNA genes are dispersed throughout the genome. Analysis of the about 100 copies of 5 S rRNA genes in *Neurospora* revealed that most constitute solitary units. Sequencing of several 5 S rRNA genes and their flanking sequences revealed considerable divergence among some gene sequences. This divergence was markedly more pronounced when the flanking sequences were compared, although characteristic T-rich stretches were present beyond the 3′-end of most genes (SELKER *et al.* 1981 a). Also, a 5 S rRNA „pseudogene", containing the first 50 nucleotides of the gene, but flanked by totally alien sequences, was identified in *Neurospora* (SELKER *et al.* 1981 b). The results of the studies with *S. pombe* resemble closely the ones obtained with *N. crassa*. Here, the solitary 5 S rRNA genes seem to be more conservative (TABATA 1981, MAO *et al.* 1982), but this may reflect the smaller number of sequenced genes. Again, T-rich sequences are found at the 3′-end of the gene, while both 5′- and 3′-flanking sequences are heterogeneous. It is of interest that in one clone a $tRNA^{Asp}$ gene was identified at about 530 np from the 3′-end of the 5 S rRNA gene (MAO *et al.* 1982).

The results obtained in the studies on dispersed 5 S rRNA genes in *N. crassa* and *S. pombe* pose several important problems regarding their evolution, organization, and function that will not be discussed here (see SELKER *et al.* 1981 a, b, MAO *et al.* 1982). It is already clear that the absence of a tandem arrangement of clustered 5 S rRNA genes is not incompatible with the survival of the species. Whether it is an advantage or disavantage in evolution remains to be elucidated in future studies.

II.4. Ribosomal Protein Genes

Until recently the study of the estimated about 85 distinct r-protein genes seemed a frightening endeavour. Nevertheless, in the last few years, several groups succeeded in identifying and cloning individual r-protein genes (or cDNA for r-protein mRNA) from *Saccharomyces* (WOOLFORD *et al.* 1979, FRIED *et al.* 1981, BOLLEN *et al.* 1981), *Drosophila* (VASLET *et al.* 1980, FABIJANSKI and PELLEGRINI 1982), *Xenopus laevis* (BOZZONI *et al.* 1981, 1982) and the mouse (MEYUHAS and PERRY 1980). These studies provided evidence that r-protein genes are present in a varying number of copies. Thus, in *Saccharomyces* one to two r-protein genes per haploid genome are found. In *Xenopus laevis* the r-protein genes are present in two to four copies and in the mouse 7 to 20 copies are present (MONK *et al.* 1981). It is still premature to draw conclusions from these results, but it seems likely that the number of r-protein genes has increased in evolution. At present, although only a small number of r-protein genes has been identified, most appear to be unlinked. For example, 15 *Saccharomyces* r-protein genes (with a single exception) were shown to be unlinked (WOOLFORD *et al.* 1979, FRIED *et al.*

1981). There is evidence that the individual r-protein genes in yeasts are located in different chromosomes (HYMAN *et al.* 1980) and it was found also that in the mouse 5 r-protein genes are located in more than one chromosome (D'EUSTACHIO *et al.* 1981). The absence of linkage between r-protein genes is somewhat unexpected since it is known that the production of r-protein mRNAs, although monocistronic, is coordinately regulated (see below). Whether this coordination is ensured at the gene level remains to be clarified. It is intriguing that one r-protein gene in *Drosophila* is flanked by repeated sequences homologous to a region of the ntS in rDNA (FABIJANSKI and PELLEGRINI 1982), but the significance of this finding is unknown. Another interesting property is that some r-protein genes contain introns as shown in the case of *Saccharomyces* (BOLLEN *et al.* 1982) and *Xenopus laevis* (AMALDI *et al.* 1982). The complete sequence of six *X. laevis* and one *Saccharomyces* r-protein genes has been determined and the primary structure of the respective r-proteins derived (AMALDI *et al.* 1982, LEER *et al.* 1982).

The still limited information on r-protein genes in eukaryotes is rapidly expanding. Although much more remains to be learned, these studies are likely to provide crucial information on the regulatory mechanisms controlling ribosome biogenesis.

II.5. Synopsis

The information on ribosomal genes in eukaryotes is rapidly growing. Much has been learned on the organization and structure of rRNA genes, including the complete sequence of one rDNA repeating unit (*S. cerevisiae*). Knowledge about r-protein genes is still at an initial stage, but many have been cloned and their detailed characterization is anticipated in the near future. At present, the main features of the genes for rRNA and r-proteins may be outlined as follows:

1. Two distinct sets of multiple rRNA genes operate in eukaryotes: a) the rRNA genes for S-rRNA, 5.8 S rRNA and L-rRNA and b) 5 S rRNA genes. In most eukaryotes these two sets are located in distinct chromosomes. Many organisms live with 100–200 rRNA genes per haploid genome, but numerous exceptions are known and in some cases (*i.e.* higher plants) the number of rRNA genes is significantly higher. The number of 5 S rRNA genes in somatic cells is usually higher than that of rRNA genes. In some lower eukaryotes rRNA and 5 S rRNA genes are equal in number and located in the same repeating unit, although transcribed separately.

2. The rRNA genes are clustered in the genome and located at the *nucleolus organizer* loci of specific "nucleolar chromosomes". The number of NOs and "nucleolar chromosomes" varies from species to species and does not seem to be related to a distinct evolutionary advantage. The

number of rRNA genes in a given NO may be subject to variations within the population from a given species. The 5 S rRNA genes are also clustered at one or more sites in the genome. However, in some lower organisms the 5 S rRNA genes are scattered throughout the genome without apparent deleterious consequences.

3. Both rRNA and 5 S rRNA genes are always organized in *repeating units,* usually arranged in tandem in the NO cluster. The rDNA repeating units contain *nontranscribed spacer* (ntS) sequences and a *transcription unit*, containing sequences corresponding to the mature rRNA species and *transcribed spacer* (tS) sequences separating the three genes in the repeating unit. The 5 S rDNA repeating unit also contains ntS sequences but not tS sequences. Some 5 S rDNA repeating units (*X. laevis* oocytes) contain "pseudogenes" of unknown significance.

4. The ntS sequences display considerable heterogeneity in size and structure, not only among species, but even within different cell types. In many cases repeated sequences are identified in the ntS segment. Some ntS sequences are also found in the chromosomes (*i.e.* in the centromere region) outside the rRNA gene clusters. These and other observations suggest that ntS sequences may be involved in meiotic (and possibly mitotic) recombination events directing the chromosomal location and the maintenance of sequence homogeneity of the multiple rDNA repeating units. The ntS may contain putative promoter, initiation, termination, and other control signals for the transcription of rRNA genes.

5. The transcription unit of all rDNA repeating units displays a common pattern of organization. Constituent sequences follow the order: (promoter)—tS_e—S-rRNA—tS_i 1—5.8 S rRNA—tS_i 2—L-rRNA. No explanations for this order are known. The sequences corresponding to mature rRNA are strongly conserved in evolution (at least the terminal segments of S-rRNA and L-rRNA genes), while the tS_e and tS_i sequences are widely divergent. This fact indicates that considerable genetic pressure exists to maintain constant the primary structure of rRNA genes. The character and the operation of postulated gene rectification mechanisms, (*i.e.* unequal crossing-over or gene conversion), maintaining identical the multiple copies of rDNA repeating units are still hypothetical. The 5 S rRNA transcription unit coincides with the gene, but it may include a few nucleotide pairs distal of the gene.

6. Eukaryotes possess the capacity for differential replication of their rRNA genes. Amplification of rRNA genes in the oocytes of many amphibia and insects results in the formation of up to several thousand extrachromosomal rDNA repeating units. Differential replication of rRNA genes is observed also in some somatic cells as shown by magnification and polytenization phenomena in *Drosophila.* Formation of extrachromosomal rDNA repeating units and the possibility of their reintegration in the

chromosomes of germ or possibly somatic cells is an attractive field for further exploration. The production of a large number of extrachromosomal rRNA genes organized as giant palindromes is another form of differential replication exploited in some lower eukaryotes (*e.g.*, *Tetrahymena, Physarum, Stylonychia*). It is not yet known whether 5 S rRNA genes are also subject to differential replication mechanisms.

7. The genes for r-proteins are present in single (lower eukaryotes?) or multiple (mammals) copies per haploid genome. Most r-protein genes studied till now are unlinked and located in different chromosomes. Some of the r-protein genes contain introns, even in *Saccharomyces*, where introns in structural genes are rare. The sequence of some r-protein genes is known, but there is not yet detailed information on their flanking sequences.

In summary, the complex organization and structure of rRNA and 5 S rRNA genes outlined above creates conditions for their adequate replication and maximally efficient transcription. Inclusion of S-rRNA, 5.8 S rRNA, and L-rRNA genes in one transcription unit ensures—even at the gene level—synchrony in the formation of the two ribosomal particles.

III. Transcription of Ribosomal Genes

Transcription of rRNA genes is the first step in ribosome biogenesis and therefore is crucial in the understanding of the molecular and cellular aspects of ribosome production. Moreover, a specialized cell structure—the nucleolus—is built at the site of the active rRNA genes. Here, I shall consider only some characteristics of the active rRNA genes in eukaryotes directly related to the elucidation of the molecular architecture and dynamics of the nucleolus. Although not made there, the product of the 5 S rRNA genes is integrated into preribosomes in the nucleolus. The transcription of 5 S rRNA genes is understood to considerable detail (KORN and GURDON 1981, BROWN 1982) and an analysis of the relevant data will be presented. Unfortunately, the information on the transcription of r-protein genes is still in an embryonic stage. We can only hope that our knowledge on the expression of these important "household genes" will soon be expanded. At present it is perhaps justified to think that r-protein genes are transcribed and expressed according to the general rules elucidated for many eukaryotic structural genes (see DARNELL 1982).

III.1. Components of the Transcription Complex

III.1.1. RNA Polymerases

Transcription is catalyzed by DNA-dependent RNA polymerases, a subject that has been extensively reviewed (JACOB 1973, CHAMBON 1975, BISWAS *et al.* 1975, ROEDER 1976, BEEBEE and BUTTERWORTH 1980, JACOB and ROSE 1980); only some features will be considered here.

The RNA polymerase reaction requires the presence of all four ribonucleoside-5'-triphosphates as substrates. The absence of any one of the four nucleotides greatly reduces the reaction rate. However, many synthetic or natural analogues can replace the respective nucleoside-5'-triphosphate and be incorporated into the polyribonucleotide chain (LANGEN 1975, SUHADOLNIK 1979). The reaction is also dependent on the presence of DNA as template. *In vivo*, as well as under stringent conditions *in vitro*, the sequence of the polyribunucleotide product is a faithful replica of the non-coding DNA strand. Therefore, transcription is asymmetric and vectorial.

In most known cases the RNA chains synthesized *in vivo* or *in vitro* start with 5'-terminal phosphates of adenosine or guanosine, indicating that purine nucleotides are preferred at the transcription initiation site; but this requirement is not as absolute as believed earlier.

Since the pioneer studies on the subject, eukaryotic RNA polymerase has been known to be a nuclear enzyme, probably associated with the chromatin fraction (WEISS 1960). An important later breakthrough was the finding that multiple forms of RNA polymerase may be present in eukaryotic cells (WIDNELL and TATA 1964, 1966). The existence of multiple RNA polymerases was confirmed by their solubilization and fractionation (ROEDER and RUTTER 1969, 1970, BLATTI *et al.* 1970 and others), as well as by their differential sensitivity to α-amanitin (STIRPE and FIUME 1967, JACOB *et al.* 1970, KEDINGER *et al.* 1970). Three major classes of eukaryotic RNA polymerases specialized in the transcription of distinct types of genes are now recognized: I—for rRNA genes; II—for nucleoplasmic genes and III—for lower molecular weight RNA (including 5 S rRNA) genes[1]. All three eukaryotic RNA polymerases are multimer enzymes containing two high molecular weight and four to six low molecular weight subunits. It appears that the polypeptide sub-units of the three different classes of RNA polymerases are encoded by different genes, but the respective genes are not yet identified and the evidence is still circumstantial. Also, the relationship between the molecular structure of RNA polymerases and their transcription specificity remains totally unknown. Selective inhibition by α-amanitin (and related amatoxins) (FIUME and WIELAND 1970) has been and still is broadly exploited in *in vivo* and *in vitro* studies of eukaryotic RNA polymerases. RNA polymerase II is inhibited selectively by very low concentrations (10^{-8} to 10^{-9} M) of α-amanitin, which binds stoichiometrically to the enzyme molecule. RNA polymerase III is also inhibited, but at markedly higher concentrations (10^{-4} to 10^{-5} M), while RNA polymerase I is fully resistant. There is some evidence that thuringiensin (an exotoxin from *B. thuringiensis*) inhibits RNA polymerase I and II by altering the enzymes (SMUCKLER and HADJIOLOV 1972, BEEBEE *et al.* 1972).

The activity of the three distinct classes of RNA polymerases has been investigated in a broad variety of eukaryotes under different physiological and pathological conditions, but the accumulated data will not be considered here and some relevant information will be discussed later. Here, I shall review briefly some data related to the cellular mechanisms of transcription of rRNA and 5 S rRNA genes.

Considerable evidence exists that active RNA polymerase I is confined to the nucleolus. Isolated nucleoli display most of the nuclear RNA poly-

---

[1] In some studies RNA polymerases I, II, and III are designated as A, B, and C, respectively.

merase I activity (YU 1980) and provide a convenient source of the enzyme (see GRUMMT 1978, MURAMATSU *et al.* 1979). Also, the nucleolar location of RNA polymerase I was evidenced by light (MAUL and HAMILTON 1967, POGO *et al.* 1967, MOORE and RINGERTZ 1973) and electron microscopic radioautography (LAVAL *et al.* 1976, GRUCA *et al.* 1978), as well as by immunofluorescence microscopy using an anti-RNA polymerase I antibody (SCHEER *et al.* 1983). Some evidence was obtained that in regenerating rat liver RNA polymerase I is associated with the fibrillar component of nucleoli (GRUCA *et al.* 1978). The intranuclear sites of RNA polymerase III are still unknown, although its loose association with nuclei would indicate a nucleoplasmic location (DUCEMAN and JACOB 1980). RNA polymerases are known to exist in template-bound and free forms (YU 1975). In rat liver about 20% of the total RNA polymerase I is in the free form and presumably remains associated with the nucleolus (YU 1980). Further elucidation of the intranuclear site of free RNA polymerase I and III seems of particular interest.

The number of RNA polymerase molecules per nucleus is also important to know in order to fully understand some of the mechanisms controlling transcription of rRNA and 5 S rRNA genes. Determination of this number in animal cells was first attempted by taking advantage of the known stoichiometry of binding of α-amanitin to RNA polymerase II and a figure of 4.6×10^4 enzyme molecules per diploid genome was obtained (COCHET-MEILHAC *et al.* 1974). Biochemical estimates of the number of transcribing RNA polymerase II molecules provided figures of $6.5–7 \times 10^3$ in liver (WEAVER *et al.* 1971, COUPAR *et al.* 1978) and 1×10^4 in chick oviduct (COX 1976) and KB cells (SUGDEN and KELLER 1973). Because the amounts of the three RNA polymerases per cell are roughly equal (YU 1980), the data for form II are informative about forms I and III as well. In the careful study of COUPAR *et al.* (1978) the number of active molecules RNA polymerase I in normal rat liver was estimated to be 2×10^4 per diploid genome. Corrected for the presence of free enzyme (see YU 1980) this will yield a figure of 2.5×10^4 RNA polymerase I molecules per diploid genome. Pending more direct estimates, *e.g.* by titration with specific antibodies (*cf.* BUHLER *et al.* 1980, HUET *et al.* 1982 a, b), this figure seems to be a sufficiently reliable landmark for liver cells and probably for other cells in mammals. Finally, it should be stressed that in a broad variety of organisms and experimental conditions studied till now, a correlation between the rates of *in vivo* pre-rRNA synthesis and RNA polymerase I acitivity has been shown (see BEEBEE and BUTTERWORTH 1980).

In summary: the specialization of eukaryotic RNA polymerases in the transcription of different types of genes is a remarkable achievement in evolution. The molecular mechanisms for discriminating among gene templates remain unknown, but many properties of RNA polymerases have been elucidated, permitting at least a semi-quantitative assessment of their participation in transcription.

III.1.2. Nucleolar rDNA and r-Chromatin

All genomic DNA (including rDNA) in interphase chromatin and mitotic chromosomes is complexed with histone and nonhistone proteins. We assume, therefore, that chromatin (rather than "naked" DNA) is the actual physical structure involved in eukaryotic transcription complexes. This assumption remains to be proved however, since electron microscopic studies suggest that active nucleolar chromatin from *Xenopus laevis* oocytes is indistinguishable from "naked" DNA fibers (LABHART and KOLLER 1982). Whatever the case (see below), isolation and analysis of the chromatin containing rRNA genes (r-chromatin) is of particular importance for understanding the transcription process.

The known characteristics of rDNA repeating units of eukaryotes were outlined in the preceding Chapter. It should be emphasized here that most rDNA repeating units are clustered within the nucleolus organizer. In interphase cells the nucleolus is formed at the site of the nucleolus organizer and therefore the active rRNA genes are a major constituent of the nucleolus. Yet, in isolated nucleoli only 20–30% of the chromatin is truly intranucleolar and a still smaller part (about 2%) of the DNA in the preparation is in the form of rDNA[2] (BACHELLERIE *et al.* 1977a). The studies on the isolation and structure of r-chromatin are still at an initial stage. Different methods yielding intranucleolar fractions that are substantially enriched in transcriptionally active rDNA have been described (BACHELLERIE *et al.* 1977a, b, BOMBIK *et al.* 1977, FABER *et al.* 1981) and that has opened the way for further analytical and functional studies on r-chromatin. However, the fractions enriched in nucleolar r-chromatin still contain a large amount of non-rDNA chromatin, making clear-cut conclusions unlikely.

The isolation of practically pure r-chromatin (containing more than 90–95% rDNA) was achieved by the use of cells containing amplified extrachromosomal rDNA: *Xenopus laevis* oocytes (HIGASHINAKAGAWA *et al.* 1977), *Tetrahymena pyriformis* (LEER *et al.* 1976, MATHIS and GOROVSKY 1976, JONES 1978a) and *Physarum polycephalum* (GRAINGER and OGLE 1978). The isolated r-chromatin contains histones and a large number of ribosomal and other non-histone proteins (HIGASHINAKAGAWA *et al.* 1977, JONES 1978b). A detailed analysis of *Tetrahymena* r-chromatin shows (JONES 1978b) that it contains histones H 2b, H 3, H 4, and Hx (equivalent

---

[2] Earlier figures usually are underestimations by a factor of 5 (mammalian cells) because it was not recognized that the rDNA repeating unit is in the range of 40 kb, not the 8 kb that is the sum of 18 S + 28 S rRNA genes.

to H 2 a) in a ratio similar to that found in total chromatin. Histone H 1 is also present in r-chromatin, but it may be less extensively phosphorylated than bulk chromatin. Also the amount of different modified histones is comparable in bulk and r-chromatin. The main distinguishing feature of *Tetrahymena* r-chromatin seems to be a histone/DNA ratio only about 40% of that found in total chromatin. Because of the difficulties inherent in chromatin isolation techniques, more reliable studies are required before we can be confident about generalizations on the composition of r-chromatin. At present, it appears that by and large the main constituents of bulk and r-chromatin are similar.

III.2. The Transcription Process

III.2.1. Topology of Primary Pre-rRNA

All eukaryotes are characterized by the presence in the nucleolus of a pool of large, primary pre-rRNA molecules containing S-rRNA, 5.8 S rRNA and L-rRNA sequences. These large pre-rRNA molecules are the first to be labeled following administration of radioactive precursors and therefore are considered to be the initial transcripts of rRNA genes (see below). Since the primary pre-rRNA molecules are a faithful replica of the non-coding strand of rRNA transcription units, the arrangement of tS and rRNA sequences in pre-rRNA reflects their topology in the respective transcription unit. The introduction of an elegant technique for the direct electron microscopic visualization of double-stranded (GC-rich) loops in pre-rRNA molecules, partly denatured with formamide-urea (WELLAUER and DAWID 1973), permitted ascertainment of the location of the L-rRNA and S-rRNA segments within pre-rRNA of different higher eukaryotes (WELLAUER and DAWID 1975, SCHIBLER *et al.* 1975, see HADJIOLOV and NIKOLAEV 1976). Precise transcription mapping of pre-rRNA within cloned rDNA repeating units extended these observations to a broad variety of eukaryotic cells. Considerable variation in the size of primary pre-rRNA, ranging from molecular weights of $1.9–2.1 \times 10^6$ in *Acetabularia* (SPRING *et al.* 1976) or 2.3×10^6 in *Sciara coprophila* (RENKAWITZ *et al.* 1979) to about 4.7×10^6 in mammalian cells (see WELLAUER and DAWID 1975, HADJIOLOV and NIKOLAEV 1976) exist among species. Aside from these large variations in size, a general pattern in the topology of primary pre-rRNA molecules (and rRNA transcription units) has remained stable in evolution (HADJIOLOV 1977) and no exceptions to this regularity have yet been uncovered (see Fig. 9).

Unlike primary transcripts of structural genes, the role of intron sequences in conferring size variations to primary pre-rRNA molecules is of limited significance. In fact, primary pre-rRNA molecules containing introns (of about 400 nucleotides) were identified by R-loop mapping only in some species of *Tetrahymena* (DIN *et al.* 1979, CECH and RIO 1979). Although intron sequences may be identified in pre-rRNA of other eukaryotes this phenomenon seems not to have been favored in the evolution of eukaryotic rRNA genes.

Identification of the polarity of primary pre-rRNA molecules was a tedious and controversial problem for a long time (see HADJIOLOV and NIKOLAEV 1976). The new techniques of gene analysis provided direct and conclusive evidence (REEDER *et al.* 1976, DAWID and WELLAUER 1976) that the following 5'- to 3'-polarity of primary pre-rRNA is typical of eukaryotes:

5'—tS_e—S-rRNA—tS_i 1—5.8 S rRNA—tS_i 2—L-rRNA—3'.

This polarity is similar to that of primary pre-rRNA in prokaryotes (see NOMURA *et al.* 1977), showing that the pattern of rRNA gene transcription sequence has remained stable in evolution.

III.2.2. Morphology of Transcribed rRNA Genes

Introduction of the chromosome-spreading technique by MILLER and his co-workers (MILLER and BEATTY 1969 a, b, see MILLER and HAMKALO 1972 a, b, FRANKE *et al.* 1979, TRENDELENBURG 1983) allowed the direct visualization of the transcription process. The original studies were carried out with amplified rRNA genes of *Triturus viridescens* and other amphibian oocytes. Since then extensive studies were performed with many different cells and organisms, but in general the best results are obtained with amplified extrachromosomal rRNA genes. Thus, actively transcribed rRNA genes are investigated in considerable detail in the oocytes of such amphibia as *Triturus* (MILLER and BEATTY 1969 a, b, SCHEER *et al.* 1973), *Rana pipiens* (TRENDELENBURG and MCKINNELL 1979), *Xenopus laevis* (MILLER and BEATTY 1969 a, b, SCHEER *et al.* 1977), *Pleurodeles waltlii* (ANGELIER and LACROIX 1975) and others. The rRNA genes in the germ or embryonic cells of numerous insects like *Dytiscus marginalis* and *Achaeta domesticus* (TRENDELENBURG 1974, TRENDELENBURG *et al.* 1973, 1976), *Drosophila* (MEYER and HENNIG 1974, GLÄTZER 1975, LAIRD and CHOOI 1976, MCKNIGHT and MILLER 1976), *Oncopeltes fasciatus* (FOE *et al.* 1976, FOE 1978) and many others also provided fruitful material for extensive chromosome-spreading studies.

Successful visualization of active rRNA genes was achieved also in many lower eukaryotes such as the green alga *Acetabularia, Chlamydomonas,* and others (TRENDELENBURG *et al.* 1974, SPRING *et al.* 1974, 1976, BERGER and SCHWEIGER 1975 a, b, WOODCOCK *et al.* 1975, BERGER *et al.* 1978) and the slime mold *Physarum polycephalum* (GRAINGER and OGLE 1978).

Application of the chromosome-spreading technique to the analysis of mammalian cells is technically more difficult and only limited information is presently available. Nevertheless, active rRNA genes have been observed in HeLa (MILLER and BAKKEN 1972) and CHO (PUVION-DUTILLEUL *et al.* 1977 a) cells, in mouse spermatids (KIERSZENBAUM and TRES 1975), and in rat hepatocytes (PUVION-DUTILLEUL *et al.* 1977 b), thus allowing us to deduce a general morphological picture of the transcription process in eukaryotes.

The general characteristics of transcribed rRNA genes may be summarized as follows:

a) Active transcription units are visualized as comprising axes of rDNA (r-chromatin) covered with densely packed lateral fibrils forming a gradient of increasing length and resulting in a typical "Christmas tree" pattern, the apexes of which coincide with the transcription initiation regions. The active transcription units are designated also *matrix units* by some authors. In most intra- or extrachromosomal rRNA genes studied, transcriptional units with identical polarity are arranged in tandem on the rDNA axes and separated by fibril-free spacers. Up to 120–150 matrix units have been identified on a single axis (*cf.* SPRING *et al.* 1976, 1978). Normally, the bulk (80–90%) of the active transcription units are of standard size, corresponding to the size of the primary pre-rRNA typical for a given eukaryote. For example, a matrix unit of 2.1 μm in *Acetabularia mediterranea* corresponds to a molecular weight for its pre-rRNA of about $1.7–2.1 \times 10^6$ (SPRING *et al.* 1976); the matrix unit in *Xenopus laevis* is 2.2–2.6 μm and the molecular weight of its pre-rRNA is in the range of $2.1–2.6 \times 10^6$ (SCHEER *et al.* 1977); in CHO cells matrix units of up to 5.2 μm correspond to a primary pre-rRNA of about 4.5×10^6 (PUVION-DUTILLEUL *et al.* 1977 a). Thus, at a first approximation the length of transcribed rDNA in matrix units is only slightly shorter than the calculated length for B-form DNA (MILLER and HAMKALO 1972 a, b; see below).

The size of spacer intercepts and the rDNA repeating units in which they reside, show greater variations, not only between different organisms, but even between units on the same rDNA axis (see FRANKE *et al.* 1976, 1977, RUNGGER and CRIPPA 1977). This observation reflects, at least partly, the length heterogeneity of ntS's determined by other techniques.

A common polarity of tandem rRNA transcription units is not found in all eukaryotes. An intriguing alternating polarity of matrix units has been observed in *Acetabularia exigua* (BERGER *et al.* 1978). Also, two matrix units of opposite polarity, starting from the central part of the linear extrachromosomal rDNA, have been observed in *Physarum* (GRAINGER and OGLE 1978) and *Tetrahymena* (CECH and RIO 1979).

b) The lateral fibrils of matrix units, appear to be pre-rRNA chains growing as transcription progresses and this was proven to be the case by electron microscopic autoradiography (ANGELIER *et al.* 1979). In all cases studied thus far the growing lateral fibrils are 5–10-fold shorter than expected if they reflect pre-rRNA chains. This observation, combined with enzymatic and specific staining studies, led to the important conclusion that the growing pre-rRNA chains are already coated with proteins (see MILLER and HAMKALO 1972a, b) and electron microscopic studies with specific antibodies provided direct evidence that structural r-proteins are selectively bound to the growing pre-rRNA chains (CHOOI 1976, CHOOI and LEIBY 1981). A constant length increment of growing pre-rRNA fibrils is observed in some organisms. In most cases, terminal granules (up to 30 nm in diameter) are formed at the free ends of the growing fibrils (*cf.* FRANKE *et al.* 1976, 1979, ANGELIER *et al.* 1979), suggesting that the tightest RNA-protein packing occurs at the 5′-end of the pre-RNA chain.

c) The growing pre-rRNP fibrils are bound to the rDNA axes by 12–14 nm diameter granules identified as RNA polymerase I (see MILLER and HAMKALO 1972a, b, FRANKE *et al.* 1976, 1979). Transcription of rRNA genes seems to proceed normally at a very high efficiency since a close to maximal packing of RNA polymerase along the rDNA axis is usually found. In a broad variety of organisms the observed maximal packing ratios are in the range of 40–50 enzyme molecules per μm axis, which indicates that 50–70% of the rDNA template is covered. These findings suggest that in fully active rRNA genes spatial hindrance by adjacent RNA polymerase molecules may be an important limiting factor in the transcription rate. The observed regular spacing of RNA polymerase along the rDNA axis suggests that in fully active rRNA genes, equal rates of initiation, elongation and termination of transcription are maintained.

d) Several experimental findings provide evidence that each rRNA transcription unit has separate promoter and termination signals. Detailed investigations of the morphology of ntS segments showed that in most cases the observed beaded structure in the spacer axis results from changes in chromatin conformation rather than from the presence of non-transcribing polymerase molecules (FRANKE *et al.* 1976, 1979, SCHEER 1978, PRUITT and GRAINGER 1981). Studies on the activation or inactivation of pre-rRNA synthesis in *Triturus alpestris* oocytes also provided evidence for the existence of independent promoters for each rRNA transcription unit (SCHEER *et al.* 1975, 1976).

Direct proof for the existence of an independent promoter for each rRNA transcription unit was obtained upon analysis of the transcription of bacterial plasmids, containing *X. laevis* rDNA, after injection into oocyte nuclei (TRENDELENBURG and GURDON 1978). Faithful transcription of rDNA was visualized in some of the injected plasmids, practically indistinguishable from the Christmas-tree pattern typical for the transription of endogenous rDNA (Fig. 15). These results demonstrate that a short ntS segment (about 2.5 knp) of the rDNA repeating unit contains enough information to initiate correctly transcription by RNA polymerase

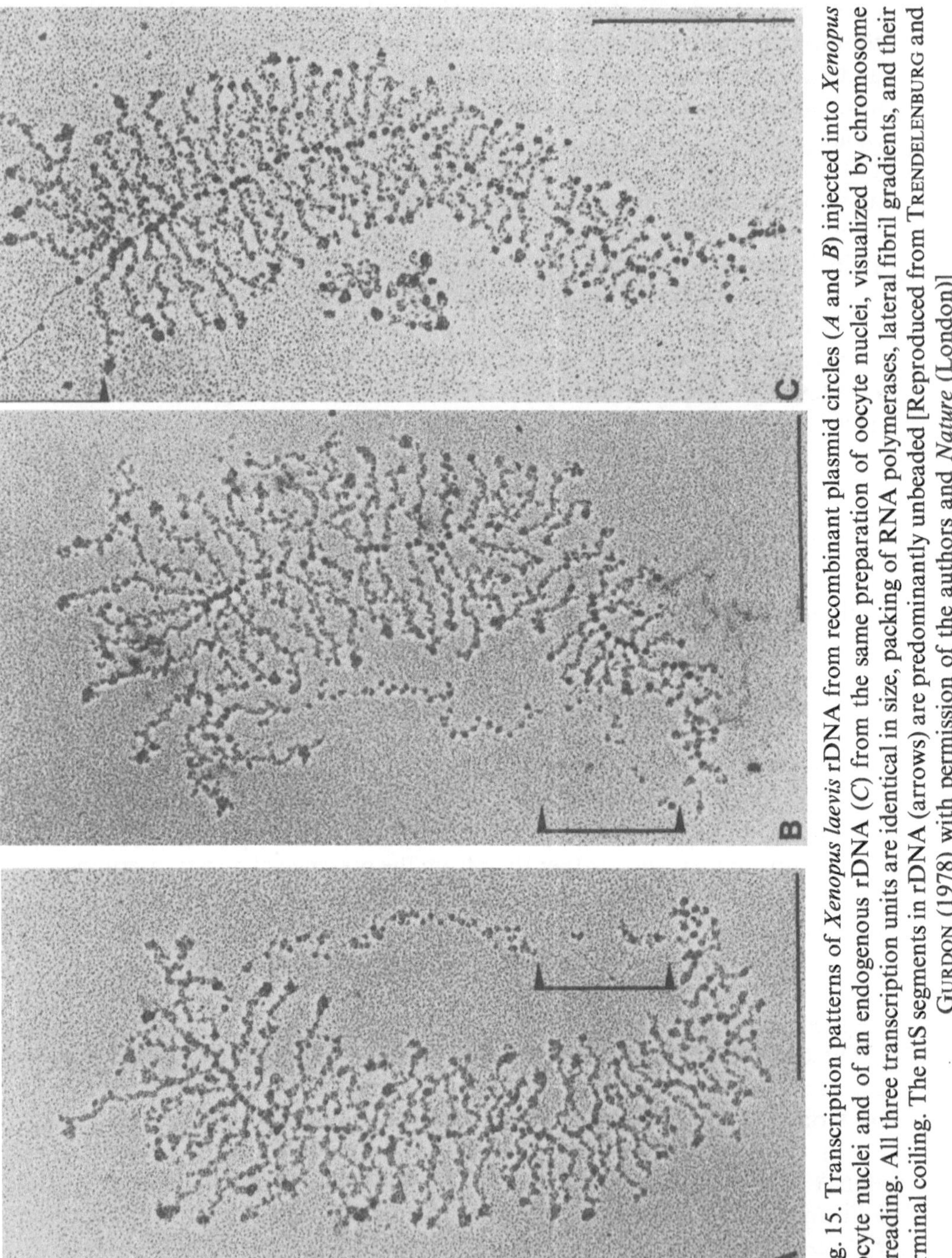

Fig. 15. Transcription patterns of *Xenopus laevis* rDNA from recombinant plasmid circles (*A* and *B*) injected into *Xenopus* oocyte nuclei and of an endogenous rDNA (*C*) from the same preparation of oocyte nuclei, visualized by chromosome spreading. All three transcription units are identical in size, packing of RNA polymerases, lateral fibril gradients, and their terminal coiling. The ntS segments in rDNA (arrows) are predominantly unbeaded [Reproduced from TRENDELENBURG and GURDON (1978) with permission of the authors and *Nature* (London)]

I (see Moss 1982) and to attain a maximal packing of enzyme molecules and pre-rRNA fibrils along the rDNA axis. Injection of amplified *Dytiscus marginals* circular rDNA into *X. laevis* oocytes nuclei displayed a similar pattern of transcription (TRENDELENBURG *et al.* 1978), thus showing that transcription start signals operate in heterologous systems as well.

The results outlined above clearly demonstrate the successful visualization of the basic steps in the transcription of rDNA, *i.e.*, initiation, elongation, and termination. It is evident that the major control elements reside in the interactions among the components of the ternary transcription complex: rDNA (r-chromatin)—RNA polymerase I—growing pre-rRNP fibrils. In this respect, the role of rDNA or r-chromatin structure seems to be the least understood and it will be considered next in more detail.

III.2.3. Transcribed and Non-Transcribed r-Chromatin

The discovery of nucleosome and higher order organization of chromatin caused a huge expansion of studies on the molecular structure of the eukaryotic genome. As a result we can anticipate that some important clues to regulatory mechanisms involved in gene expression will emerge from studies on the structure of transcriptionally active r-chromatin. The basic findings on chromatin in general have been expertly reviewed (MATHIS *et al.* 1980, WEISBROD 1982), so that I shall limit myself to the available information on r-chromatin.

The nucleosome organization of inactive rDNA repeating units, including both transcription units and ntS segments is now firmly established (SCHEER 1978, FOE 1978). The "beads on a string" image of non-transcribed r-chromatin is apparently the same as in bulk chromatin, suggesting a similar structure of nucleosomes. A study on rDNA circles in *Dytiscus marginalis* oocytes provides a direct approach to the structure of inactive r-chromatin (SCHEER and ZENTGRAF 1978); inactive rDNA organized in nucleosomes displays a packing ratio (μm B-form DNA/μm chromatin) of 2.2–2.4[3]. Further compaction of the nucleosome chain into supranucleosomal globules (diameter 21–34 nm, packing ratio = 11) is also found in inactive rDNA circles. Supranucleosomal structures with a packing ratio of above 20 are observed also in ntS regions of inactive chromatin in *Xenopus laevis* oocytes (PRUITT and GRAINGER 1981). Typical nucleosome organization of non-transcribed r-chromatin, with rDNA packing ratios of 2.0–2.3 is seen also in *Drosophila melanogaster* (MCKNIGHT *et al.* 1978) and *Oncopeltes fasciatus* (FOE 1978) embryos. A packing ratio of 2.1 is obtained

---

[3] The linear density of B-form DNA (0.337 nm between adjacent base pairs) is 1 μm = 1.96×10^6 Da = 2,976 np. However, electron microscopy reveals the linear density of double-stranded DNA to be 1 μm = 2.08×10^6 Da = 3,142 np (STÜBER and BUJARD 1977). This implies a "packing ratio" for DNA (related to B-form DNA) of 1.06.

in inactive chromatin circles formed upon injection of rDNA containing plasmids into oocyte nuclei (TRENDELENBURG and GURDON 1978). Together with data on the composition of isolated r-chromatin, the above findings indicate that the chromatin containing non-transcribed rDNA is organized in nucleosomes and supranucleosomal structures similar to those found in bulk chromatin (WEISBROD 1982). Whether some quantitative or qualitative differences exist in the histone and nonhistone complement of nucleosomes from inactive ribosomal and bulk chromatin cannot be decided at present. Recent observations that a conserved histone variant (hv 1, from the H2a-like group) is enriched in *Tetrahymena* macronuclei and in nucleoli of mammalian cells (ALLIS *et al.* 1982) and that nonhistone proteins HMG-T 1 and HMG-T 2 (analogous to HMG-1 and HMG-2) are concentrated in nucleolar areas of trout cells (BHULLAR *et al.* 1981) provide attractive hints for future studies on r-chromatin constituents.

The DNA in nucleosomes surrounds a histone core and both strands are fixed by strong ionic interactions. Therefore, transcription requires unfolding of nucleosomes. Observations of transcribed rRNA genes clearly show that this is the case of r-chromatin. Measurements of the length of active transcription units yield a rDNA packing ratio of only 1.1 to 1.4 (see MCKNIGHT *et al.* 1978, FOE 1978, FRANKE *et al.* 1976, 1978). Such a low packing ratio requires an almost complete unfolding of the nucleosome (see RICHARDS *et al.* 1978). Because transcribed rRNA genes are covered by RNA polymerases, the observed shortening of rDNA could in fact be caused by local polymerase-induced unwinding (MILLER and HAMKALO 1972 b, MCKNIGHT *et al.* 1978). Therefore, one may ask: what is the actual template in transcription, naked rDNA or r-chromatin? A conclusive answer is not yet possible. Several observations support the view that transcribed rDNA remains associated with histones and nonhistone proteins. These include: 1. the thickness of the active r-chromatin axes is 40–70 Å (rather than the 20 Å of B-form DNA) and r-chromatin stains with phosphotungstic acid, a reagent for basic proteins (FOE *et al.* 1976, FOE 1978); 2. digestion with micrococcal nuclease of r-chromatin containing presumably fully active rRNA genes in *Tetrahymena* (MATHIS and GOROVSKY 1976, 1978), *Physarum polycephalum* (JOHNSON *et al.* 1978 a, b, BUTLER *et al.* 1978) or *Xenopus laevis* (REEVES 1978 a, b) yields nucleosome-size (about 140 np) rDNA fragments. This indicates that active rDNA is still associated with histones in a spatial arrangement similar to that in bulk chromatin; 3. complete unfolding of bulk chromatin nucleosomes does not result in the removal of core histones (ZAYETZ *et al.* 1981); 4. Antibodies to histone H2a are shown to interact with the rDNA axes in matrix units (REEDER *et al.* 1978); 5. Ag-NOR proteins (nonhistone acidic proteins) are found associated with the rDNA axes of *Pleurodeles* matrix units (ANGELIER *et al.* 1982).

It should be pointed out that the above evidence is not conclusive and alternative interpretations are not ruled out. Thus, the results of an electron microscopic study of nucleolar chromatin from *X. laevis* oocytes prepared under varying ionic conditions led to the conclusion that transcribed rRNA genes are depleted of histones and probably most nonhistone proteins (LABHART and KOLLER 1982). Also, possible artifacts are not excluded in some of the experiments outlined above. For example, after micrococcal nuclease digestion nucleosome-sized rDNA fragments were obtained from the ntS, but not from rRNA transcription units of *Tetrahymena* r-chromatin (BORCHSENIUS *et al.* 1981). These and similar findings leave open the possibility that "naked" rDNA is present within actively transcribed rRNA genes.

If RNA polymerase I transcribes along a fully unfolded r-chromatin axis, the histones may remain attached by their highly basic N-terminal segments to the sugar-phosphate backbone of the DNA chains (WEINTRAUB *et al.* 1976. RICHARDS *et al.* 1978, ZAYETZ *et al.* 1981). That transcription of histone-coated DNA is still possible is demonstrated by model experiments with prokaryotic (WILLIAMSON and FELSENFELD 1978, WASYLYK *et al.* 1979) and eukaryotic (KARAGYOZOV *et al.* 1978; WASYLYK and CHAMBON 1979) RNA polymerases.

The mechanisms causing unfolding to whatever degree of nucleosomes and suptranucleosomal structures during activation of r-chromatin remain unknown, although several possibilities could be envisaged (see MATHIS *et al.* 1980, WEISBROD 1982). It is now clear that many probes for active chromatin (*e.g.* DNase I sensitivity) do, in fact, reflect a *potentially active* state of a given gene rather than its actual involvement in transcription. This is clearly demonstrated in the case of extrachromosomal rRNA genes of *Physarum polycephalum*: both fully transcribed and mitotic (inactive) rRNA genes are similarly digested by DNAse I (STADLER *et al.* 1978). Consequently, the extrachromosomal rRNA genes in *Physarum* may be considered to be in an *unlocked* state, independent of whether they are actually transcribed or not. The *locked* and *unlocked* states of genes within chromatin is certainly an attractive area for further exploration. Conceivably unlocking of rRNA genes may be related to decondensation and unfolding of r-chromatin. Indeed, an important observation in studies on the morphology of rRNA transcription units is that unfolding of r-chromatin is not directly coupled with transcription. Thus, fully extended r-chromatin fibers are observed at developmental or experimental stages preceding or following transcription of rRNA genes (FOE *et al.* 1976, LAIRD *et al.* 1976, FOE 1978, SCHEER 1978, FRANKE *et al.* 1979, DERENZINI *et al.* 1983). Unlocked r-chromatin is conceivably more sensitive to nucleases and other disruptive treatments as has been shown to be the case in several studies on the state of rRNA genes in a variety of eukaryotic cells and organisms (REEVES and JONES 1976, STALDER *et al.* 1978, MATHIS and GOROVSKY 1978, HIGASHI *et al.* 1978, GIRI and GOROVSKY 1980, BIRD *et al.* 1981 a, CHIU *et al.* 1982). More recently, the presence of DNase I hypersensitive sites within or adjacent to actively transcribed genes has been demonstrated in different experimental systems (see ELGIN 1981, WEISBROD

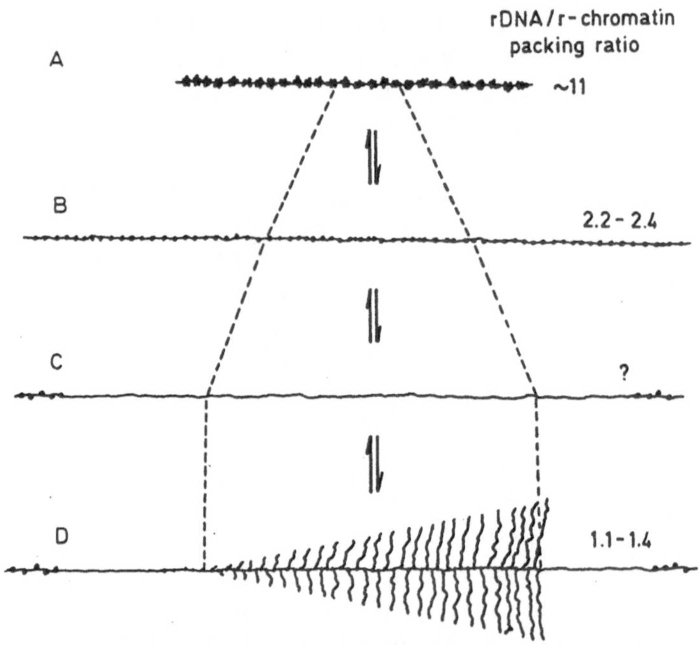

Fig. 16. Successive stages in the activation of r-chromatin. The broken lines indicate the changes in the length of r-chromatin from a single rRNA transcription unit. The packing ratio of rDNA in r-chromatin is approximate, based on data obtained with different cells and organisms. The lettering (*A* to *D*) denotes the different states of r-chromatin as described in the text

1982). Such sites were identified in extrachromosomal r-chromatin of *Tetrahymena pyriformis* (BORCHSENIUS *et al.* 1981), mouse liver (BIRD *et al.* 1981 b) and *X. laevis* × *X. borealis* hybrids (MACLEOD and BIRD 1982). That DNase I hypersensitivity is indeed related to *in vivo* transcribed rRNA genes is clearly demonstrated in the studies with *Xenopus* hybrids. It was shown that only the *X. laevis* rRNA genes, which are predominantly expressed in the hybrids (HONJO and REEDER 1973, CASSIDY and BLACKLER 1974), are hypersensitive to DNase I. These studies showed also that rRNA gene expression and DNase I hypersensitivity are not related to rDNA hypomethylation, unlike the case of mouse liver rRNA genes, where a positive correlation was found (BIRD *et al.* 1981 b). Thus it is still premature to generalize that the active state of rRNA genes is related to hypomethylation of rDNA (BIRD and SOUTHERN 1978). Also, the role of histone and/or nonhistone proteins in locking-unlocking of r-chromatin remains to be explored further.

In conclusion, non-transcribed r-chromatin appears to be tightly packed in nucleosomal and supranucleosomal structures (Fig. 16). Activation of rRNA genes may be tentatively envisaged as proceeding in several consecutive stages. The initial stage involves unfolding of supranucleosomal globules or other structures (*A*) followed by the unfolding of nucleosomes (*B*). It is likely that unlocking of rRNA genes is related to disruption of the higher order structure of r-chromatin, rather than to nucleosome unfolding. Fully or partly unfolded r-chromatin is committed for transcription, but still non-transcribed (*C*). At this stage, its protein constituents may be altered resulting in a higher sensitivity to nucleases or other agents. At a next stage, interactions of RNA polymerase I with promoter sites results in the switching on of transcription (*D*). The process prior to *D* may be related to the creation of DNase I hypersensitive sites within or adjacent to rRNA transcription units. The overall process is reversible and supplies the cell with versatile mechanisms for transcriptional control of ribosome formation. However, the locked-unlocked state of r-chromatin may be more stably fixed during the life cycle of the cell and related to the organization of the genome (see HANCOCK 1982, HANCOCK and BOULIKAS 1982).

III.2.4. Primary Transcripts and Primary Pre-rRNA

As mentioned above, formation of mature rRNA in all eukaryotes starts from a discrete pool of large pre-rRNA molecules containing both S-rRNA and L-rRNA sequences. Thus, *primary pre-rRNA* may be defined as the largest pre-rRNA species forming a distinct pool in the nucleolus. The *primary rRNA transcript* may be defined as the RNA chain corresponding to the full length of the rRNA transcription unit. It is possible that the primary transcript and the primary pre-rRNA are not identical, *i.e.*, the primary transcript may contain sequences removed *before* the accumulation of primary pre-rRNA. That processing takes place before the end of transcription has been shown for bacteria. In *E. coli* separate precursors to L-rRNA and S-rRNA are normally cleaved from the growing pre-rRNA chain. An intact 30 S pre-rRNA (containing S-rRNA, L-rRNA, and 5 S rRNA segments) can be found only in RNase III$^-$ mutants (see NIKOLAEV *et al.* 1975). Short-term labeling in eukaryotes invariably shows (Fig. 17) that initially more than 90% of the label is in primary pre-rRNA, with no detectable labeling of free S-rRNA. Therefore, we conclude that primary pre-rRNA is not normally processed before transcription termination (HADJIOLOV *et al.* 1978).

An exception to this rule may occur in *Dictyostelium discoideum* in which chromosome-spreading reveals unusual matrix units that suggest that processing may take place during transcription (GRAINGER and MAIZELS 1980). Pending direct biochemical evidence, the absence of processing of pre-rRNA during transcription may be considered as a general property of

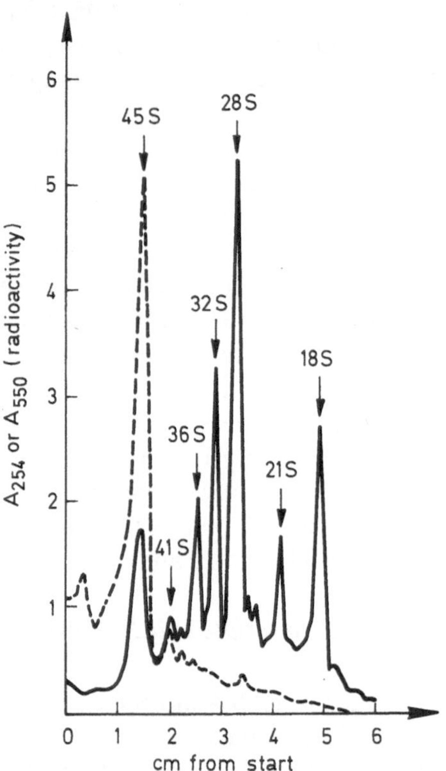

Fig. 17. Short-term *in vivo* labeling of rat liver nucleolar RNA. The animals are injected intraperitoneally with 50 μGi per rat of [^{14}C]orotate. The *in vivo* labeling is for 10 min. Nucleolar RNA is isolated and fractionated by urea/agar gel electrophoresis. ———, absorbance at 260 nm; ----, densitometric tracing at 550 nm of the radioautogram. Aisde from the small gradient of labeled, but unfinished pre-rRNA molecules, the bulk of the label is in 45 S pre-rRNA. [From K. V. HADJIOLOVA, unpublished results)]

eukaryotes. If the primary transcript is shown to be processed before the formation of primary pre-rRNA, three possibilities may be envisaged: 1. partial or complete removal of 5′-end tS_e sequences; 2. removal of 3′-end sequences located beyond the L-rRNA segment and 3. splicing of intron sequences transcribed from intron$^+$ rRNA genes.

Until now the only transcribed intron$^+$ genes that have been found are in some strains of *Tetrahymena*. Two types of a 35 S primary pre-rRNA are identified by R-loop mapping: with and without intron sequences (DIN *et al.* 1979, CECH and RIO 1979). These results suggest that splicing of the intron sequence takes place shortly after or during transcription, but in any case *before* further processing of the primary pre-rRNA molecule. The sequences at the intron/exon junctions have nothing in common with the respective consensus sequences in structural genes and therefore require distinct splicing mechanisms (WILD and SOMMER 1980). Recently, the intriguing observation was made that splicing of *Tetrahymena* primary pre-rRNA may be autocatalytic, *i.e.*, the precise intron excision, covalent cyclization of the intron, and ligation of the two exons may result from a splicing activity residing solely in the structure of the RNA (KRUGER *et al.* 1982, CECH *et al.* 1983). Since most rRNA genes and primary pre-rRNA do not contain introns, this exotic activity of rRNA does not seem to be related to the problem discussed here.

Introduction of the chromosome-spreading technique permitted a direct approach to probing the identity between primary transcript and primary pre-rRNA. Since the time of the initial studies (see MILLER and HAMKALO 1972a) many investigators have compared the length of rRNA transcription units and the size of primary pre-rRNA in various cells and organisms. In most cases the size of primary pre-rRNA corresponds to the length of the transcription unit, assuming that rDNA is B-form. That this is not a coincidence is shown by the extrachromosomal genes of *Dytiscus marginalis* oocytes, where the contour length of isolated circular rDNA coincides with that of actively transcribed circles (TRENDELENBURG *et al.* 1976). However, at the molecular level important limitations exist. These include: 1. the growing transcript is coated with proteins and direct measurement of its length is impossible; 2. B-form DNA appears to be 1.1–1.4-fold compacted, but more precise estimates of packing ratios are impossible; 3. short transcripts (up to 300 nucleotides) may not be visualized and 4. errors in the range of $0.1–0.2 \times 10^6$ are possible in estimates on the molecular mass of the rather large primary pre-rRNA molecules. These limitations leave open the possibility that the primary transcript is longer than primary pre-rRNA.

Biochemical analyses of short-term labeled RNA reveal in many cases some heterogeneity of primary pre-rRNA and the existence of molecules larger than the bulk of primary pre-rRNA (see LOENING 1975, HADJIOLOV and NIKOLAEV 1976). For example, three distinct primary pre-rRNA peaks were identified in a mouse cell line (TIOLLAIS *et al.* 1971), two of them differing by about 700 nucleotides (GALIBERT *et al.* 1975). Unexpectedly large pre-rRNA molecules were found in cultured *Acer pseudoplatanus* L. cells (MIASSOD *et al.* 1973, COX and TURNOCK 1973). Pronounced heterogeneity of primary pre-rRNA was found also in rat liver, assigned to differences in their tS_e segment (DABEVA *et al.* 1976). The heterogeneity of primary pre-rRNA may be correlated with the length heterogeneity of rRNA transcription units frequently observed in spread chromosome preparations (see MILLER and BEATTY 1969 a, SCHEER *et al.* 1973, FRANKE *et al.* 1976). Thus, a detailed analysis on *Xenopus laevis* oocytes (SCHEER *et al.* 1973, 1977) showed that the bulk of rRNA transcription units are homogeneous (2.2–2.6 μm), but longer ones (3.4–4.2 μm) are consistently found. Since the size of the rRNA gene region is constant, the observed heterogeneity could reflect: 1. initiation at different sites and variations in the tS_e length or 2. processing of part of the tS_e sequence during transcription.

Distinguishing between these two possibilities has been the subject of considerable controversy (discussed in LOENING 1875, FRANKE *et al.* 1976, HADJIOLOV and NIKOLAEV 1976, RUNGGER and CRIPPA 1977) and will not be considered in detail here.

Direct biochemical proof for the absence of 5′-end processing may be provided by the identification of 5′-terminal triphosphates (pppN) in primary pre-rRNA. The initial attempts to do this with different animal cells yielded negative results (SLACK and LOENING 1974, KOMINAMI and MURAMATSU 1977, NAZAR 1977). However, evidence for the presence of 5′-terminal pppAp was obtained by showing that primary pre-rRNA from *Xenopus laevis* is a capping enzyme substrate (REEDER *et al.* 1977). Later, a similar approach identified pppGp at the 5′-end of primary pre-rRNA from mouse (BACH *et al.* 1981 b) and rat (FINANCSEK *et al.* 1982 b) tissues. Direct demonstration of 5′-terminal triphosphates in the primary pre-rRNA from several eukaryotes was also achieved. Thus, pppAp and pppGp in *Saccharomyces cerevisiae* (HADJIOLOV *et al.* 1978, NIKOLAEV *et al.* 1979), pppAp in *Tetrahymena pyriformis* (NILES 1978) and pppAp in *Drosophila melanogaster* (LEVIS and PENMAN 1978) primary pre-rRNA have been found.

The identification of 5′-terminal triphosphates in primary pre-rRNA shows that in most eukaryotes processing of 5′-end tS_e sequences *during* transcription does not take place. These results show also that processing of the primary rDNA transcript cannot start normally before the transcription of its L-rRNA segment is completed (HADJIOLOV *et al.* 1978). Accordingly, the observed heterogeneity of primary pre-rRNA is explainable by the presence of multiple transcription initiation sites rather than by 5′-end processing during transcription.

The matter has been studied in more detail in *Xenopus laevis*. It was shown that the putative promoter site sequences are reduplicated within the ntS (MOSS and BIRNSTIEL 1979, BOSELEY *et al.* 1979). Transcription within ntS of some rDNA repeating units was observed previously by chromosome-spreading (SCHEER *et al.* 1973, FRANKE *et al.* 1976) or RNA-DNA hybridization (RUNGGER *et al.* 1979). In a recent chromosome-spreading

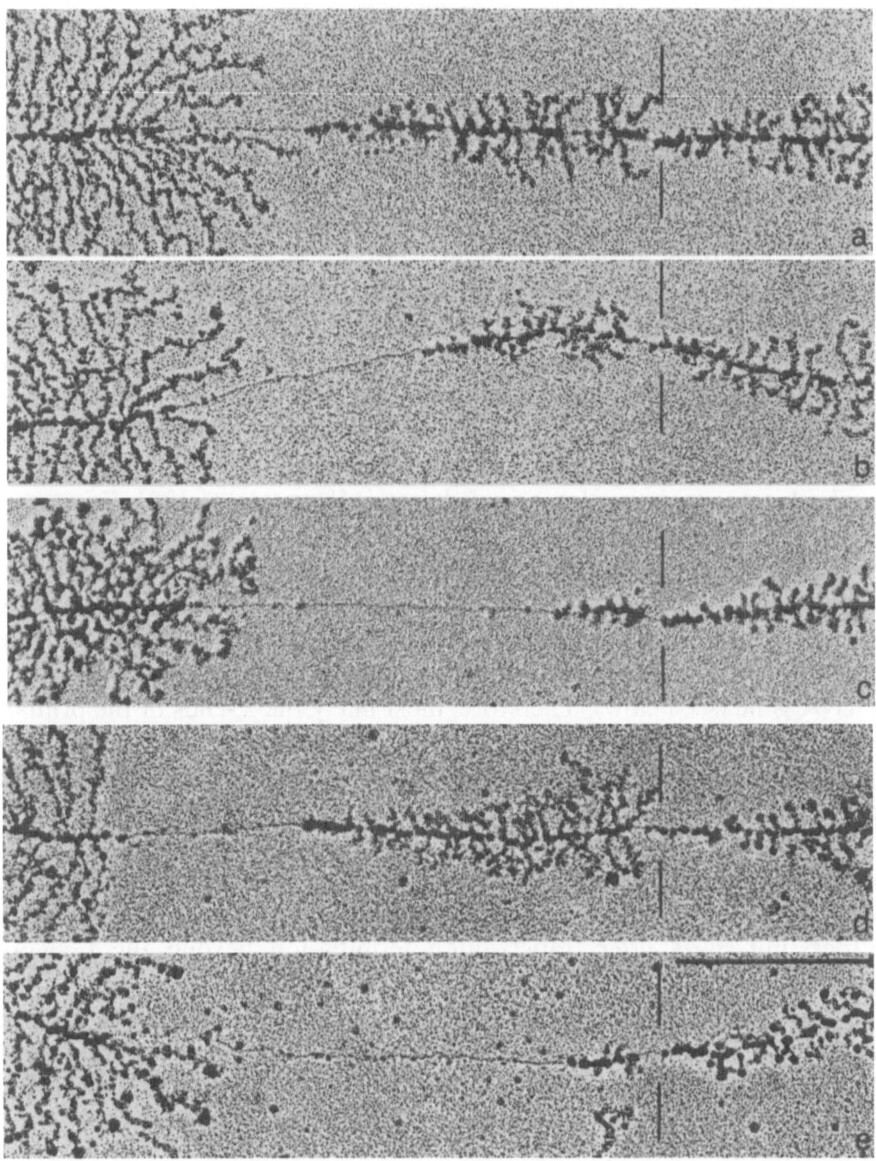

Fig. 18. Electron micrographs of spread preparations from *Xenopus laevis* oocyte chromatin showing different patterns of transcription in ntS. The most frequent patterns observed are shown in *a* and *d*. Start regions of *X. laevis* 40 S pre-rRNA transcription are denoted by vertical lines. Note the absence of RNA polymerase particles and nucleosomes in the non-transcribed part of ntS. Transcription of ntS terminates in front of the 40 S pre-rRNA transcription start site. [From TRENDELENBURG (1981). Reproduced with permission of the author and *Biology of the Cell*)]

study, ntS transcription was studied precisely (Fig. 18) and it was shown that: a) this is a relatively rare event, the "true" promoter being prefered; b) initiation within ntS occurs at positions containing reduplicated promoter sequences and c) most ntS transcripts terminate close to the site of the "true" promoter (TRENDELENBURG 1981). These findings with *Xenopus laevis* support the view that length heterogeneity among primary pre-rRNA molecules may be due to spurious initiation within ntS and failure to terminate in front of the favored intiation site. Such phenomena could be enhanced by treatments altering RNA processing, as exemplified by the action of 5-fluorouridine (RUNGGER *et al.* 1978). Technical problems should not be underestimated. Thus, it was shown that length heterogeneity among transcripts studied by the chromosome-spreading technique may arise by local stretching forces and be misleading (GLÄTZER 1980). On the other hand, the 5′-end tS_e segment in primary pre-rRNA may be less protected and subject to easier nuclease degradation (MISHIMA *et al.* 1981, ROTHBLUM *et al.* 1982) thus contributing to primary pre-rRNA heterogeneity.

The possibility that primary rDNA transcripts contain some 3′-end tS_e sequences (beyond the L-rRNA segment) is more difficult to assess. Most studies on transcription mapping of primary pre-rRNA show that termination takes place at or a few nucleotides beyond the 3′-end of the L-rRNA gene (see below). Therefore, most likely the 3′-ends of the primary transcript and pre-rRNA coincide. This of course does not rule out the possibility that some oversize transcripts may be generated by occasional read-through events due to faulty transcription termination.

In summary, presently available evidence permits us to conclude that:

a) The primary transcript and primary pre-rRNA are identical and therefore processing *during* transcription does not normally take place. Processing starts *after* transcription termination beyond the L-rRNA segment.

b) Occasional heterogeneity of primary pre-rRNA may be due to spurious initiation within ntS, faulty termination, or degradation of short tS_e segments. These phenomena may be species specific and reflect peculiarities in the primary structure of rRNA transcription units.

III.2.5. Transcription Initiation and Termination

Recent techniques of cloning, sequencing, and transcription mapping of rDNA repeating units has led to the location of transcription initiation and termination sites and resulted in a better understanding of these processes.

III.2.5.1. Initiation

By definition transcription initiation takes place at the ntS/tS_e junction. The precise mapping of initiation sites depends critically on the identification of the primary transcript or at least of its 5'-end. As pointed out above, this has been achieved mainly by direct or indirect identification of 5'-terminal triphosphates in primary pre-rRNA. Accordingly, transcription initiation sites were mapped within the rDNA of *Saccharomyces cerevisiae* (KLEMENZ and GEIDUSHEK 1980, BAYEV *et al.* 1980), *Tetrahymena pyriformis* (NILES *et al.* 1981 a, SAIGA *et al.* 1982), *Dictyostelium discoideum* (HOSHIKAWA *et al.* 1983), *Drosophila melanogaster* (LONG *et al.* 1981 a), *Xenopus laevis* (SOLLNER-WEBB and REEDER 1979, BOSELEY *et al.* 1979, MOSS and BIRNSTIEL 1979), *X. clivii* and *X. borealis* (BACH *et al.* 1981 a) and three mammalian species, including mouse (MILLER and SOLLNER-WEBB 1981, GRUMMT 1981 a, MISHIMA *et al.* 1981), rat (FINANCSEK *et al.* 1982 a, ROTHBLUM *et al.* 1982, HARRINGTON and CHIKARAISHI 1983) and man (FINANCSEK *et al.* 1982 b, MIESFELD and ARNHEIM 1982). It is noteworthy that due to difficulties in the isolation of intact primary pre-rRNA from mammalian species, unambiguous identification of the transcription initiation site was achieved only after correlation with results obtained by *in vitro* transcription. The sequences around some presently identified transcription initiation sites are presented in Fig. 19. It seems that *in vivo* the bulk of primary rDNA transcripts starts at *one* defined rDNA sequence. Additional minor initiation sites were identified in yeast (BAYEV *et al.* 1980) and *Tetrahymena* (NILES *et al.* 1981 a), located at 30 and 21 np respectively upstream from the major one. This finding suggests the recognition of transcription initiation sequences by RNA polymerase I. However, comparison of the known sequences does not reveal any *common features* either in the preceeding ntS region or in the tSe segment, a situation markedly distinct from the case of eukaryotic structural genes, where fairly conserved sequences serving as RNA polymerase II promoter signals are found (see BREATHNACH and CHAMBON 1981). Apparently, the rate of sequence divergence in the initiation site domain does not differ markedly from that of ntS and tS_e in general. This pattern is clearly seen when the respective sequences in three *Xenopus* species are compared (BACH *et al.* 1981 a). Comparison of the initiation domain sequences of three mammalian species reveals the presence of a conserved 19 np sequence immediately after the initiation point (FINANCSEK *et al.* 1982 b), which might be implicated in transcription start control. Again, this sequence has little in common with the ones in other eukaryotes. These observations suggest three possibilities: a) the promoter of RNA polymerase I is species specific and each enzyme recognizes (directly or through initiation control proteins) its own promoter sequence; b) conserved initiation signal sequences are far away (more than ± 200 np) from the transcription start; or c) the promoter signal is not related to rDNA sequence, but depends on the specific conformation of r-chromatin within the initiation region.

The putative promoter domain is AT-rich in most cases and several rather

Species	Sequence (−40 … +20; ↓ = transcription start site)
Sacch. cerevisiae	TCATGGAGTACAAGTGTGAGGAAAAGTAGTTGGGAGGTACTTCATGCGAAAGCAGTTGAAGACAAGTTC
Drosophila melanogaster	TGAGGACAGCGGGTTCAAAAACTACTATAGGTAGGCAGTGGTTGCCGACCTCGC
Xenopus borealis	CACGGGGCCTCCCCTCCCCCGGGTACTGCTCCGGCAGGAAGGTAGGGACTGAGGTACTAATCACCCTG
Xenopus laevis	CGCCTCCATGCTACGCTTTTTTGGCATGTGCGGGCAGGAAGGTAGGGGAAGACCGGCCCTCGGCGCGAC
Mus musculus	TTGTGATCTTTTCTATCTGTTCCTATTGGACCTGGAGATAGGTACTGACACGCTGTCCTTTCCCTATT
Rattus norvaegicus	ACTTTTCATCTTTGCTATCTGTCCTATTGTACTGGAGATATATGCTGACACGCTGTCCTTTGACTCTTT
Homo sapiens	ATCTTTCGCTCCGAGTCGGCAATTTTGGGCCGCCGGGTTATATGCTGACACGCTGTCCTCTGGCGACCT

Fig. 19. Sequence of rDNA at the ntS/tS_e junction of different eukaryotes. The vertical arrow indicates the transcription start site. Homologous sequences are boxed. Data for rDNA sequence and transcription start site: *S. cerevisiae* (Bayev *et al.* 1980); *D. melanogaster* (Long *et al.* 1981 a); *X. laevis* (Sollner-Webb and Reeder 1979); *X. borealis* (Bach *et al.* 1981 a); *Mus musculus* (Grummt 1981 a; Mishima *et al.* 1981), *Rattus norvegicus* (Financsek *et al.* 1982 a); *Homo sapiens* (Financsek *et al.* 1982 b)

long (> 6 np) stretches of T's (non-coding strand) are encountered in the ntS of yeasts (BAYEV *et al.* 1980), *Tetrahymena* (NILES *et al.* 1981 a), *Dictyostelium* (HOSHIKAWA *et al.* 1983) and some other species, while similar T-stretches are also abundant in the tS_e of yeasts (BAYEV *et al.* 1980), rat (ROTHBLUM *et al.* 1982), mouse (URANO *et al.* 1980, BACH *et al.* 1981 b) and other eukaryotes, but not in the *X. laevis* tS_e (MOSS *et al.* 1980, MADEN *et al.* 1982). Whether some of these T-stretches play a role as initiation signals remains to be proven. It is plausible that some may be involved in RNA polymerase I binding, since in the case of *S. cerevisiae,* for example, it was shown that the enzyme binds (GABRIELSEN and ØYEN 1982) and initiates *in vitro* (SAWADOGO *et al.* 1981) far upstream in the ntS, probably within regions characterized by the presence of very long T-stretches.

Further insight on RNA polymerase I promoter sequences was obtained by studies on transcription in *X. laevis* oocytes with different plasmids carrying ntS/tS_e junction sequences (BAKKEN *et al.* 1982, MOSS 1982). The active promoter was found to encompass a — 145 to + 16 np segment of the rDNA sequence surrounding the transcription initiation site. The precise boundaries of promoter sequences remain to be specified, but the major control elements clearly are located in the ntS, about 50–100 np upstream from the transcription initiation site. Similar studies with mouse rDNA provided evidence that a principal initiation control region is located at — 39 to — 12 np upstream from the initiation site (GRUMMT 1982). Several authors obtained results showing that rDNA transcription is species specific (GRUMMT 1981 b, GRUMMT *et al.* 1982, MIESFELD and ARNHEIM 1982, but see WILKINSON and SOLLNER-WEBB 1982) and it may be expected that distinct promoter sequences will be recognized by the RNA polymerase I complex. That additional factors, besides rDNA sequences, may be important is strongly suggested by results showing that the same ntS/tS_e sequences are present in transcribed (intron$^-$) and non-transcribed (intron$^+$) *Drosophila* rDNA (LONG *et al.* 1981 a). That r-chromatin structure plays a role in transcription initiation is supported by nuclease digestion studies on isolated yeast nuclei; the results show that r-chromatin in the ntS, near the initiation site, has a structure distinct from that within the transcription unit (LOHR 1983).

Thus, the identity of RNA polymerase I promoter signals requires further specification. It is likely that the signal sequences are located upstream in the ntS and that species specific protein factors are involved in the precise recognition of transcription initiation sites.

III.2.5.2. Termination

At present transcription termination sites are precisely mapped for *Saccharomyces carlsbergensis* (VELDMAN *et al.* 1980 b), *S. cerevisiae* (BAYEV *et al.* 1981), *Tetrahymena thermophila* (DIN *et al.* 1982), *T. pyriformis* (NILES

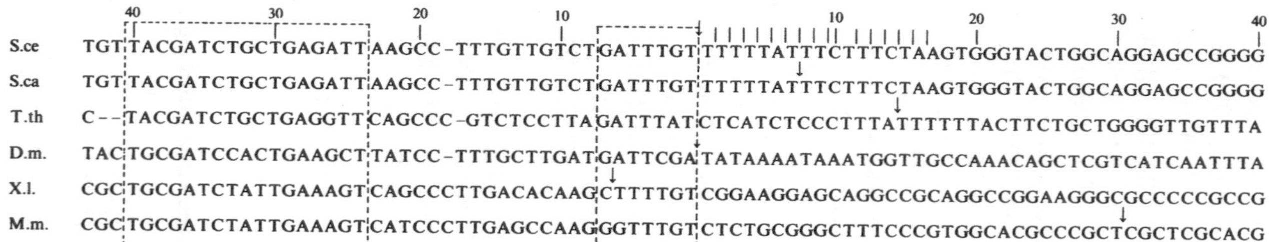

Fig. 20. Sequence of rDNA at the transcription termination site (the L-rRNA/ntS junction). Homologous sequences at the 3′-termini of L-rRNA are boxed with dotted lines. The vertical arrows indicate the position of identified transcription termination sites. The vertical lines above the *S.ce.* sequence indicate minor termination sites in this organism. *S.ce. Saccharomyces cerevisiae* (Bayev *et al.* 1981); *S.ca. Saccharomyces carlsbergensis* (Veldman *et al.* 1980b); *T.th. Tetrahymena thermophila* (Din *et al.* 1982); *D.m. Drosophila melanogaster* (Mandal and Dawid 1981); *X.l. Xenopus laevis* (Sollner-Webb and Reeder 1979); *M.m. Mus musculus* (Kominami *et al.* 1982)

et al. 1981 b), *Drosophila melanogaster* (MANDAL and DAWID 1981), *Xenopus laevis* (SOLLNER-WEBB and REEDER 1979) and the mouse (KOMINAMI *et al.* 1982). Data about other eukaryotes will soon be forthcoming, but even at this stage (Fig. 20) several important conclusions may be outlined: 1. the 3′-end segment of L-rRNA is conserved including the terminal heptanucleotide (GAUUUGU$_{OH}$). In contrast, wide sequence divergence is found in the adjacent ntS; 2. In several cases (*S. cerevisiae, Drosophila, X. laevis*) transcription terminates at the 3′-end of L-rRNA. In others, termination takes place a few nucleotides within the ntS (up to 30 in the mouse). *Xenopus laevis* termination seems to take place at 6 nucleotides within the L-rRNA 3′-end conserved sequence, a finding which deserves verification by direct L-rRNA sequencing; 3. no common structural features may be uncovered in the termination site adjacent ntS. In *Saccharomyces* and *Tetrahymena* T-stretches flank the putative termination site, while in mouse a short inverted repeat is present.

In looking for transcription termination signals, it seems reasonable to think that they reside primarily in the structure of L-rRNA. Double-stranded structures involving the 3′-end of L-rRNA may be constructed either by L-rRNA folding (*e.g.* VELDMAN *et al.* 1981 b) or by its interaction with 5.8 S rRNA (*e.g.* KELLY and COX 1981, VELDMAN *et al.* 1981 a, GEORGIEV *et al.* 1981, 1983). In both cases the emerging double-stranded structure could be a suitable target for a selective endonuclease cleavage. An interesting case is provided by termination in *S. cerevisiae* (BAYEV *et al.* 1981). Here, most primary pre-rRNA molecules seem to terminate at the 3′-end of L-rRNA. However, some longer transcripts are also found, covering a T-stretch of up to 15 np. Thus, it seems that the termination signal is provided by the secondary structure at the 3′-end of L-rRNA and the transcript is released by endonuclease action. Occasional read-through transcripts (bearing unpaired 3′-end U-residues) may be trimmed by exonuclease or endonuclease action. A similar trimming mechanism could be envisaged for longer transcripts encountered in other eukaryotes. That the primary pre-rRNA transcript is detached by an endonuclease cleavage rather than by release of RNA polymerase I is shown by an electron microscopic analysis of transcription termination in *Xenopus laevis* (TRENDELENBURG 1982). It is clearly demonstrated that RNA polymerase I remains attached to template r-chromatin after release of the transcript, the enzyme molecules covering 150–200 np in the ntS. The mechanism of the subsequent RNA polymerase I release remains unknown, but it is interesting to note that rather long T-stretches are found within this region of the *Saccharomyces* (BAYEV *et al.* 1981) and *Tetrahymena* (NILES *et al.* 1981 b, DIN *et al.* 1982) ntS.

In conclusion, it seems that the major termination signal resides in the specific 3′-end structure of L-rRNA. Termination appears to take place by an endonuclease cleavage of the transcript (at U-residues in most cases), rather than by RNA polymerase I release, the latter event taking place at some distance in the ntS. Trimming of possible extra sequences in primary pre-rRNA appears to take place soon after transcription termination.

III.2.6. Transcription in vitro

Analysis of the transcription of rRNA genes in isolated nuclei or nucleoli, with endogenous template and enzyme, has been studied extensively under a variety of experimental conditions. Evidence for selective transcription of rRNA genes in such *in vitro* systems was obtained by characterization of the reaction product by nucleotide composition, nearest-neighbour frequency, finger printing, methylation of nucleotides, and hybridization to rDNA (see references in HADJIOLOV and MILCHEV 1974, GRUMMT 1978). Yet, proof for the fidelity of transcription, *i.e.*, for correct initiation, elongation and termination, is not easily achieved in such systems. Analysis of the size of the product labeled *in vitro* revealed the presence of a gradient of labeled RNA molecules with a maximal size (in mammalian systems) of about 45 S, corresponding to the *in vivo* primary pre-rRNA (YOUNGER and GELBOIN 1970, ZYLBER and PENMAN 1971, HADJIOLOV and MILCHEV 1974, UDVARDY and SEIFART 1976, MATSUI *et al.* 1977, ONISHI and MURAMATSU 1978, GRUMMT 1978). However, time-course analyses show that the reaction levels-off rapidly (*i.e.* REEDER and ROEDER 1972, MARZLUFF *et al.* 1973, HADJIOLOV and MILCHEV 1974, UDVARDY and SEIFART 1976, COUPAR *et al.* 1978). Also, addition of initiation inhibitors like heparin (HADJIOLOV and MILCHEV 1974, FERENCZ and SEIFART 1975, COUPAR and CHESTERTON 1977) or Sarkosyl[4] (GREEN *et al.* 1975; SAMAL *et al.* 1978, BALLAL *et al.* 1980) do not alter appreciably RNA polymerase I activity. All these results indicate that mainly elongation (and possibly termination) of preexisting rDNA transcripts takes place in isolated nuclei and nucleoli, while initiation and reinitiation are virtually abolished.

The relative stability of rDNA transcript elongation in isolated nuclei and nucleoli is demonstrated by observations showing that removal of more than 90% of the nuclear or nucleolar proteins by Sarkosyl (GREEN *et al.* 1975) does not destroy transcription complexes (SCHEER 1978) and when the latter are isolated they still retain most of their RNA polymerase I activity (SAMAL *et al.* 1978, BALLAL *et al.* 1980, KARAGYOZOV and HADJIOLOV 1981). However, evaluation of the RNA elongation rate in isolated rat liver nuclei shows that it is about 0.15 nucleotides/sec (COUPAR *et al.* 1978), which is about 80-fold lower than the estimated rates of *in vivo* rRNA synthesis

---

[4] Sarkosyl is a trade name for N-dodecylsarcosine, Na salt [CH_3-$(CH_2)_{10}$-CO-N(CH_3)-CH_2-CO-ONa], a product of Ciba-Geigy, Basel, Switzerland.

(KARAGYOZOV *et al.* 1980). Although estimated *in vitro* elongation rates may be artifactually lowered, these considerations indicate that, while markedly more stable than initiation, elongation in isolated nuclei is at least quantitatively altered and estimated rates seem to be 1–2 orders of magnitude lower than the ones attained *in vivo*. Whether these observations reflect the need for a continuous supply of cytoplasmic factor(s) or the effects of transcription complexes damage during isolation of nuclei and nucleoli remains to be clarified and these experimental systems need to be perfected further before they may be considered to represent a faithful model of the situation *in vivo*. Recently, it was reported that RNA polymerase I activity in premeabilized HeLa cells is 4-fold higher than in isolated nuclei (WYDRO *et al.* 1980), offering an attractive alternative model for studying transcription *in vitro*.

As pointed out above, initiation and reinitiation of rDNA transcription in isolated nuclei is an extremely vulnerable process, yet this process does not seem to be completely abolished during isolation of nuclei and nucleoli. Several authors have reported α-amanitin-resistant incorporation of γ-labeled nucleoside-5′-triphosphates into RNA of isolated nuclei (BUSIELO and DI GIROLAMO 1975, SUN *et al.* 1979, GROSS and RINGLER 1979, SAIGA and HIGASHINAKAGAWA 1979), which suggests that some transcription initiation does take place *in vitro*. However, it has been known for some time that RNA polymerase I can initiate at nicks, ends, or denaturated regions of the DNA template, thereby acting as pseudopromoters. Faithful initiation of transcription, therefore, can be proven only by a precise sequence analysis of the *in vitro* transcript. Such studies have been undertaken recently and the results have shown that accurate rDNA transcription initiation *in vitro* does occur. Accurate transcription of rRNA genes was observed after microinjection into *Xenopus* oocyte nuclei of plasmids carrying *Xenopus laevis* rDNA (TRENDELENBURG and GURDON 1978) or of amplified *Dytiscus* rDNA circles (TRENDELENBURG *et al.* 1978). Other studies with this *in vitro* system demonstrated that accurate *in vitro* initiation of transcription occurs at the *in vivo* site (BAKKEN *et al.* 1982, MOSS 1982). While precise transcription of exogenous rDNA has now been convincingly proven, the *in vitro* systems are not yet quantitatively satisfactory since only one molecule of pre-rRNA per 2–5 molecules of injected rDNA seems to be synthesized (SOLLNER-WEBB and MCKNIGHT 1982, MOSS 1982).

Others also have described cell-free systems capable of accurate initiation of transcription on endogenous rDNA. Homogenates of manually isolated *X. laevis* nuclei have been shown to sustain initiation and reinitiation by endogenous RNA polymerase I on internal DNA templates and to elongate the transcripts at a rate of 2–5 nucleotides/sec (HIPSKIND and REEDER 1980). Electron microscopic analysis confirmed that in the cell-free extract apparently correct initiation takes place at about one fifth of the *in vivo* frequency. Evidence was also obtained that *in vitro* elongation of pre-initiated transcripts proceeds at a rate of 2.2 nucleotides/sec (McKnight *et al.* 1980). With additional improvement of the *Xenopus* cell-free system a homogenate of *X. borealis* oocyte nuclei was found to initiate transcription of exongenous *X. laevis* cloned rDNA at the *in vivo* initiation site (Wilkinson and Sollner-Webb 1982). It was shown also that about one pre-rRNA molecule is made per 50 molecules of input rDNA. Several authors have described also cell-free systems from mouse (Miller and Sollner-Webb 1981, Grummt 1981a, Mishima *et al.* 1981) and rat (Rothblum *et al.* 1982) tumor cells capable of accurate transcription initiation on added rDNA containing plasmids. These cell-free systems are all analogous to the RNA polymerase II system described previously (Weil *et al.* 1979). Essentially, all contain the 100,000 × g supernatant from a total cell lysate and rely on the presence of released endogenous RNA polymerase I. Although these systems are still poorly characterized enzymologically, their capacity for accurate transcription initiation is convincingly proven. Thus, we can anticipate that additional information on mechanisms controlling transcription initiation will soon be forthcoming.

Analysis of *in vitro* transcription termination on rDNA has received much less attention than initiation. Isolated *Tetrahymena* nucleoli have been shown to terminate transcription at the correct site if a protein factor is added (Leer *et al.* 1979).

Generally, the study of systems capable of accurate *in vitro* transcription of rRNA genes by RNA polymerase I is lagging behind the advances realized with RNA polymerase II and III cell-free systems. Nevertheless, the gap is rapidly narrowing and both correct initiation and termination have been demonstrated in some *in vitro* systems. For the time being efficient RNA polymerase I systems are still rather complex, suggesting the contribution of numerous additional factors. The ultimate goal of accurate transcription of rDNA (or r-chromatin) by purified RNA polymerase I is still remote, but the required tools are now available and the crucial experiments appear to be perfectly feasible.

III.3. Transcription of 5 S rRNA Genes

In most eukaryotes the clusters of 5 S rRNA and rRNA genes are unlinked. However, the contribution of both gene sets is necessary for the production of ribosomes. Therefore, elucidation of mechanisms controlling the expression of 5 S rRNA genes will help us understand some facets in the regulation of ribosome biogenesis, a subject of particular fascination in *Xenopus* (and perhaps in other eukaryotes) in which three structurally distinct sets of 5 S rDNA repeating units are differentially expressed in oocytes and somatic cells (see II.3.). As pointed out earlier, the 5 S rRNA transcription units coincide almost exactly with the gene, with only a few extra nucleotides being present in the precursors to 5 S rRNA synthesized *in vivo* (RUBIN and HOGNESS 1975, JACQ *et al.* 1977). This, and the known sequence of many 5 S rRNA, makes identification of transcripts much easier than those of rDNA. On the other hand, direct visualization of 5 S rRNA transcription has not been possible and it is only recently that some structures in spread chromatin have been proposed to correspond to RNA polymerase III transcription units (SCHEER 1982). Despite the difficulties in studying the structural features of chromatin containing 5 S rRNA genes, transcription of these genes has been elucidated in considerable detail aided primarily by the involvement of a specific enzyme—RNA polymerase III, which happens to be more easily solubilized than the other two RNA polymerases (see ROEDER 1976).

Several studies *in vivo* have identified 5′-terminal triphosphates in 5 S rRNA (see ERDMANN *et al.* 1983), thus showing that 5 S rRNA transcription starts at the 5′-terminal nucleotide of mature 5 S rRNA. The precursor to 5 S rRNA contains a few additional nucleotides at its 3′-end; its maturation *in vivo* is extremely fast, with a half-life of less than one minute being reported for *Drosophila* (LEVIS 1978).

The use of *in vitro* transcription systems has provided a wealth of information about 5 S rRNA synthesis. Studies with isolated nuclei revealed that the *in vitro* formation of 5 S rRNA proceeds linearily for extended time periods, indicative of efficient reinitiation (WEINMANN and ROEDER 1974, MARZLUFF *et al.* 1974, 1975, UDVARDY and SEIFART 1976). Precise analysis showed conclusively that discrete 5 S rRNA molecules having the correct size and sequence are formed in isolated nuclei (YAMAMOTO and SEIFART 1977), some of the molecules bearing extra 3′-end nucleotides as in 5 S pre-rRNA made *in vivo* (YAMAMOTO and SEIFART 1978). With further improvo-ment of the *in vitro* systems faithful transcription of 5 S rRNA genes was shown to take place in isolated chromatin, markedly and selectively enhanced by the addition of purified RNA polymerase III (PARKER *et al.* 1977, YAMAMOTO *et al.* 1977). Also, transcription of isolated human chromatin is faithfully directed by heterologous mammalian, but not by yeast, RNA polymerase III (KRAUSE and SEIFART 1981), thus showing some species specificity in template recognition by the enzyme. Faithful transcription also was obtained for 5 S rDNA injected into *Xenopus* oocyte nuclei (BROWN and GURDON 1977, 1978). However, cloned 5 S rDNA was randomly transcribed in isolated chròmatin systems (*cf.* PARKER *et al.* 1977) thus opening the way for the analysis of chromatin-associated or soluble factors involved in accurate initiation and termination of 5 S rDNA transcription. Efficient cell-free systems, based on the 100,000 g supernatant of whole cell or nuclear homogenates have been improved and shown to permit accurate transcription of cloned 5 S rDNA by exogenous RNA polymerase III. Such cell-free systems were isolated from *Xenopus* oocytes (BIRKENHEIMER *et al.* 1978, NG *et al.* 1979, WORMINGTON *et al.* 1981) and cultured mammalian or amphibian cells (WU 1978, WEIL *et al.* 1979, GRUISSEM *et al.* 1981).

The use of improved cell-free systems has permitted the dissection of initiation and termination signals in 5 S rDNA. It was established unexpectedly that in the Xbs (*X. borealis* somatic) 5 S rDNA a transcription initiation control region is located *within* the gene spanning from nucleotide

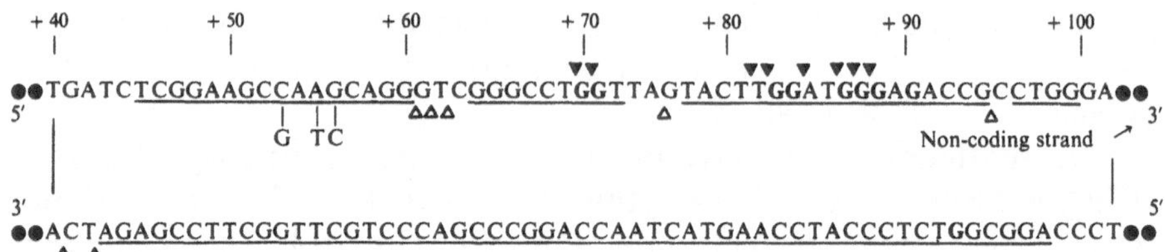

Fig. 21. Interaction of the positive transcription factor TF III A with the internal control region in the *Xenopus* 5 S rRNA gene. The somatic *Xenopus borealis* 5 S rDNA sequences are shown with the numbering starting at the 5′-end of the gene. The sequences protected against DNase I by TFIIIA binding are underlined. Open triangles—non-protected sites. The solid triangles indicate residues making contact with TFIIIA. The boldface G's interfere with TFIIIA binding when methylated. The substitutions indicate differences between Xbs and Xlo 5 S rDNA [From SAKONJU and BROWN (1982). Reproduced with permission by Dr. D. D. BROWN and *Cell*]

residues 50/55 to 80/83 (SAKONJU *et al.* 1980). Moreover, accurate initiation of transcription in cell-free systems depends on the presence of a 5 S rRNA gene-specific protein (Mr 40,000) called TF III A (ENGELKE *et al.* 1980, PELHAM and BROWN 1980). This factor interacts directly with 5 S rDNA (ENGELKE *et al.* 1980) and determines the formation of a stable transcription complex with RNA polymerase III (BOGENHAGEN *et al.* 1982). The presence of TF III A permits reconstitution of transcriptionally active chromatin containing 5 S rDNA (GOTTESFELD and BLOOMER 1982). The contact points of TF III A with 5 S rDNA have been analyzed in detail (Fig. 21) and the strongest contact points have been shown to be on the non-coding strand, indicating that the TF III A-rDNA complex may not be dissociated during transcription (SAKONJU and BROWN 1982). The tissue and species specificity of TF III A remains to be elucidated, but TF III A seems at least qualitatively, to recognize 5 S rRNA genes from different organisms, which probably is related to the evolutionary stability of the control sequence within the gene. Some data suggest the participation of additional control factors in 5 S rDNA transcription (GRUISSEM *et al.* 1981, SHASTRY *et al.* 1982).

Transcription termination sequences in 5 S rDNA also were analyzed (Bogenhagen and Brown 1981). Apparently the presence of a T-stretch in the non-coding strand at the end of the gene is a major termination signal. Indeed, rather long T-stretches are found in 5 S rDNA at the gene/ntS junction, a notorious example being *S. cerevisiae* with more than 30 Ts in that region (Maxam *et al.* 1977, Valenzuela *et al.* 1977). Transcription of extra 3′-end sequences may be variable and related to the distance between the 3′-end of the gene and the T-stretch that serves as a termination signal.

III.4. Transcription of r-Protein Genes

Knowledge about r-protein genes is still very limited. Little can be said of the mechanism of their transcription, although they seem to share most features with nucleoplasmic structural genes (see Darnell 1982). Some interesting aspects of r-protein gene transcription have been observed in recent studies. For example, splicing of pre-mRNA seems to play an important role, even in organisms in which it had been thought to be a rarity. Thus, in *S. cerevisiae* many temperature sensitive mutants (*rna* mutants) defective in the synthesis of mature r-protein mRNA have been isolated and it has turned out that in many of these *rna* mutants splicing of r-protein pre-mRNA was selectively abolished at the restrictive temperature (Rosbash *et al.* 1981, Fried and Warner 1982, Larkin and Woolford 1983). Whether splicing of r-protein pre-mRNA plays a special regulatory role in ribosome biogenesis remains to be clarified.

Another interesting example of r-protein gene transcription is related to the onset of transcription during *Xenopus laevis* embryogenesis. By using cloned r-protein DNA probes, Pierandrei-Amaldi *et al.* (1982) established that transcription of r-protein genes in O_{nu} mutants starts at the same stage as in wild type embryos, thereby showing that transcription of rRNA and r-protein genes are independent of each other. However, the accumulated r-protein mRNA was not translated and no r-proteins were detected in the O_{nu} mutants. This intriguing observation suggests the participation of important post transcriptional control mechanisms in the expression of r-protein genes (see Chapter VI).

Although scanty, the above examples reveal that many important facets of ribosome biogenesis will be understood when r-protein gene expression is further elucidated.

III.5. Synopsis

Transcription of ribosomal genes is the first step in ribosome biogenesis. The whole process is centered on the transcription of rRNA genes in the nucleolus organizer, but the simultaneous transcription of 5 S rRNA and r-protein genes is required in most cells and organisms. The following basic features characterize the present knowledge on the transcription of ribosomal genes.

1. Transcription is catalyzed by DNA-dependent RNA polymerases. In eukaryotes, three types of RNA polymerases exist, specialized in the transcription of different sets of genes: RNA polymerase I—for rRNA genes; II—for nucleoplasmic, including r-protein, genes; and III—for 5 S rRNA and tRNA genes. The RNA polymerases are multicomponent enzymes, built up of two large subunits (M_r, 190 and 195 kDa) and a variable number of smaller subunits. Different forms of each type of RNA polymerases differing by small subunit constituents, and possibly function, exist. Equal amounts of RNA polymerases I, II, and III seem to be present in animal cells, with estimates in the range of 2 to 5×10^4 molecules per cell for each type. RNA polymerase I is confined mainly to the nucleolus, while RNA polymerases II and III are nucleoplasmic. The mechanisms directing the intranuclear compartmentalization of RNA polymerases are unknown.

2. The rRNA genes in eukaryotes are complexed with histones and non-histone proteins to constitute r-chromatin. r-Chromatin appears to be the physical structure transcribed by RNA polymerase I. Non-transcribed r-chromatin has a nucleosomal and suptranucleosomal structure similar to that of bulk chromatin. It is likely that the protein constituents of nucleolar r-chromatin differ (in quantity and/or quality) from those present in bulk chromatin, but conclusive evidence is not yet available. The protein constituents of transcribed r-chromatin are also unknown. It is certain that the transcribed r-chromatin fiber is not folded in nucleosomes. Also, several studies provide evidence that the presence and/or spatial arrangement of non-histone and possibly histone proteins is altered in transcribed r-chromatin segments. The structure of non-transcribed and transcribed chromatin containing r-protein genes is totally unknown.

3. Multiple RNA polymerase I molecules transcribe in tandem each rRNA transcription unit. The process can be visualized by chromosome-spreading and observation of the "exposed" transcription complexes, each including: template r-chromatin (rDNA)—RNA polymerase I—growing pre-rRNA chain. Growing pre-rRNA is coated with proteins to form pre-rRNP fibrils. In most cases a maximal loading of rDNA with RNA polymerases and fibrils (40–50 per μm rDNA axis) is observed, thus showing a maximal frequency of initiation events limited mainly by steric hindrance. The r-chromatin in active transcription units is unfolded, with a packing ratio (μm B-form DNA/μm chromatin) in the range of 1.1. to 1.4, thereby providing independent evidence that nucleosomes are not present along the transcribed rDNA axis. Gene spreading studies on cells in different states of transcriptional activity reveal that each rRNA transcription unit possesses separate initiation and termination signals.

4. Initiation of transcription (both *in vivo* and *in vitro*) occurs at a major

start site. Additional minor start sites are present in the rDNA repeating units of some organisms. The sequences at the ntS/tS$_e$ boundaries are evolutionarily divergent and characteristic structural features have not been uncovered. There is evidence that in animal cells major transcription start signals encompass about 150 np of the ntS and 20 np of the tS$_e$ segment in rDNA, but relevant signal sequences are not yet identified. It is possible that a specific local structure of r-chromatin is also involved in specifying the transcription start site. Studies *in vitro* provide evidence that RNA polymerase I (or transcription initiation factors) displays some species specificity.

Elongation of pre-rRNA chains proceeds by displacement of RNA polymerase I along the rRNA transcription unit. The estimated *in vivo* elongation rate of pre-rRNA chains is in the range of 20 to 40 nucleotides per second. Therefore, completion of a primary pre-rRNA molecule in mammalian cells (about 12.5×10^3 nucleotides) would require 5 to 10 minutes. Initiation is markedly more vulnerable to isolation procedures than is elongation. Elongation may continue to occur in isolated nuclei (although at a markedly reduced rate), but initiation and reinitiation is virtually abolished. The factors needed for transcription initiation remain unknown.

Transcription termination occurs at the 3′-terminus of the L-rRNA gene, or a few nucleotides downstream from this terminus. The ntS sequences at the L-rRNA/ntS boundary are widely divergent. A T-tract is present at the transcription termination site of many organisms suggesting that an endonuclease cleavage (of a UpU bond) may release the completed pre-rRNA. The release of RNA polymerase I seems to take place further downstream (100–200 np) in the ntS segment.

5. The primary pre-rRNA in all eukaryotes has the following topology: 5′: tS$_e$—S-rRNA—tS$_i$ 1—5.8 S rRNA—tS$_i$ 2—L-rRNA: 3′. In many cases, pppNp structures have been identified at the 5′-end of primary pre-rRNA, showing that this molecule is identical to the primary rDNA transcript. The primary pre-rRNA molecules in eukaryotes constitute a small, but significant pool, apparently because processing of primary pre-rRNA normally does not occur *during* transcription. The small amount of heterogeneity of primary pre-rRNA size observed in some eukaryotes may be due to: (i) transcription initiation at secondary start sites or (ii) nuclease degradation of short tS$_e$ segments. Splicing of intron$^+$ pre-rRNA molecules is observed only in some strains of *Tetrahymena*. It takes place shortly after transcription and the mechanisms involved are apparently different from those responsible for pre-mRNA splicing.

6. Transcription of 5 S rRNA genes (both *in vivo* and *in vitro*) by RNA polymerase III is understood to considerable detail. Initiation starts at the 5′-terminal nucleotide of the mature 5 S rRNA. It is specifically directed by

a positive control protein factor (TF III A in *Xenopus* oocytes) that binds to the middle of the 5 S rRNA gene. The presence or absence of this (and possibly other) protein factors may ensure both faithful transcription and on-off switching of 5 S rRNA genes. Transcription termination occurs a few nucleotides downstream from the 5 S rRNA gene. In several eukaryotes, a T-tract is present near the gene/ntS boundary, suggesting that an endonuclease cleavage at a UpU bond releases the 5 S rRNA transcript.

IV. Maturation of Preribosomes

Preribosomes may be defined as discrete, free ribonucleoprotein particles containing pre-rRNA. The assembly of preribosomes starts during transcription of rRNA genes and the growing pre-rRNA chain is already coated with ribosomal and non-ribosomal proteins. The selective interaction of individual r-proteins with the respective segments in pre-rRNA clearly shows (Chooi and Leiby 1981) that its protein coating involves specific protein-RNA and protein-protein interactions. Thus, assembly of preribosomes takes place simultaneously with transcription to release upon termination an already defined particle which may be designated as the *primary preribosome* (containing primary pre-rRNA).

The maturation of the primary preribosome is a complex, sequential process and its best understood facet is the processing or maturation of its constituent primary pre-rRNA. As discussed earlier, maturation of pre-rRNA in eukaryotes starts normally from a rather homogeneous pool of primary pre-rRNA molecules containing the sequences for S-rRNA, 5.8 S rRNA and L-rRNA. Therefore, the polynucleotide chain of primary pre-rRNA should be complete and the respecitive preribosome should have the "correct" structure in order to enter the sequence of reactions involved in its processing (Hadjiolov *et al.* 1978). These features of primary pre-rRNAs and preribosomes stress the point that they are prepared for processing in many respects during transcription of rRNA genes and that processing can be considered as the second stage in ribosome biogenesis.

Since the maturation of pre-rRNA is understood to considerable detail (reviews in Perry 1976, 1981, Hadjiolov and Nikolaev 1976, Hadjiolov 1980), I shall discuss first the available evidence on the maturation of pre-rRNA in order to evaluate subsequently the much more limited information on the structure and maturation of preribosomes.

IV.1. Structure of Primary Pre-rRNA

IV.1.1. Size and Primary Structure

The mature rRNA species in eukaryotes are characterized by: 1. size variations of L-rRNA in the Mr range of 1.2 to 1.7 × 10^6 Da (about 3,400 to

5,000 nucleotides) and 2. size homogeneity of S-rRNA with Mr of approximately 0.7×10^6 Da (about 2,000 nucleotides) (see ATTARDI and AMALDI 1970). The size of primary pre-rRNA molecules is proportionately larger in evolutionarily more advanced species, seen mainly in a pronounced divergence of both tS_e and tS_i sequences (see Fig. 9).

The nucleotide composition and primary structure of rRNA species in eukaryotes reveals several characteristic features of these molecules (see Cox 1977 for earlier data): 1. considerable evolutionary divergence exists in the sequence of both L-rRNA and S-rRNA; 2. conserved and non-conserved sequences in L-rRNA and possibly S-rRNA follow a more or less mosaic pattern (see II.2.5.e), the 5′- and 3′-terminal segments showing a high degree of homology among species. An exceedingly high conservation is encountered in some internal L-rRNA sequences (GOURSE and GERBI 1980) (see Fig. 10); 3. large (100–500 np) GC-rich segments (up to 85% GC) are typical for L-rRNA of higher eukaryotes (see HADJIOLOV and NIKOLAEV 1976) and 4. 5.8 S and 5 S rRNA sequences are also conserved in evolution. Following mutations, compensatory base changes occur to make possible the preservation of characteristic secondary structures (see DELIHAS and ANDERSEN 1982).

The non-conserved tS_e and tS_i sequences in primary pre-rRNA are much less studied. Generally, they display a marked divergence in evolution, with little or no homology among species. They contain no methylated nucleotides and probably no pseudouridine (see MADEN *et al.* 1977). Long U-tracts are found in some species as exemplified by the tS_e in the mouse (URANO *et al.* 1980, BACH *et al.* 1981) and rat (ROTHBLUM *et al.* 1982) primary pre-rRNA. On the other hand, tS_e and tS_i segments in animals are extremely GC-rich in some regions, as for example the tS_i of *Xenopus laevis* (HALL and MADEN 1980), the mouse (GOLDMAN *et al.* 1983, MICHOT *et al.* 1983) and the rat (SUBRAHMANYAM *et al.* 1982, HADJIOLOVA *et al.* 1984b). All these and other features clearly show that considerable changes in tS_e and tS_i sequences have taken place in the evolution of eukaryotes.

IV.1.2. Modifications

Extensive modifications of nucleotides are typical for eukaryotic pre-rRNA. These include: 1. methylation of the base or the 2′-OH group in the ribose and 2. conversion of uridine into pseudouridine. The evidence for these modifications has been reviewed (MADEN *et al.* 1977) and will be considered here only briefly. Most of the methylated nucleotides and pseudouridine are already present in primary pre-rRNA, the modifications having been made at specific sites within the S-rRNA, 5.8 S rRNA, and L-rRNA segments of the primary pre-rRNA chain. These observations show that modification is a highly selective process, largely taking place during or shortly after transcription. Detailed studies with HeLa cells of the oligonucleotides containing methylated nucleotides demonstrated that methylations of L-rRNA and S-rRNA sequences follow a specific pattern (MADEN *et al.* 1974, MADEN and SALIM 1974, MADEN 1982), with the number of pseudouridine and 2′-O-methyl-nucleoside residues in L-rRNA, S-rRNA and 5.8 S rRNA, as well as in primary pre-rRNA, being similar in some (see MADEN *et al.* 1977), but not all (HUGHES and MADEN 1978), vertebrate species. This observation suggests that the two types of modification may be interdependent. In animal cells there are about 200 modification sites in primary pre-rRNA—100 for methylation (about 90% being 2′-O-methyl groups) and 100 for pseudouridylation (KHAN *et al.* 1978). In lower eukaryotes the number of modified nucleosides is markedly less. *Saccharomyces* L-rRNA possesses 43 methyl groups (86% 2′-O-methyl) and 37 pseudouridines (KLOOTWIJK and PLANTA 1973 a, b, BRAND *et al.* 1979). The methylated nucleotides appear to be clustered along the rRNA chains. For example, in *Xenopus laevis* S-rRNA most (60%) of the 2′-O-methyl- nucleotides are found in the first 700 nucleotides and in cluster in the region of nucleotides 1,240 to 1,400. In contrast, practically all base methylations are at the 3′-end of the S-rRNA (MADEN 1980, SALIM and MADEN 1981, MADEN 1982). Clustering of methylated nucleosides is encountered also in *Saccharomyces* L-rRNA (VELDMAN *et al.* 1981 b). In general the methylation sites are within the most evolutionarily conserved L-rRNA and S-rRNA segments (MADEN 1982). The highly specific modification pattern of L-rRNA and S-rRNA sequences in primary pre-rRNA suggests that methylation and pseudouridylation play an important role in the assembly and processing of preribosomes, as well as in ribosome structure and function.

The timing of modification events implies that modification enzymes should be in the nucleolus and associated with transcription complexes. Unfortunately, the modifying enzymes are poorly characterized, mainly because of difficulties in obtaining unmodified polyribonucleotide substrates (LIAU and HURLBERT 1975, LIAU *et al.* 1976). Nevertheless, some enzyme systems catalyzing methylation (OBORA *et al.* 1982, BROWN *et al.* 1982) or pseudouridylation (GREEN *et al.* 1982) of different RNA species are partially characterized. In particular, a nucleolar methylase (methylase I) has been purified and was shown to use S-adenosyl methionine as a substrate for methylating undermethylated mammalian pre-rRNA (OBORA *et al.* 1982). These studies open the way for a more precise elucidation of the mechanisms of pre-rRNA and rRNA modification and its role in ribosome biogenesis.

IV.1.3. Conformation

The available information on the conformation of primary pre-rRNA is limited. Studies of its denaturation spectra reveal a structure in solution closer to that of L-rRNA than of S-rRNA (HADJIOLOV and COX 1973). Electron microscopy of pre-rRNA molecules partly denatured in formamide/urea shows characteristic double-stranded loops along the polynucleotide chain (WELLAUER and DAWID 1973, see WELLAUER and DAWID 1975). In all vertebrates studied until now such loops (100–500 np long) are observed mainly in tS_e, tS_i, and L-rRNA segments of pre-rRNA (WELLAUER and DAWID 1975, SCHIBLER *et al.* 1975, DABEVA *et al.* 1976). These loops most likely correspond to the GC-rich segments identified earlier in animal rRNA (see HADJIOLOV and NIKOLAEV 1976, COX 1977). Such large GC-rich segments are not found in lower eukaryotes and seem to be a late acquisition in evolution with as yet unknown significance.

The ever increasing number of completely described S-rRNA and L-rRNA sequences from different eukaryotes and an understanding of the sequence conservation of S-rRNA and L-rRNA has allowed a relatively precise evaluation of possible secondary structure interactions to be made. As a result, models have been proposed for the secondary structure of the eukaryotic rRNA species (STIEGLER *et al.* 1981 a, VELDMAN *et al.* 1981 b, MADEN 1982, HADJIOLOV *et al.* 1984), based on earlier ones for 16 S and 23 S rRNA of *E. coli* (*i.e.* NOLLER *et al.* 1981, GLOTZ *et al.* 1981, STIEGLER *et al.* 1981 b, BRANLANT *et al.* 1981). Several structural domains (4 in S-rRNA and 7 in L-rRNA) showing considerable conservation of secondary structure may be delineated. Generally, it may be inferred that: 1. Considerable structural homology has been preserved through evolution, resulting in analogous secondary structure organization of rRNA among species; 2. sequence conservation is more pronounced within single-stranded regions of the proposed structures than in 2–3° regions; 3. homologous double-stranded loops may be built as a result of numerous base-pair changes compensating for mutations; 4. increase in length (for L-rRNA in particular) resulting from the appearance of new stretches all along the rRNA molecule, has given rise to additional double-stranded loops within a domain.

The models proposed for the secondary structures in S-rRNA and L-rRNA are, of course, only a first step toward an understanding of the conformation of pre-rRNA and rRNA within the preribosome and the ribosome. Many long-range interactions, *e.g.* between the 5′- and 3′-terminal segments in L-rRNA, are possible and may play an important role in ribosome processing and function.

IV.2. Pre-rRNA Maturation Pathways

IV.2.1. General Considerations

Because pre-rRNA and rRNA constitute the backbone of the respective ribonucleoprotein particles, pre-rRNA processing is an important aspect of the overall process of ribosome biogenesis. The conversion of primary pre rRNA into mature rRNA is a multistep process, occurring mainly in the nucleolus, but some late steps take place in the nucleoplasm, and even in the cytoplasm. Processing of pre-rRNA is relatively fast and only minute amounts of primary and intermediate pre-rRNA species are present in the cell under steady-state conditions (Table 3). Consequently, the correct

Table 3. *Absolute pool sizes of pre-rRNA and rRNA molecules in rat liver nuclei*[a]

rRNA species (S)	$10^{-4}\times$ number of molecules per nucleus
45	1.2 ± 0.2 (1.0)[b]
41	0.9 ± 0.2
36	1.2 ± 0.2
32	2.9 ± 0.4 (4.0)
21	2.5 ± 0.5[c]
28	9.6 ± 1.5 (7.0)
18	8.7 ± 1.7 (6.0)

[a] The amounts of pre-rRNA and rRNA molecules are determined by A_{260} measurements of RNA extracted from detergent-purified nuclei and fractionated by gel electrophoresis (DABEVA *et al.* 1978). The sum of all nuclear pre-rRNA and rRNA molecules is about 3.5% of the rRNA molecules in the liver cell.

[b] The values in brackets are for HeLa cell nuclear pre-rRNA and rRNA measured by methods similar to those used for liver (WEINBERG and PENMAN 1968).

[c] This value is likely to be an overestimation owing to interference by RNA species different from 21 S pre-rRNA.

analysis of pre-rRNA maturation pathways obviously depends critically on the techniques used for the identification and analysis of large pre-rRNA molecules. The introduction of gel-electrophoresis techniques for fractionation of large RNA molecules (TSANEV 1965, HADJIOLOV *et al.* 1966, LOENING 1967) permitted the identification of several intermediate pre-rRNA species. Degradation of RNA during isolation of nuclei or nucleoli, interference by hnRNA, formation of aggregates, etc. are hazards that require carefully controlled experimental conditions. Several methods have been used to identify separate pre-rRNA components: estimation of molecular mass, hybridization with rDNA, comparative methylation patterns or fingerprints of enzyme hydrolyzates, electron microscopic secondary structure mapping after partial denaturation, mapping on rDNA restriction fragments, etc. The combination of these techniques permitted the unequivocal identification of primary and intermediate pre-rRNA in several eukaryotic cells and organisms. Yet, most of these techniques permit only the outline of *possible* pre-rRNA maturation pathways. The actual precursor-product relationships require detailed labeling kinetics and pulse-chase studies. Because of the very rapid transformations involved and numerous methodical imperfections, only in a limited number of cases have the actual pre-rRNA processing steps been documented sufficiently. Consequently, one must keep in mind, the tentative character of proposed pre-rRNA maturation pathways.

IV.2.2. Common Pattern of Pre-rRNA Maturation

Studies on the structure and topology of primary pre-rRNA reveal a similar organization in all eukaryotes, suggesting that similar mechanisms may operate in its processing. Most of the early information on pre-rRNA maturation pathways in different cells and organisms has been reviewed (HADJIOLOV and NIKOLAEV 1976) and will not be considered in detail here. Well documented schemes for pre-rRNA maturation are available for unicellular organisms (*Saccharomyces*, *Tetrahymena*), plants (*Phaseolus vulgaris*, *Acer pseudoplatanus*), insects (*Drosophila*), lower vertebrates (*Xenopus laevis*), and several mammalian organisms. Invariably, discrete intermediate pre-rRNA species have been shown to form, indicating that initially processing involves a succession of endonuclease attacks (see WINICOV and PERRY 1975, PERRY 1976, HADJIOLOV and NIKOLAEV 1976). The endonuclease attacks may follow a rigid sequential pattern so that each processing step induces conformational changes that trigger the next one. As a consequence, a single pre-rRNA maturation pathway is predominant for a given cell type. Taking into consideration the available information from a variety of eukaryotic cells and organisms, it was proposed (PERRY 1976, HADJIOLOV and NIKOLAEV 1976) that a common pre-rRNA processing pathway is shared by all eukaryotes (Fig. 22).

The proposed scheme is still tentative; the evidence supporting the occurrence of each step, as well as possible mechanisms, will be discussed below.

The *first step* generates an intermediate molecule containing both S-rRNA and L-rRNA segments and its early formation in vertebrate (mainly mammalian) species has been thoroughly documented by a variety of experimental approaches (see HADJIOLOV and NIKOLAEV 1976), including secondary structure mapping of partially denaturated molecules (WELLAUER and DAWID 1975, SCHIBLER *et al.* 1975, DABEVA *et al.* 1976). That the tS_e segment is split by endonuclease attack was shown for HeLa cells by electron microscopic identification of its characteristic secondary structure retained after tS_e release from the primary pre-rRNA (WELLAUER and DAWID 1973). An analogous intermediate pre-rRNA was found also in cultured *Drosophila* cells, although in this case only a very small (about 0.2×10^6 Da) tS_e segment is discarded (LEVIS and PENMAN 1978). Whether an initial endonucleolytic removal of the entire tS_e segment takes place also in lower eukaryotes is not known. In *S. cerevisiae*, a putative intermediate smaller than the primary pre-rRNA was observed (KLEMENZ and GEIDUSHEK 1981), but not consistently (*i.e.* NIKOLAEV *et al.* 1979, VENKOV and VASILEVA 1979, VELDMAN *et al.* 1981 a) and whether its formation is compulsory is uncertain. More conclusive evidence was reported for *Tetrahymena* where small amounts of a pre-rRNA, devoid of the tS_e segment only, were identified (KISTER *et al.* 1983).

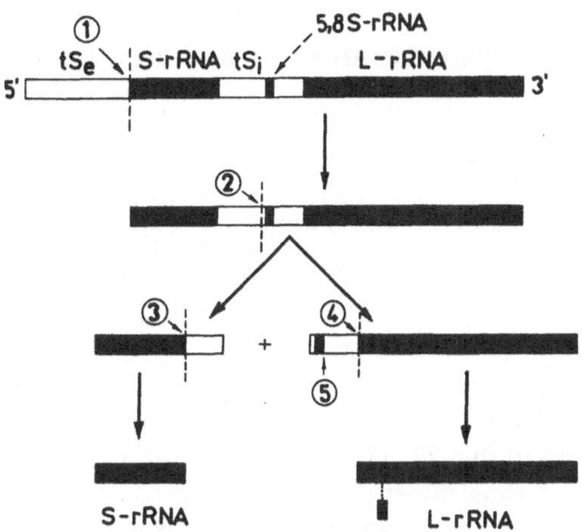

Fig. 22. Common pattern of pre-rRNA maturation in eukaryotes. The putative major endonuclease attack sites are indicated by small arrows and numbered 1 to 5 as discussed in the text. The S-rRNA, 5.8 S rRNA, and L-rRNA segments in pre-rRNA, corresponding to the respective mature rRNA are shown in black. tS_e external transcribed spacer: tS_i internal transcribed spacer. The mature L-rRNA and 5.8 S rRNA are hydrogen-bonded

As discussed earlier, tS_e sequences in eukaryotes are widely divergent, a finding suggesting that a variety of mechanisms might exist for the removal of the tS_e segment. For example, a marked size heterogeneity was observed in the pool of primary pre-rRNA molecules of the mouse (TIOLLAIS *et al.* 1971, GALIBERT *et al.* 1975) and the rat (DABEVA *et al.* 1976). But the recognition of long (> 40) stretches of U-residues in the tS_e segment (total length about 4,000 nucleotides) near the 5′-end of primary pre-rRNA led to the suggestion that they may specify early processing sites (MILLER and SOLLNER-WEBB 1981, GRUMMT 1981 a, MISHIMA *et al.* 1981, ROTHBLUM *et al.* 1982). The tS_e segment in *S. cerevisiae* also is extremely rich in U-tracts (SKRYABIN *et al.* 1979, BAYEV *et al.* 1981) raising the possibility of its rapid degradation by a series of endonucleolytic attacks. These findings suggest that degradation of tS_e segments may vary among species and may involve more than one nuclease attacks. The correct shaping of the 5′-end of the S-rRNA segment is unknown. At present, the sequences at the tS_e/S-rRNA junction are known only for *Saccharomyces* (SKRYABIN *et al.* 1979, BAYEV *et al.* 1981), *Xenopus laevis* (SALIM and MADEN 1980, MADEN *et al.* 1982) and the rat (CASSIDY *et al.* 1982). Both the 5′-end and 3′-end segments of S-rRNA have been almost absolutely conserved in evolution (see RUBTSOV *et al.* 1980, HALL and MADEN 1980, CASSIDY *et al.* 1982, SUBRAHMANYAM *et al.* 1982, MADEN *et al.* 1982). A model for the long range base-pairing of 5′ and 3′ termini of S-rRNA is shown in Fig. 23. It is plausible that the conformation of S-rRNA resulting from such an interaction plays a signal role for endonuclease cleavages in site *1* and *3* (see below).

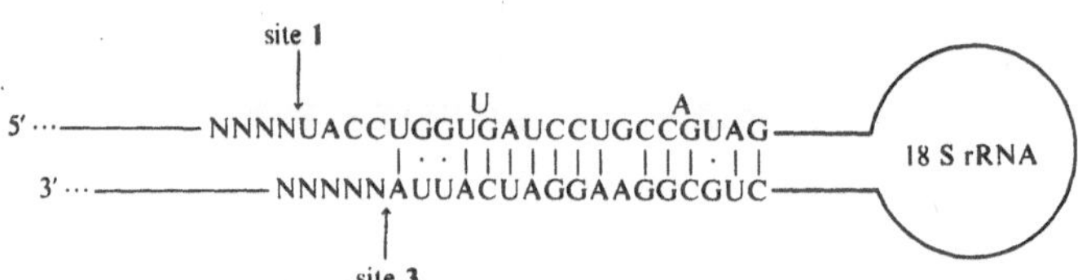

Fig. 23. Possible base-pairing between the strongly conserved 5′- and 3′-termini of eukaryotic S-rRNA (18 S rRNA). The numbering of endonuclease processing sites (arrows) is the same as in Fig. 22. The residues denoted by N are non-conserved in the eukaryotic species thus far studied. Only limited complementarity of these sequences, flanking the 5′- and 3′-termini of S-rRNA may be discerend in some species

2. The *second step* in pre-rRNA processing results in the formation of intermediate pre-rRNA for both L-rRNA and S-rRNA. The formation of intermediate precursors to L-rRNA in eukaryotes was already established in the early studies on pre-rRNA processing (SCHERRER and DARNELL 1962) and their existence has been proven for all eukaryotes studied up till now (see ATTARDI and AMALDI 1970, MADEN 1971, HADJIOLOV and NIKOLAEV 1976). The precise site of the endonuclease cleavage that generates precursors to L-rRNA is not known, but presumably should be in tS_i 1 since conclusive evidence has been obtained that in all cases the pre-rRNA contains the 5.8 S rRNA sequence. In several cases, as in *Saccharomyces* (VELDMAN *et al.* 1981 a), *Tetrahymena* (ENGBERG *et al.* 1980), mouse (BOWMAN *et al.* 1981, 1983), rat (DUDOV *et al.* 1983, HADJIOLOVA *et al.* 1984 b) and human (KHAN and MADEN 1976) cells, convincing evidence was obtained that the site 2 endonuclease cut in tS_i 1 is at or near the 5′-end of the 5.8 S rRNA segment.

The formation of an intermediate precursor to S-rRNA was observed in several eukaryotes, the best documented cases including *Saccharomyces* (UDEM and WARNER 1972, TRAPMAN and PLANTA 1975, see VELDMAN *et al.*

1981 a), *Tetrahymena* (POUSADA *et al.* 1975, ENGBERG *et al.* 1980), *Musca domestica* (HALL and CUMMINGS 1975), mouse and rat liver (EGAWA *et al.* 1971, FUJISAWA *et al.* 1973, DABEVA *et al.* 1976, 1978, BOWMAN *et al.* 1981) and HeLa cells (WEINBERG and PENMAN 1970, MADEN *et al.* 1972, WELLAUER and DAWID 1973). In some of these studies (*c.f., e.g.*, BOWMAN *et al.* 1981, 1983) evidence was obtained showing that at least in mammalian cells the precursor to S-rRNA contains tS_i 1 sequences and therefore is generated by a site 2 endonuclease cut. However, in a broad variety of eukaryotes, including *Drosophila melanogaster* (LEVIS and PENMAN 1978), *Xenopus laevis* (LOENING *et al.* 1969, WELLAUER and DAWID 1974), and mouse L cells (WELLAUER *et al.* 1974 a), a precursor to S-rRNA could not be detected. Instead, several intermediate precursors to L-rRNA were identified, the largest containing almost the entire tS_i sequence (*cf., e.g.*, WELLAUER and DAWID 1974, WELLAUER *et al.* 1974 a). These results indicate that in these organisms an endonuclease split at site 3 in fact precedes the one at site 2. Given the possibility for the formation of double-stranded structures involving 5′- and 3′-termini of S-rRNA (see Fig. 23), an earlier split at site 3 is easily visualized. Alternatively, the existence in most eukaryotes of precursors to S-rRNA strongly suggests that in these cases the formation of the envisaged double-stranded structure is sufficient to specify a site 1 nuclease cut, while the site 3 nuclease attack must await the split at site 2 and some putative secondary structure rearrangements involving tS_i 1 sequences. It is noteworthy that the order of endonuclease attacks at site 2 and 3 may alternate even in the same cell type, as shown by the example of mouse cells (WELLAUER *et al.* 1974 a, BOWMAN *et al.* 1981, 1983), thereby generating different intermediate pre-rRNA molecules.

4. The endonuclease cuts at sites *4* and *5* are considered in more detail, because they participate in the formation of L-rRNA. The existence of a relatively large pool of L-rRNA precursors in the nucleolus has been ascertained to be the situation in all eukaryotes. The recognized variations in size for this pre-rRNA among different species reflects differences in constituent tS_i 2 and L-rRNA segments. Thus, in *Saccharomyces* this pre-rRNA has a molecular mass of 1.3×10^6 Da, while in mammalian cells its M_r is about 2.1×10^6 Da. Conclusive evidence has been obtained that this pre-rRNA contains the sequences for both L-rRNA and 5.8 S rRNA (MADEN and ROBERTSON 1974, NAZAR *et al.* 1975), as well as tS_i 2 sequences (BOWMAN *et al.* 1981, 1983, DUDOV *et al.* 1983). Therefore, the formation of 5.8 S rRNA hydrogen bonded to L-rRNA is obviously a rather complicated process, involving a minimum of two or three endonuclease cuts that shape the ends of both L-rRNA and 5.8 S rRNA. A 12 S rRNA has been identified in mouse (HADJIOLOVA *et al.* 1973, HADJIOLOV *et al.* 1974 a, BOWMAN *et al.* 1981), rat (DUDOV *et al.* 1983), and human (KHAN and MADEN 1976) cells and contains both 5.8 S rRNA and almost the complete tS_i 2 segment.

Therefore, at least in mammalian species, endonuclease cut at site 4 seems to be an early processing step, shaping the 5′-end of L-rRNA. The formation of the 5.8 S rRNA 3′-end seems to result from a subsequent cleavage at site 5. In fact, it may be preceded by cleavages within the tS_i 2 segment, since shorter than 12 S RNA precursors to 5.8 S rRNA have been identified in *Saccharomyces carlsbergensis* (TRAPMAN *et al.* 1975 a) and rat liver (REDDY *et al.* 1983).

Recent sequencing studies on the tS_i regions of some eukaryotes permits some insight into the structures responsible for the processing at sites 4 and

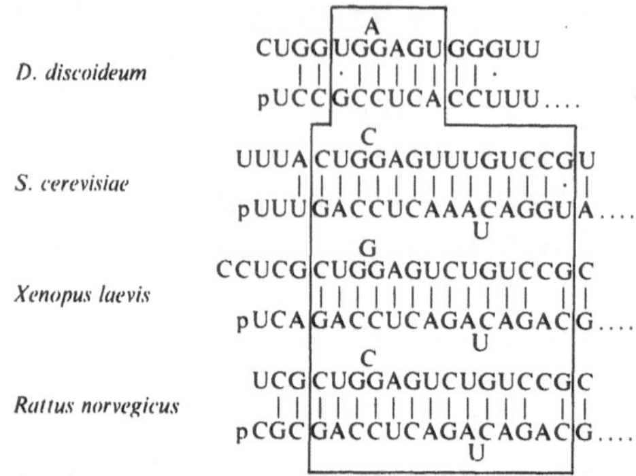

Fig. 24. Possible base-pairing between the 3′-termini of 5.8 S rRNA (upper row) and the 5′-termini of L-rRNA (lower row). Homologous sequences are boxed. Note the presence of several compensatory base substitutions preserving the possibility for base-pairing. For sources of sequence data see the text

5, as well as at site 2. These results suggest a role for possible interactions of the 5.8 S rRNA segment with L-rRNA and therefore deserve attention.

Converging experimental evidence from various eukaryotes shows that both 5′- and 3′-end sequences of 5.8 S rRNA are involved in the interaction with L-rRNA (PACE *et al.* 1977, PAVLAKIS *et al.* 1979, NAZAR and SITZ 1980, SITZ *et al.* 1981, WALKER *et al.* 1982, PETERS *et al.* 1982, OLSEN and SOGIN 1982). There is now general agreement that the 3′-end of 5.8 S rRNA interacts with the 5′-end of L-rRNA (VELDMAN *et al.* 1981 b, GEORGIEV *et al.* 1981, SUBRAHMANYAM *et al.* 1982, MICHOT *et al.* 1982) and the structural basis for this assumption is illustrated in Fig. 24. It is noteworthy that numerous *compensatory mutations* seem to occur, preserving the complementarity between the 5.8 S and L-rRNA. The site of interaction of the 5′-end of 5.8 S rRNA is less clearly defined. However, models for a rather strong base pairing with L-rRNA sequences in "domain I" (nucleotides 330–417 in *Saccharomyces*) were proposed for *Saccharomyces* (VELDMAN *et al.* 1981 b) and the mouse (MICHOT *et al.* 1982 a). An interaction with 3′-end segments in L-rRNA has been proposed as an alternative (GEORGIEV *et al.*

1981, 1983) and is supported by the results of reassociation studies with *Drosophila* (ISHIKAWA 1979) and *Neurospora crassa* (KELLY and COX 1981) 5.8 S and L-rRNA species. In any case, there is agreement that the 5′-end of 5.8 S rRNA is base-paired to sequences in L-rRNA. Accordingly, it may be proposed that endonuclease attack sites 4 and 5 are specified by double-

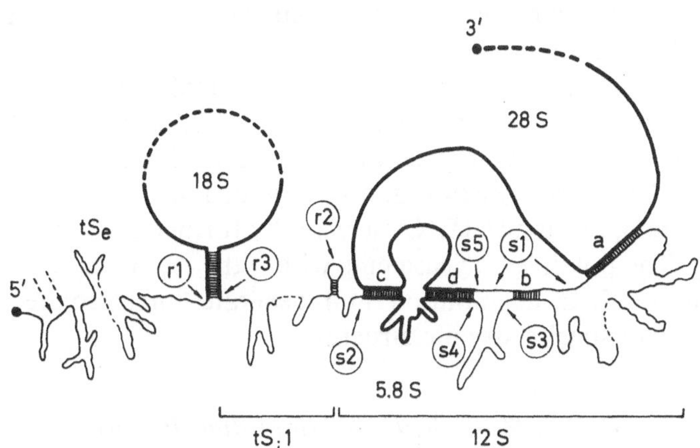

Fig. 25. A model for possible base-pairing interactions involved in the processing of mammalian 45 S pre-rRNA. The segments corresponding to mature rRNA are represented by a thick line. It is proposed that processing occurs in two main phases: rapid (*r*) and slow (*s*). The rapid phase generates 18 S rRNA and 32 S pre-rRNA. The slow phase generates 28 S rRNA and 5.8 S rRNA starting from 32 S pre-rRNA. tS_e—external transcribed spacer; tS_i1—internal transcribed spacer spanning the sequence from the 3′-end of 18 S rRNA to the 5′-end of 5.8 S rRNA. 12 S denotes a precursor to 5.8 S rRNA, corresponding to the 5′-part of 32 S pre-RNA and containing 5.8 S rRNA and tS_i2 sequences. The arrows indicate endonuclease attack sites and their numbering reflects the preferred order of cleavage during the rapid (r 1 to r 3) and slow (s 1 to s 5) phases of pre-rRNA processing. Base-paired segments are expanded. [According to HADJIOLOVA *et al.* (1984 b). Reproduced with permission by *Biochem. J.*]

stranded structures involving the segments in pre-rRNA corresponding to mature rRNA. A possible model for secondary structure interactions defining the endonuclease attack sites involved in primary pre-rRNA processing is proposed in Fig. 25 (HADJIOLOVA *et al.* 1984 b).

The following features of this model deserve further comments:

a) All endonuclease attack sites are defined by base-pairing between relatively conserved regions within mature rRNA species. This makes superfluous the search for concensus sequences at tS_e/rRNA and tS_i/rRNA junctions. In fact, such sequences do not seem to exist, since wide divergence is observed at sites immediately flanking the mature rRNA segments.

b) The 5.8 S rRNA plays a central role in pre-rRNA processing. It is immaterial whether the site of interaction of its 5′-end sequence is at the 3′-end segment or the "domain I" region of L-rRNA, since in both cases the site 2 attack will delineate the 5′-end of 5.8 S rRNA. If the interaction with a 3′-end segment in L-rRNA is indeed taking place, this could supply an additional termination signal and give a plausible explanation of why processing of primary pre-rRNA in eukaryotes does not start before transcription termination.

c) The proposed endonuclease cuts may in fact take place at sites *near*, not necessarily *at* the double-stranded structures. In such cases they will be specified by species-specific structures (*e.g.*, U-stretches) within tS_e and tS_i sequences. Again, the existing double-stranded structures will direct the precise shaping of mature rRNA termini by further exo- or endonuclease trimming. The possible nuclease cuts within the tS_e or tS_i segments could contribute to their degradation and generate intermediate pre-rRNA molecules with species-specific structure.

IV.2.3. Multiplicity of Maturation Pathways

Until recently it was agreed that a single pre-rRNA maturation pathway is the case for a given cell type. Studies on the kinetics of labeling of pre-rRNA and rRNA species in *Saccharomyces* (UDEM and WARNER 1972, TRAPMAN and PLANTA 1975) provided evidence in support of this concept, at least for the case of a simple eukaryote.

The possibility that processing of pre-rRNA may take place along parallel pathways occurring in the same cell was raised by observations showing the simultaneous existence of several "minor" pre-rRNA components. In fact, an increased amount of such "aberrant" intermediates was observed in virus-infected HeLa cells (WEINBERG and PENMAN 1970) and comparison of pre-rRNA species in cultured mouse L cells (WELLAUER *et al.* 1974 a) and mouse liver (HADJIOLOV *et al.* 1974 a) suggested the prevalence of different pathways probably resulting from an inverted sequence of endonuclease attacks at sites 2 and 3 in the pre-rRNA molecule (WINICOV 1976; see Fig. 22). Studies with cultured human lymphocytes, showed the simultaneous formation of 32 S (nuclease cut at sites 2 → 3) and 36 S (3 → 2) pre-rRNA; phytohaemagglutinin stimulation caused a 4-fold increase only in the rate of processing through 32 S pre-rRNA (PURTELL and ANTHONY 1975). Further evidence for the simultaneous operation of two pre-rRNA processing pathways was obtained with a temperature-sensitive mutant of BHK cell line 422 E (TONIOLO *et al.* 1973). At the permissive temperature these cells generate L-rRNA from a 32 S pre-rRNA, but at the restrictive temperature the temporal order of endonuclease attack at sites 2 and 3 is altered and 36 S pre-rRNA is produced (WINICOV 1976, TONIOLO and

BASILICO 1976). The simultaneous existence of 36 S, 32 S, and 20 S pre-rRNA in rat liver further demonstrates the variability in nuclease attacks at sites 2 and 3 and the consequent simultaneous processing of pre-rRNA along at least two parallel pathways (DABEVA *et al.* 1976).

These observations raised the question as to what is the actual precursor-

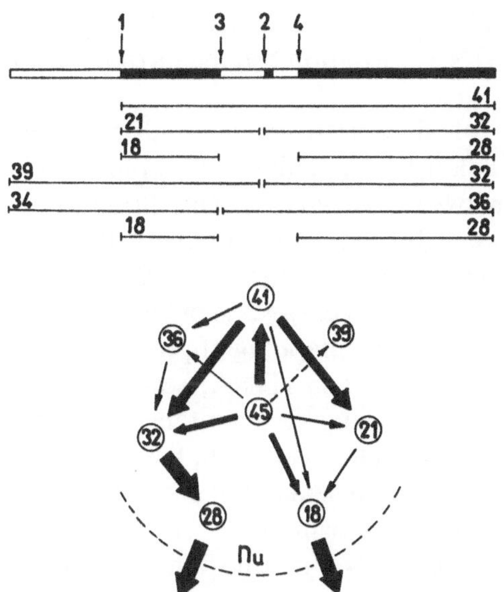

Fig. 26. Multiplicity of primary pre-rRNA processing pathways in rat liver. *Above*—Endonuclease attack sites and major pre-rRNA species generated during processing (the numbers denote S values of respective pre-rRNA and rRNA). *Below*—Possible conversions during primary pre-rRNA processing. The thickness of the arrows reflects the differences in transfer rates between the pools of nucleolar pre-rRNA and rRNA as derived from labeling kinetics studies. Note that 28 S rRNA can be generated only from one precursor—32 S pre-rRNA. [According to DUDOV *et al.* (1978) and DUDOV and DABEVA (1982)]

product relationship in pre-rRNA maturation. The flexibility in the sequence of endonuclease attacks invites the analysis of pre-rRNA processing as a stochastic process, *i.e.*, a process whereby attacks at each site in a given pre-rRNA may be possbile. Such an analysis was attempted with rat liver, a population of cells operating under steady-state conditions (DUDOV *et al.* 1978). Detailed tracer kinetics experiments were carried out and an algebraic approach to computer analysis of the relations among a large number of interconnected pools was developed. The analysis permits the selection of a most probable model for existing connections among pre-rRNA and rRNA pools (Fig. 26). The following basic features may be deduced: 1. considerable flexibility exists in the sequence of endonuclease attacks at the early stages of pre-rRNA processing, resulting in the simultaneous occurrence of several processing pathways and 2. the cleavage site 4 leading to the formation of mature L-rRNA is protected until the generation of its immediate pre-rRNA, thereby specifying what may be a crucial checkpoint before the release of mature ribosomes. Independent proof for the existence of multiple pre-rRNA processing pathways due to altered order of nuclease attack comes from studies of hybridization of pre-rRNA species to rDNA restriction fragments of mouse L cells (BOWMAN *et al.* 1981, 1983).

The possibility of the existence of multiple pre-rRNA processing pathways in other eukaryotes remains to be explored. The smaller size of tS_e and tS_i segments in lower eukaryotes creates difficulties in the identification of intermediate pre-rRNA species, but, nevertheless, it seems that the multiplicity of pre-rRNA processing pathways originating from variations in the order of endonuclease cleavages is of wide importance in eukaryotes *e.g.*, in detailed studies on the mapping of pre-rRNA and rRNA molecules within *Tetrahymena thermophila* rDNA, several intermediate pre-rRNA species were identified, revealing the simultaneous occurrence of alternative processing pathways (ENGBERG *et al.* 1980, KISTER *et al.* 1983). On the other hand, it seems that more rigid sequential patterns of pre-rRNA processing operate in some cells and organisms, *e.g.*, *S. cerevisiae* temperature-sensitive mutants, defective in pre-rRNA maturation, failed to show a switching of alternative processing pathways at the restrictive temperature (ANDREW *et al.* 1976, GORENSTEIN and WARNER 1977, VENKOV and VASILEVA 1979).

The physiological significance of multiple pre-rRNA processing pathways remains to be clarified. It appears that at high rates of transcription and output of primary pre-rRNA, the processing endonuclease cleavages are also accelerated. Simultaneous splitting at two processing sites is also enhanced and results in a direct formation of rRNA and decreased amounts of intermediate pre-rRNA. Thus, in regenerating rat liver, the bulk of primary pre-rRNA yields S-rRNA directly (simultaneous cleavage at site 1 and 3) and 32 S pre-rRNA, the immediate precursor to L-rRNA (DUDOV and DABEVA 1983). Thus, a compulsory split at site 2, before sites 4 and 5 (see Fig. 22) seems to be a basic requirement in pre-rRNA processing. This is easily understood if we envision that the complex secondary structure interactions of the 5.8 S rRNA segment make the formation of L-rRNA possible (see Fig. 25). It is plausible that the endonuclease attacks in pre-rRNA processing are critically dependent on the "correct" protein coating of the primary pre-rRNA chain and, in fact, drastic changes in the amount of different intermediate pre-rRNAs were observed upon cycloheximide block of protein synthesis (STOYANOVA and HADJIOLOV 1979).

In summary, the endonuclease actions involved in pre-rRNA processing appear to operate in a stochastic pattern, generating a variety of intermediate pre-rRNAs and multiple processing pathways. Direct formation of S-rRNA from pre-rRNA is possible, while the formation of L-rRNA can result only through at least two subsequent cleavage steps.

IV.2.4. Enzyme Mechanisms

That the nucleolytic hydrolysis of phosphodiester bonds in pre-rRNA is highly selective is now evident. If we consider the scheme of primary pre-rRNA processing (see Fig. 22), it is clear that the accurate endonuclease cleavage of only 5 or 6 (out of 8,000 to 12,000) phosphodiester bonds would be sufficient to generate mature S-rRNA, 5.8 S rRNA, and L-rRNA, but that is certainly an oversimplification; in any case, the homogeneous termini in mature rRNA reveal that their generation and shaping proceeds with a very high precision. The enzyme mechanisms involved in pre-rRNA processing are still obscure, but they certainly include endonuclease, and possibly, exonuclease actions. As outlined above, the conformational interactions between the segments of putative mature rRNA within the primary pre-rRNA seem to play a major role in defining a specific cleavage site. Additionally, the protein coating of the primary pre-rRNA likely plays an important, yet unknown, role in directing nuclease action. Hence we can envisage different processing possibilities, ranging from strictly specific nucleases for each cleavage site to fully nonspecific nucleases acting on cleavage sites determined by the unique structure of pre-rRNA and preribosomes.

Analysis of terminal nucleotides and sequences in eukaryotic rRNA have provided information on the enzymes involved in their formation. The results obtained have been tabulated (see Cox 1977) and show that generally the terminal nucleotides in S-rRNA, and to a lesser extent in 5.8 S rRNA and L-rRNA, have been conserved in evolution. Thus, it seems that in most cases 5′-ends are generated by a cleavage of NpU or NpC bonds. The 3′-end of S-rRNA results from the split of a ApN bond, while hydrolysis of a UpN bond yields the 3′-end of 5.8 S rRNA. Unfortunately, the generalizations cannot be extended to adjacent spacer nucleotides as evident from examination of the sequences at the junctions between tS_e and tS_i sequences and mature rRNA. Thus, it appears that either the endonuclease cuts are not specified by the character of partner nucleotides in the phosphodiester bond, or that they occur at specific neighbor sites. In the latter case further trimming of terminal nucleotides may occur, as suggested for the 5′-end of L-rRNA (Kominami *et al.* 1978, Hadjiolova *et al.* 1984 a). In any case, the nature of the phosphodiester bond at a given cleavage site obviously cannot, in itself, determine the high specificity of nuclease attack typical of pre-rRNA processing.

We do not yet know whether the structure of pre-rRNA alone is sufficient to define a given nuclease cleavage site, although such a possibility is supported by the accurate *in vitro* processing of *E. coli* pre-rRNA by ribonuclease III, an enzyme specific for double-stranded RNA (see NIKOLAEV *et al.* 1975), and by the accurate processing of "naked" primary pre-rRNA to L-rRNA and S-rRNA in *Saccharomyces cerevisiae* (WALTSCHEWA *et al.* 1983). The possible involvement of double-stranded structures in pre-rRNA in specifying endonuclease cleavage sites is supported by several observations. Hydrolysis with *E. coli* RNase III of purified 45 S pre-rRNA yields discrete RNA fragments, some with electrophoretic mobilities close to those of pre-rRNA products formed *in vivo* (GOTOH *et al.* 1974, TORELLI *et al.* 1977). Several authors have described the isolation and purification of endonucleases specific for double-stranded RNA from total homogenates (SHANMUGAM 1978, RECH *et al.* 1980) or isolated nuclei and nucleoli (SAHA and SCHLESINGER 1978, GRUMMT *et al.* 1979 b) of animal cells. Some of these enzymes introduce apparently selective nicks in 45 S pre-rRNA and generate pre-rRNAs and RNAs similar to those found *in vivo*. Analogous results were reported with endonucleases isolated from animal cell nuclei or nucleoli whose pre-rRNA conformational requirements were not specified (PRESTAYKO *et al.* 1972, 1973, WINICOV and PERRY 1974, 1975, GORCHAKOVA and SIDORENKO 1976, DENOYA *et al.* 1981). HALL and MADEN (1980) correctly pointed out that, given the broad variations in tS_i sequences, the formation of double-stranded structures specifying processing enzyme attack sites similar to the ones documented in *E. coli* (BRAM *et al.* 1980) are unlikely to exist in eukaryotic pre-rRNA. Thus, primary pre-rRNA processing could be the result of endonucleases attacking single-stranded RNA (HADJIOLOVA *et al.* 1984 b). Isolation and purification of such single-strand specific enzymes from animal cell nucleoli was described recently (EICHLER and TATAR 1980, EICHLER and EALES 1982). Nevertheless, while many endonucleases do exist in nucleoli, their involvement in primary pre-rRNA processing remains conjectural. Possibly the precise specification of endonuclease cleavage sites is determined by the conformation of the entire preribosome. Such a view is supported by findings on the *in vitro* processing with partially purified endonucleases of preribosomes (containing primary pre-rRNA) from HeLa (MIRAULT and SCHERRER 1972, KWAN *et al.* 1974), mouse L (WINICOV and PERRY 1974), Ehrlich ascites (GRUMMT *et al.* 1979 b) and Novikoff hepatoma (PRESTAYKO *et al.* 1973) cells. In all cases, *in vitro* processing with preribosomes was observed to be closer to the *in vivo* process then when purified primary pre-rRNA was used as a substrate. Again, further perfection of these *in vitro* systems is needed before they can be exploited fully in elucidating the mechanisms of pre-rRNA processing.

IV.3. Preribosomes: Structure and Maturation

The coating with proteins of the nascent pre-rRNA proceeds simultaneously with the RNA chain growth. Selective protein-RNA and protein-protein interactions involving individual r-proteins seem to take place during chain growth (CHOOI and LEIBY 1981), an indication that ribosome assembly starts during transcription. Thus, when the primary pre-rRNA is released from the rDNA (or r-chromatin) template, it is already assembled with ribosomal and non-ribosomal proteins following an apparently specific pattern. Actually, all pre-rRNAs in the nucleolus are associated with proteins in the form of preribosomes. Two types of preribosome particles were identified initially in HeLa cell nuclei: "80 S" containing 45 S pre-rRNA and "55 S" containing 32 S pre-rRNA (WARNER and SOEIRO 1967). The presence of dithiotreitol in the medium during isolation seems to be of critical importance for the successful extraction of these particles and has been widely used in subsequent studies. Because the "55 S" particles also contain 5 S rRNA (KNIGHT and DARNELL 1967), the latter is recognized as joining the preribosome at an early stage of ribosome biogenesis. Analogous types of particles were later identified and characterized in *Saccharomyces* (TRAPMAN *et al.* 1975b), *Tetrahymena* (LEICK 1969), amphibian oocytes (RODGERS 1968), mouse L cells (LIAU and PERRY 1969), rat liver (NARAYAN and BIRNSTIEL 1969), NOVIKOFF hepatoma (PRESTAYKO *et al.* 1972) and several other eukaryotic organisms.

The "80 S" preribosome is now firmly established as the precursor of the "55 S" preribosome, which in turn matures to produce the large ribosomal subparticle (see WARNER 1974, CRAIG 1974). In most studies, separate particles containing other intermediate pre-rRNAs could not be identified. However, in two of the preribosomal particles isolated from mouse L cell nucleoli, substantial amounts of 36 S pre-rRNA, in addition to 45 S and 32 S pre-rRNA, were identified (LIAU and PERRY 1969), whereas a discrete particle carrying 41 S pre-rRNA was reported for HeLa cells (MIRAULT and SCHERRER 1971). It appears that the "80 S" and "55 S" particles are in fact, processing particles, containing small amounts of other pre-rRNAs, different from the major component. Conceivably, minor changes during primary pre-rRNA processing (*e.g.*, removal of tS_e sequences) may not result in the formation of an identifiable intermediate particle.

The cleavage of pre-rRNA at site 2 (see Fig. 22), which generates separate precursors to L-rRNA and S-rRNA, is accompanied by sufficient change in constituents and/or conformation of the preribosome to allow unequivocal identification of at least one of the products—the "55 S" preribosome—as a discrete particle. Such a particle has been invariably found in all eukaryotes, thus far studied and constitutes, in fact, by far the largest pool of nucleolar preribosomes (see below). The other expected product (containing the precursor to S-rRNA) is not easily identified in the nucleolus. This may be due to the fact that such a particle is expected to be in the 40 S zone of density gradients, a zone which contains considerable amounts of hnRNP. Leakage during isolation or an exceedingly fast transfer to the nucleoplasm and cytoplasm may be also envisioned. In *Saccharomyces*, the last maturation step that yields S-rRNA takes place mainly in the cytoplasm where the respective preribosomes have been identified (UDEM and WARNER 1973, TRAPMAN *et al.* 1975 b). However, in *Tetrahymena* and other lower eukaryotes, formation of S-rRNA seems to take place both in the nucleolus and the cytoplasm (LEICK 1969, POUSADA *et al.* 1975). It may be that the cellular site of S-rRNA formation is related to the overall rate of ribosome synthesis. Thus, starvation-induced inhibition of protein synthesis in *Tetrahymena* results in the accumulation in nuclei of RNP particles containing both a precursor to and mature S-rRNA (RODRIGUES-POUSADA *et al.* 1979). A similar shift in the cellular site of S-rRNA formation may occur in a *S. cerevisiae* mutant carrying a defect in the transport and processing of the 20 S precursor to S-rRNA (CARTER and CANNON 1980). Isolation of HeLa nucleoli under mild conditions allows identification of a small pool of "40 S" particles containing 20 S pre-rRNA (MIRAULT and SCHERRER 1971). Recently, the "40 S" precursor particles in animal cell nucleoli were unambiguously identified by showing both the presence of 21 S pre-rRNA and immunochemically characterized S-protein constituents (TODOROV and HADJIOLOV 1981, TODOROV *et al.* 1983).

A rapidly labeled ribonucleoprotein particle sedimenting at about 30 S has been isolated from nucleoli of different animal cells. This particle was reported to contain 45 S pre-rRNA and was labeled faster with radioactive RNA precursors than the 45 S pre-rRNA in "80 S" particles suggesting that it may be an early-stage preribosome (BACHELLERIE *et al.* 1971, 1975). An analogous particle was observed also in *Tetrahymena* nuclear extracts (RODRIGUES-POUSADA *et al.* 1979). These observations are rather intriguing since they suggest considerable structural reorganization in protein constituents and conformation of the primary preribosome before it starts along its maturation pathway. Unfortunately, the protein complement of these particles has not yet been characterized.

All in all, taking into account the considerable methodical difficulties in the isolation and purification of ribonucleoprotein particles as well as the minute amounts of nucleolar preribosomes, we can consider that at present only the "80 S", "55 S", and to a lesser extent "40 S" preribosomes have been identified and partially characterized.

The protein complement of preribosomes was studied in several laboratories. The total amount of proteins recovered in preribosomes constitutes only a relatively small portion of the total proteins in isolated nucleoli, but quantitative estimates are difficult since they largely reflect the purity of the isolated nucleolar fraction (PHILLIPS and MCCONKEY 1976, ROTHBLUM *et al.* 1977).

The "80 S" preribosomes constitute 10–20% of all nucleolar preribosomes and their isolation and characterization is more difficult. In mammalian cells, they contain mainly 45 S pre-rRNA, but variable amounts of other intermediate pre-rRNAs are consistently found. Whether these minor pre-rRNAs exist *in vivo* or they are contaminants or products of processing taking place during isolation is not known with certainty. The proteins of the "80 S" preribosomes are still poorly characterized, despite the fact that direct analyses of the proteins from isolated "80 S" preribosomes of HeLa (KUTER and RODGERS 1976, LASTICK 1980), mouse leukemia (AUGER-BUENDIA and LONGUET 1978, AUGER-BUENDIA *et al.* 1978, 1979), and rat liver (FUJISAWA *et al.* 1979) cells have been carried out. Although considerable differences between the different studies exist, some basic features seem to emerge. Thus, a large number of the structural L-proteins is already present in "80 S" preribosomes. In leukemia cells the authors identified 31 (out of 40) L-proteins, while 24 (out of 35) were recorded in rat liver. The "80 S" preribosome seems to contain a smaller number of the structural S-proteins, but still, almost half of the proteins, 15 out of 31 in leukemia cells and 13 out of 25 in liver cells, found in cytoplasmic ribosomes are already present in nucleolar "80 S" preribosomes. In a recent immunochemical study 11 out of 17 S-proteins tested were detected in "80 S" preribosomes from erythroleukemia cells (TODOROV *et al.* 1983). All studies converge to show that a large number of structural r-proteins (both L- and S-) associate with primary pre-rRNA during or shortly after transcription.

The "55 S" preribosome contains mainly 32 S pre-rRNA and is therefore a direct nucleolar precursor to the large ribosomal subparticle. It constitutes about 70–80% of the nucleolar population of preribosomes and its constituent proteins have been studied in considerable detail in HeLa cells (KUMAR and WARNER 1972, KUMAR and SUBRAHMANIAN 1975, KUTER and RODGERS 1976), mouse leukemia cells (AUGER-BUENDIA and LONGUET 1978), regenerating rat liver (TSURUGI *et al.* 1973), and *Saccharomyces carlsbergensis* (KRUISWIJK *et al.* 1978). Unexpectedly, most studies agree in showing that only a small number of additional structural r-proteins apparently join the preribosome during the "80 S" → "55 S" conversion. Thus, in mouse leukemia cells, the L-proteins of "80 S" and "55 S" preribosomes are virtually identical, a single protein being added during the transition. A similar situation exists in rat liver, in which, only 2 L-proteins seem to join the preribosome during generation of "55 S" particles (FUJISAWA *et al.* 1979). Thus, it appears that in mammalian cells the conversion of 45 S pre-rRNA into 32 S pre-rRNA is accompanied by few changes in protein composition.

A similar situation seems to exist during the generation of nucleolar "40 S" preribosomes from the "80 S" precursor. Only two additional proteins were added during formation of nucleolar "40 S" preribosomes in erythroleukemia cells (TODOROV *et al.* 1983) and apparently similar minor changes occur in rat liver (FUJISAWA *et al.* 1979).

In contrast, the conversion of nucleolar "55 S" and "40 S" preribosomes into nucleolar and cytoplasmic large and small ribosomal subparticles is accompanied by many changes in r-protein constituents. For example, at least 12 L-proteins are added during the formation of the large ribosomal subparticle in HeLa cells (KUMAR and SUBRAMANIAN 1975, AIELLO *et al.* 1977) and about 10 L-proteins join the ribosome in mouse leukemia cells (AUGER-BUENDIA and LONGUET 1978). Even a larger proportion of S-proteins (about 50%) are added during the formation of the cytoplasmic small ribosomal subparticle (AUGER-BUENDIA and LONGUET 1978, FUJISAWA *et al.* 1979, TODOROV *et al.* 1983). Due to methodical difficulties it is still impossible to specify the exact cellular site of these late additions of structural r-proteins, but it may occur in both the nucleolus and in the cytoplasm. Several authors have reported that some of the structural L- and S-proteins are, in fact, added at a cytoplasmic stage of ribosome maturation (*e.g.*, AUGER-BUENDIA *et al.* 1979, KUMAR and SUBRAMANIAN 1975, LASTICK 1980) and these reports are supported by observations showing that the nascent small and large ribosomal subparticles in the cytoplasm are characterized by buoyant densities distinct from those of mature subparticles (SAMESHIMA and IZAWA 1975, VAN VENROOIJ *et al.* 1976, VAN VENROOIJ and JANSSEN 1976).

The studies on the successive stages in the maturation of preribosomes are complemented by the observation that several L- and S-proteins are recycled in the cytoplasm and their addition to the ribosome does not depend on the formation of new ribosomes. Evidence for the recycling of structural r-proteins in the cytoplasm was obtained in studies with *S. cerevisiae* (WARNER and UDEM 1972), *Drosophila* (E. BERGER 1977), mouse L cells (AUGER-BUENDIA *et al.* 1979, CAZILLIS and HOUSSAIS 1981), and HeLa cells (KUMAR and SUBRAMANIAN 1975, LASTICK and MCCONKEY 1976, AIELLO *et al.* 1977, LASTICK 1980). The process of recycling of some r-proteins may be correlated with several post-transcriptional modifications (phosphorylation, methylation etc.) that may alter their structure and possibly function. The *in vivo* phosphorylation of protein S 6 has attracted considerable interest, since it is apparently related to important cellular control mechanisms (see WOOL 1979, THOMAS *et al.* 1979). The role of that *in vivo* modification of eukaryotic ribosomes is not yet clarified and will not be considered here.

Summarizing the still limited information available, a distinct pattern of sequential addition of structural S- and L-proteins during the maturation of preribosomes in mammalian cells may be outlined (Fig. 27). The sequential addition of r-proteins is recognized from detailed labeling kinetics studies with different animal cells (KUMAR and SUBRAMANIAN 1975, LASTICK and McCONKEY 1976, AUGER-BUENDIA *et al.* 1979, LASTICK 1980). It is likely that the sequential addition of r-proteins is characteristic also for the maturation of preribosomes in lower eukaryotes.

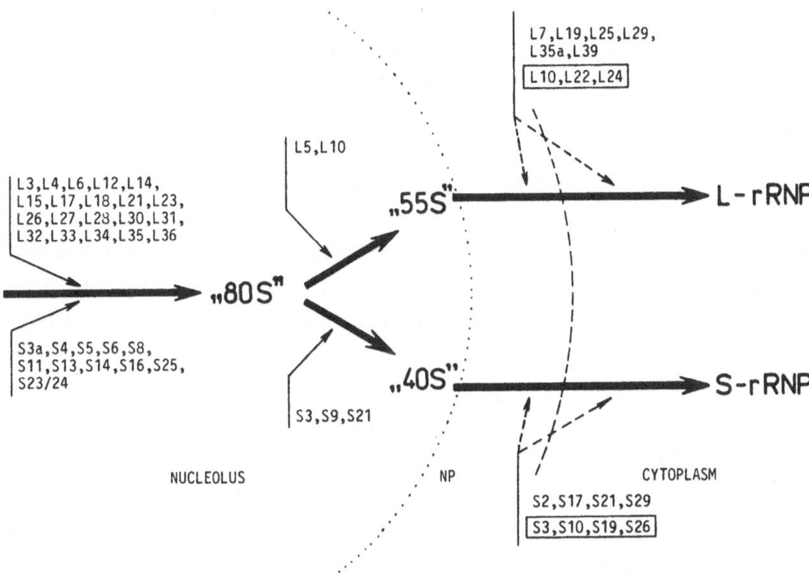

Fig. 27. Sequential addition of ribosomal proteins during the assembly and the maturation of the "80 S" primary preribosome in mammalian cells. The sites of addition of individual L- and S-proteins (nomenclature according to McCONKEY *et al.* 1979) are according to data from AUGER-BUENDIA and LONGUET (1978), FUJISAWA *et al.* (1979), LASTICK (1980), and TODOROV *et al.* (1983). The boxed proteins in the cytoplasmic compartment were found to exchange with cytoplasmic ribosomes. *NP* nucleoplasm. Due to differences in reported studies, the assignment of r-proteins to a given site of assembly is not definitive

The small changes in the protein constituents of preribosomes during the intermediate stages of preribosome maturation support the concept that accurate pre-rRNA processing is specified already during transcription. In this respect another aspect of r-protein interactions with rRNA seem to be of particular importance. Recent studies have shown, at least for rat liver, the specific interaction of proteins L 16, L 8, L 19, and S 13 with 5.8 S rRNA (ULBRICH *et al.* 1981) and L 6, L 7, and L 19 with 5 S rRNA (ULBRICH *et al.* 1980, 1982). It is noteworthy that most of these r-proteins seem to join the preribosome at the last stages of its final maturation, *i.e.*, they are not present in nucleolar "80 S", "55 S", and probably "40 S" preribosomes

(LASTICK 1980). This is an unexpected observation, although not yet proven directly, in light of the likely role of 5.8 S rRNA in pre-rRNA processing (see IV.2.2.). The absence of 5.8 S rRNA binding proteins in "80 S" and "55 S" preribosomes suggests that pre-rRNA processing requires the presence of a "naked" 5.8 S rRNA segment. The nuclease cuts that shape the 5′- and 3′-termini of 5.8 S rRNA (during the conversion of 32 S pre-rRNA into L-rRNA) may generate a structure that permits association with 5.8 S rRNA specific proteins. The 5 S rRNA situation may be of even greater interest and importance in preribosome maturation, since it seems that in amphibian oocytes (see BROWN 1982) and in mammalian cells (RINKE and STEITZ 1982) this rRNA (or its precursor) exists in nuclei in the form of free ribonucleoprotein particles associated with specific transcription-control proteins (see BROWN 1982). On the other hand, 5 S rRNA enters the preribosome at an early stage (see above), but apparently without the three r-proteins that it binds selectively. The exact 5 S rRNA: protein interactions within the preribosome remain unknown, but it seems that their elucidation could unravel important control mechanisms in ribosome biogenesis.

Since the early studies on nucleolar preribosomes it has been known that they contain substantial amounts of non-ribosomal proteins (KUMAR and WARNER 1972, SOEIRO and BASILE 1973). These non-ribosomal proteins are larger than most r-proteins. They do not exist in cytoplasmic ribosomes, are labeled at a markedly slower rate than r-proteins, and appear to be recycled in the nucleolus. In later studies the non-ribosomal proteins of both "80 S" and "55 S" preribosomes were characterized further. Ten to thirty non-ribosomal proteins have been reported by different investigators to be associated with mammalian preribosomes and apparently some of them are assembled during the conversion of "80 S" into "55 S" preribosomes (WARNER 1974, PRESTAYKO *et al.* 1974, KUMAR and SUBRAMANIAN 1975, ROTHBLUM *et al.* 1977, AUGER-BUENDIA and LONGUET 1978, FUJISAWA *et al.* 1979). From the above we can see that the advance in our knowledge of non-ribosomal proteins in preribosomes has been disappointingly slow. We know little of their origin and stoichiometry and know even less about their role in preribosome maturation. That the non-ribosomal proteins have such a role seems plausible and deserves further investigation.

Two groups of investigators have attempted to gain some insight into preribosome maturation by electron microscopic studies of isolated "80 S" and "55 S" (or analogous) particles from rat liver (MATSUURA *et al.* 1974) and HeLa cells (JOHNSON and KUMAR 1977). In accord with earlier studies (*e.g.*, NARAYAN and BIRNSTIEL 1969), preribosomes seem to have a less compact structure than mature ribosomes with some preribosomal particles showing a more extended rod-like form. Ribonuclease-sensitive fibrils are seen to protrude from the preribosome core (JOHNSON and KUMAR 1977), which also is suggestive of a looser association of proteins to pre-rRNA. Because of the rather drastic preparative techniques used in such studies, correlations with the electron microscopic images of structures observed in nuclei seems precarious and has not been attempted.

IV.4. Synopsis

The maturation of preribosomes in eukaryotes starts after transcription termination and release of the primary pre-rRNP from the rDNA template. Maturation involves several steps resulting in the formation of nascent small and large ribosomal particles. Almost all maturation steps take place in the nucleolus and end with the release of nascent particles and their transport to the cytoplasm. The following basic features may now be outlined:

1. Maturation starts normally after transcription termination and formation of the primary preribosome. The primary preribosome contains primary pre-rRNA (including tS_e—S-rRNA—tS_i 1—5.8 S rRNA—tS_i 2—L-rRNA segments). The rRNA segments are chemically modified (pseudouridylation and methylation) and the whole primary pre-rRNA has a characteristic secondary structure. The pre-rRNA is associated with ribosomal and non-ribosomal proteins, most likely at specific sites within the structure of the primary preribosome. The correct maturation of the primary preribosome depends critically on its primary structure and conformation.

2. The maturation of primary pre-rRNA involves several endonuclease cleavages. The endonuclease attack sites are specified by base-pairing interactions between the strongly conserved terminal segments of rRNA, as well as between the evolutionarily divergent spacer sequences. Endonuclease cleavages follow a preferred sequential pattern that results in the maturation of primary pre-rRNA along a major processing pathway characteristic for the species. These include: (i) removal of the tS_e segment; (ii) generation of precursors to S-rRNA and L-rRNA; (iii) generation of S-rRNA; (iv) generation of L-rRNA, base-paired with 5.8 S rRNA. The early processing steps may occur in a stochastic manner, resulting in the formation of different intermediate pre-rRNA and multiple processing pathways. The final shaping of S-rRNA is possibly defined by base-pairing between its 5′ and 3′ termini. In some lower eukaryotes it may take place in the cytoplasm. The final shaping of L-rRNA is directed by its base-pairing interaction with the 5′ and 3′ termini of the 5.8 S rRNA segment. The endonucleases catalyzing pre-rRNA processing may prefer single stranded tracts in spacer segments, but their specificity and structural requirements remain to be clarified.

3. The addition of r-proteins proceeds in two main stages during ribosome

formation: (i) during transcription and (ii) at the last stages of S-rRNA and L-rRNA formation. About half of the L-proteins and one third of the S-proteins present in mature ribosomes seem to be assembled during or shortly after transcription. The addition of r-proteins at the last nucleolar stages of preribosome maturation likely plays an important role in forming the nascent large and small ribosomal particles ready to be released from the nucleolus. The formation of the nascent large ribosomal particle (containing L-rRNA, 5.8 S rRNA, and 5 S rRNA) is slower than that of the small particle and seems to involve complex RNA : RNA, RNA : protein and protein : protein interactions.

Many facets of the process of preribosome maturation in the nucleolus remain obscure. Undoubtedly, the process involves complex reorganizations in the sequence and conformation of pre-rRNA and rRNA molecules and their highly selective interaction with r-proteins. Transcription of all mature r-RNA segments within a common primary pre-rRNA molecule guarantees the synchronous formation of the two ribosomal particles. However, more stringent structural requirements appear to be involved in the formation of the large ribosomal particle that result in a longer intranuclear maturation. The structural characteristics determining the readiness of a nascent ribosomal particle (small or large) to be released from the nucleolus are unknown. Further analysis of preribosome maturation and the molecular mechanisms resulting in the formation of nascent ribosomal particles will certainly clarify important aspects of the structure and function of the nucleolus.

V. Molecular Architecture of the Nucleolus

V.1. Introduction

The nucleolus was observed soon after the nucleus was recognized as being ubiquitous in eukaryotic cells. One of the first observations made when cells were examined with the electron microscope was that the nucleolus is the only prominent cellular organelle not delimited by a lipoprotein membrane. Yet, nucleoli are more or less sharply delineated from the nucleoplasm and usually possess a roughly spherical shape. In most cells, the nucleolus is partially surrounded by a layer of condensed chromatin, designated nucleolus-associated chromatin. Only in some primitive eukaryotes does the nucleolus have a notably different structure, *e.g.*, in *Saccharomyces* the nucleolus is seen as a "dense crescent", a structure that occupies a large part of the nucleus (MOLENAAR *et al.* 1970, SILLEVIS-SMITH *et al.* 1972). Nucleoli disappear during cell division and are reformed in telophase at the chromosomal nucleolus organizer sites where the rRNA genes are located. In some cells, a large number of spherical nucleoli are formed in connection with amplified extrachromosomal rRNA genes (see II.2.3.).

As the main cellular center of ribosome biogenesis the nucleolus is recognized as the site of the following major structures and functions discussed in preceding Chapters: a) active and inactive rDNA repeating units assembled with proteins into r-chromatin; b) transcription of rRNA genes and formation of primary preribosomes; c) maturation of preribosomes to form large and small nascent ribosomal particles. The nucleolus is thus an uniquely dynamic structure in which gene expression can be visualized electron microscopically and in which changes in ultrastructure under different physiological and pathological conditions for a broad variety of eukaryotic cells and organisms have been thoroughly investigated (see BERNHARD and GRANBOULAN 1968, BUSCH and SMETANA 1970, SMETANA and BUSCH 1974, BOUTEILLE *et al.* 1982). The purpose of the following discussion is to outline the qualitative and quantitative macromolecular features of various nucleolar structures.

V.2. Nucleolus Organizer

V.2.1. Chromosomes

Observations of the nucleolus throughout the mitotic cycle demonstrate that it reforms at telophase in association with the "secondary constriction" of certain chromosomes (HEITZ 1931, 1933); the number and size of nucleoli appear to correspond to the number of "secondary constrictions". Secondary constrictions are distinct from "primary constrictions" (more often called "centromeres" or "kinetochores")—the attachment site of mitotic spindle microtubules. Cytogenetic studies with *Zea mays* have shown that the material in the secondary constriction and the adjacent heterochromatin of chromosome 6 is responsible for the formation of the nucleolus and this chromosome segment has come to be called the *nucleolus organizer* (NO) or *nucleolus organizer region* (NOR) (MCCLINTOCK 1934, see BUSCH and SMETANA 1970). These two terms are considered synonymous, I shall use the designation *nucleolus organizer*.

Numerous investigative approaches, including observations of deletion and translocation NO mutants in plants and animals, *in situ* rRNA : DNA, and rDNA : DNA hybridization, have proven conclusively that nucleolus organizers are the sites of the rRNA gene clusters (see II.2.2.). The site of the NO and the "secondary constriction" usually coincide, but the correlation is not absolute—some cytologically defined "secondary constrictions" may not contain rRNA genes and *vice versa* (*cf.* HSU *et al.* 1975, PARDUE *et al.* 1970, PARDUE and HSU 1975, BATISTONI *et al.* 1978; DUHAMEL-MAESTRACCI and SIMARD 1980). It seems that the "secondary constriction" part of the NO corresponds to the site of active rRNA genes in the nucleolus of the previous interphase. The adjacent heterochromatic part of the NO also may contain rRNA genes possibly inactive during interphase (GIVENS and PHILIPS 1976, PHILIPS 1978). This situation seems to be typical for plant species characterized by a content of rRNA genes markedly higher than in most eukaryotes a large number of which are not expressed, and in the case of wheat have been found to be clustered in the nucleoplasmic heterochromatin (FLAVELL and MARTINI 1982). These observations suggest that r-chromatin within the NO of mitotic or meiotic chromosomes may exist in two distinct conformations. That the "secondary constriction" part of the NO r-chromatin does indeed have a distinct structure is strongly supported by the electron microscopic observation of the NO in metaphase chromosomes (HSU *et al.* 1965, 1967, ESPONDA and GIMENEZ-MARTIN 1972, GOESSENS and LEPOINT 1974). In both plant and animal cells the NO is composed of fine fibrillar material of low electron density. The fibrils in the NO are only about 5 nm in diameter and markedly thinner than the fibrils in the adjoining condensed chromatin. The molecular structure of r-chromatin in the NO is unknown, but it is conceivable that even in the mitotic chromosome it is in an unlocked (potentially active) state (see III.2.3.). The potentially active state of r-chromatin seems to be related to the presence of at least one NO specific non-histone protein, the so-called Ag-NOR protein. The NOs of plant and animal cells have been shown to be selectively "stained" by a specially devised Ag-staining procedure (GOODPASTURE and BLOOM 1975, HOWELL *et al.* 1975) and that results in Ag binding to a specific acidic, non-histone protein (HOWELL 1977, OLERT *et al.* 1979, SCHWARZACHER *et al.* 1979). The presence of this Ag-NOR protein is related to the potentially active state of r-chromatin in NO as shown by a study of mouse-human hybrid cells that do not synthesize human rRNA (ELICEIRI and GREEN 1969). The NOs in the human "nucleolar" chromosomes of such hybrids were negative following the Ag-staining reaction, while in the respective mouse chromosomes they were positive thus showing that the Ag-NOR protein is present only in the unlocked r-chromatin (MILLER *et al.* 1976, DEV *et al.* 1979).

V.2.2. Interphase Nuclei

Because the NO is equivalent to r-chromatin, we can imagine that the r-chromatin of interphase nuclei may exist in at least three forms: (i) nucleolar, transcribed by RNA polymerase I; (ii) nucleolar, non-transcribed; and (iii) nucleoplasmic in heterochromatin, most likely non-transcribed. Leaving aside (iii) we can ask: what are the morphological counterparts of forms (i) and (ii) in interphase nucleoli? Form (i) obviously is the counterpart of the "Christmas tree" pattern seen by chromosome-spreading and will be considered later. Here, I shall discuss the morphology of form (ii) r-chromatin. After more than a decade of intensive investigations and often controversial interpretations, interphase NO now seems to correspond to nucleolar structures observed by many investigators and designated by at least 20 different terms (see GOESSENS and LEPOINT 1979), most widely "*fibrillar centers*", the term first proposed by RECHER *et al.* (1969). The "fibrillar centers" of nucleoli are observed in many, but not all, plant and animal cells, and appear as small rounded areas containing fibrils (about 5 nm of diameter) and displaying a low electron density. The evidence favoring the proposal that the "fibrillar centers" correspond to non-transcribed nucleolar r-chromatin (or nucleolus organizer) (CHOUINARD 1971, GOESSENS and LEPOINT 1979) has been reviewed (GOESSENS and LEPOINT 1979, STAHL 1982) and the main arguments may be summarized as follows:

a) The structure of the "fibrillar centers" is similar to the structure of the r-chromatin in the "secondary constriction" part of the NO (GOESSENS and LEPOINT 1974) (Fig. 28).

b) In several cases the early stages of nucleologenesis have been thoroughly investigated and it has been possible to trace the relationship between the chromosomal NO and the emerging nucleolus. The nucleolus is formed invariably in conjunction with the chromosomal NO, the latter structure evolving into intranucleolar fibrillar centers (see STAHL 1982).

c) The fibrillar centers contain a small amount of DNA (GOESSENS 1976, MIRRE and STAHL 1978 b), which at least in the case of meiotic human spermatocytes, has been shown by *in situ* hybridization to be rDNA (ARROUA *et al.* 1982).

d) The r-chromatin in fibrillar centers is apparently non-transcribed. Autoradiographic studies show invariably that RNA is synthesized in the vicinity, but not within the fibrillar center structures (MIRRE and STAHL 1978 a, LEPOINT and GOESSENS 1978, HERNANDEZ-VERDUN and BOUTEILLE 1979, FAKAN and PUVION 1980, MIRRE and KNIBIEHLER 1981).

e) The fibrillar centers in nucleoli display a positive Ag-staining and therefore contain the Ag-NOR protein specific for NO r-chromatin (HERNANDEZ-VERDUN *et al.* 1978, 1980, DASKAL *et al.* 1980, PEBUSQUE and SEITE 1981, GOESSENS and LEPOINT 1982).

Thus, the "fibrillar centers" in interphase nucleoli appear to correspond to the chromosomal nucleolus organizer, but it should be stressed that they correspond only to *the non-transcribed part of the r-chromatin* in the nucleolus organizer. Accordingly, the proportion of nucleoli occupied by fibrillar centers is expected to vary in relation to the actual number of transcribed rRNA genes. An inverse correlation between the number and size of fibrillar centers and growth rate has been documented for a variety of cells (LOVE and SORIANO 1971, SMETANA *et al.* 1970, JORDAN and McGOVERN 1981). These studies support the notion that fibrillar centers correspond to the non-transcribed r-chromatin, although some questions remain unanswered and deserve further comment.

Studies of fibrillar centers revealed, somewhat unexpectedly, that in some cells their number exceeds by far the number of NOs (JORDAN and McGOVERN 1981, MIRRE and STAHL 1981) making for a quantitative mismatch. Moreover, the relatively high amounts of proteins present in fibrillar centers has been considered as evidence for some functions of these structures besides being the depositories of non-transcribed rRNA genes (JORDAN and McGOVERN 1981, MIRRE and KNIBIEHLER 1982). The *in situ* hybridization evidence for the presence of rRNA genes has thus far been

Fig. 28. Ultrastructure of NOR and fibrillar center during the mitotic cycle of Ehrlich tumor cells. *a* Chromosomes in anaphase. The arrow denotes the NOR. × 9,500. *b* Part of an anaphase chromosome with NOR. × 33,000. *c* A nucleolus in late telophase, × 91,000. *d* An interphase nucleolus with one prominent fibrillar center, × 22,000. Courtesy by Dr. G. GOESSENS [Figs. *b* and *d* are reproduced from GOESSENS and LEPOINT (1974) and GOESSENS (1976) with permission by the authors and *Experimental Cell Research*]

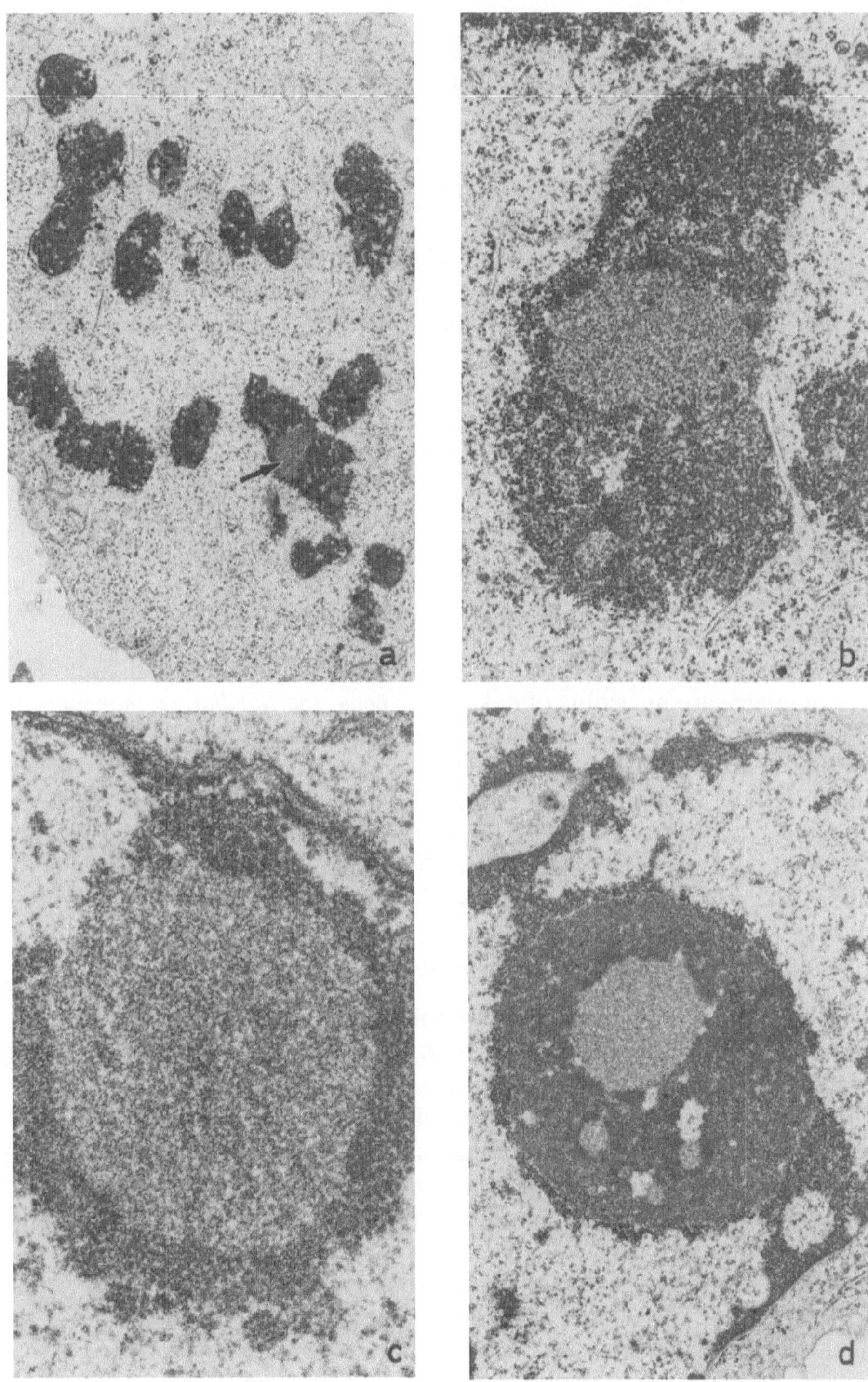
a
b
c
d

limited to some fibrillar centers in germ cells at the resolution of the light microscope (Knibiehler *et al.* 1977, Arroua *et al.* 1982). Whether all fibrillar centers in interphase nucleoli contain rDNA and to what extent ntS and rRNA genes are represented in these structures remains to be seen. It may be that fibrillar centers contain a relatively small amount of DNA, particularly rDNA, including mainly the ntS sequences of the rDNA repeating unit, an interpretation in line with the observation that in cultured mouse fibrolasts the number of fibrillar centers is close to the number of rRNA genes (Jordan and McGovern 1981). Still, the bulk of the material in fibrillar centers may well be contributed by some rDNA-associated proteins of unknown function.

These and other unresolved questions still await clarification. Nevertheless, from the available evidence we can conclude that non-transcribed r-chromatin is indeed present in nucleoli of some eukaryotic cells and that this r-chromatin is confined to the nucleolar fibrillar centers. Because Ag-NOR protein(s) are present in nucleolar fibrillar centers, this non-transcribed r-chromatin may be in an unlocked state.

V.3. Fibrillar and Granular Components

Two major components, containing ribonucleoproteins, constitute the active interphase nucleolus of eukaryotic cells: the *fibrillar component* and the *granular component*. Recent stereological studies on cultured human diploid fibroblasts (Jordan and McGovern 1981) and Ehrlich ascites tumor cells (Lepoint and Goessens 1982) show that 90% of the nucleolar volume is accounted for by the fibrillar (about 15%) and granular (about 75%) components. Therefore, identyfying the molecular counterparts of these two components will largely elucidate the basic structure of the nucleolus.

V.3.1. The Fibrillar Component

A filamentous structure in nucleoli, called *nucleolonema,* was first observed by Ag-staining of cells and attempts to trace its role in the organization of the nucleolus followed (Estable and Sotelo 1951, Estable 1966). Ultrastructural studies of nucleoli in different cells and tissues revealed many structural features that may be grouped in two main types: (i) the *compact nucleolus* and (ii) the *reticulated nucleolus* (nucleolus with nucleolonema) (see Smetana and Busch 1974, Stahl 1982). Both types of nucleoli contain fibrillar and granular components although in different proportions and spatial arrangements. Intermediate and transitional forms are often found, so that any generalization about the structure of the nucleolus risks being an oversimplification. Also, the relative amount of the fibrillar component varies with cell type, physiological state, and the stage of the cell cycle. The fibrillar component is constituted by more or less continuous filaments with a reported diameter in the range of 4–10 nm. It is the first distinct structure formed during nucleologenesis (see BOUTEILLE *et al.* 1982) and its continuity with the chromosomal NO or interphase fibrillar centers has been thoroughly documented in some cases (*i.e.* MIRRE and STAHL 1981, STAHL 1982). It seems also that there is a continuity between the nucleolus-associated chromatin and the fibrillar components of both compact and reticulated nucleoli (*cf.* BERNHARD 1966, SMETANA and BUSCH 1974, ANTEUNIS *et al.* 1979).

An important breakthrough in our understanding of the molecular architecture of the nucleolus was achieved by the introduction of the chromosome-spreading technique that provided an opportunity to visualize both the ntS and transcribed rRNA transcription units. The molecular structures underlying the observed "Christmas tree" pattern are now understood in much detail (see III.2.2.). However, because the highly contorted and compact state of nucleolar chromatin prevents visualization of active transcription units in ultrathin sections, we may ask: what structures observed in the nucleolus are the counterparts of the structures seen in spread chromosomes?

Numerous studies with amplified extrachromosomal nucleoli of amphibian oocytes have enabled investigators to trace the relation between the electron microscopic images seen in spread-chromosome preparations and ultrathin sections of nucleoli. The studies of MILLER and BEATTY (1969 a) showed that extrachromosomal nucleoli in *Triturus* oocytes are constituted by a central fibrous core and a granular cortex that correspond to the usual fibrillar and granular components seen in ultrathin sections of somatic cell nucleoli. These authors showed that upon spreading the fibrillar core is constituted almost entirely of a continuous array of tandemly arranged ntS segments and transcribed rRNA transcription units. This correspondence was confirmed and further investigated in several studies of amphibian or insect oocytes (TRENDELENBURG 1974, SCHEER *et al.* 1976, TRENDELENBURG and MCKINNELL 1979), as well of *Acetabularia* (SPRING *et al.* 1974) and a few other eukaryotic cells (see FRANKE *et al.* 1979). A detailed examination of partially spread whole extrachromosomal nucleoli from *Rana pipiens* oocytes, using a procedure for providing an enhanced contrast of transcribed rDNA axes, showed that the fibrillar components of the nucleolus seen in ultrathin sections correspond to the spread image of (Fig. 29): a) transcribed r-chromatin and b) non- transcribed r-chromatin in different stages of condensation (TRENDELENBURG and MCKINNELL 1979).

The molecular counterpart of the fibrillar component in interphase nucleoli of somatic cells is more difficult to assess. In these cases a direct correlation between ultrathin and chromosome-spreading images is practically impossible because of the low proportion of r-chromatin in total nucleolar chromatin. Short-term labeling experiments with ^{3}H-labeled RNA precursors, followed by electron microscopic autoradiography showed that the fibrillar component is labeled before the granular component suggesting a precursor-product relationship between the two

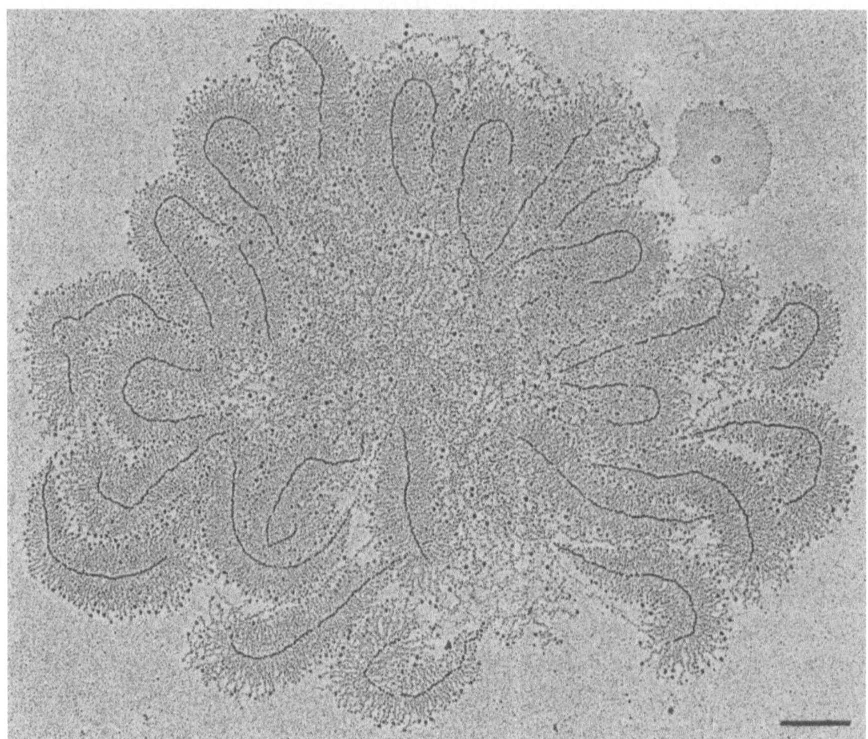

Fig. 29. Ultrastructure of partially inactivated nucleolus from full-grown *Rana pipiens* oocytes. The transcriptionally active matrix units are located peripherally in this case and the rDNA axis is prominent. The non-transcribed chromatin is in the central region and associated with some matrix units where RNA polymerase I granules may be discerned. Electron micrograph of spread nucleolus, × 15,000. [From TRENDELENBURG and MCKINNELL (1979). Reproduced with permission by Dr. M. TRENDELENBURG and *Differentiation*]

(GRANBOULAN and GRANBOULAN 1965, KARASAKI 1965). This relationship was supported by experiments that used pulse-chase procedures upon actinomycin D blockage of transcription (GEUSKENS and BERNHARD 1966). Very short (2–5 minutes) ^{3}H-uridine pulse labeling experiments showed that the label appears initially in the fibrillar component adjacent to extranucleolar and intranucleolar chromatin or to fibrillar centers (FAKAN

1978, FAKAN and PUVION 1980). Biochemical fractionation of fibrillar and granular components, with electron microscopic control of the fractions, confirmed the earlier labeling of the fibrillar component (DASKAL *et al.* 1974, BACHELLERIE *et al.* 1975). In a combined biochemical and auto-

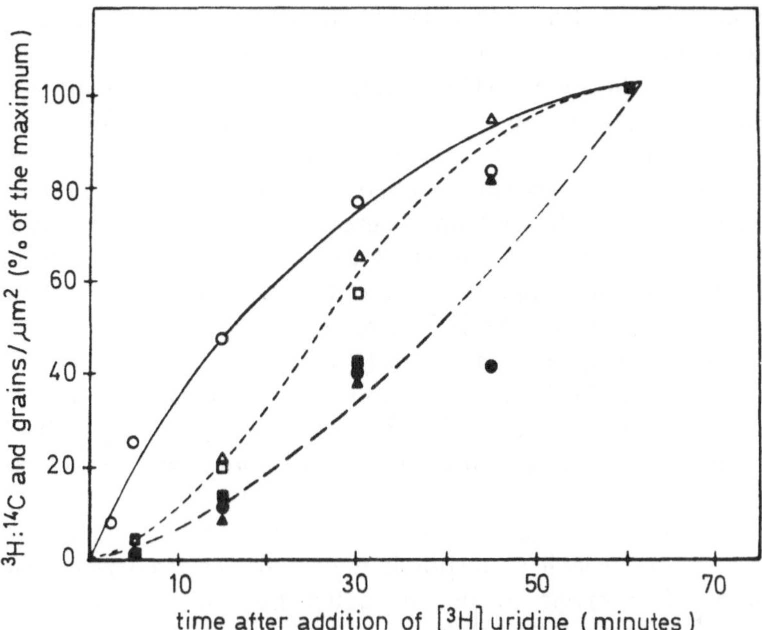

Fig. 30. Kinetics of labeling of nucleolar fibrillar and granular components, preribosomes, and pre-rRNA in cultured CHO cells. The labeling with [^{3}H]-uridine in the fibrillar (○) and the granular (●) components is estimated by autoradiography. The labeling of nucleolar "80 S" preribosomes (□), "55 S" preribosomes (■), 45 S pre-rRNA (△), and 32 S pre-rRNA (▲) is estimated by the ^{3}H/^{14}C ratio in cells, labeled with [^{14}C]-uridine for 24 h and with [^{3}H]-uridine for the indicated times. [From ROYAL and SIMARD (1975). Reproduced with permission by the authors and *The Journal of Cell Biology*]

radiographic study of CHO cells, the fibrillar component was found to be labeled *before* newly completed 45 S pre-rRNA molecules, while labeling of both of those entities preceded that of the granular component and 32 S pre-rRNA (Fig. 30) (ROYAL and SIMARD 1975).

All these studies provide evidence that the fibrillar component corresponds either to: (i) transcribed rRNA genes or (ii) primary preribosomes (containing 45 S pre-rRNA in mammalian cells) or (iii) both molecular entities, but no concensus among investigators has yet been achieved. Until recently, most investigators accepted the view that the fibrillar component corresponds to primary preribosomes (containing 45 S pre-rRNA), a view apparently rooted in the fact that the early autoradiographic studies were carried out before the advent of the chromosome-spreading technique and at a time when completed 45 S pre-rRNA molecules were considered to be the first detectable product of active rRNA genes. It was only gradually realized that the presence in animal cells of about 100–200 transcribed rRNA transcription units cannot escape visualization. More recently, several authors have adopted the interpretation that active rRNA transcription units are confined to the nucleolar fibrillar component (FRANKE *et al.* 1979, DIMOVA *et al.* 1979, FAKAN 1980, FAKAN and PUVION 1980). In particular, the nucleolonema in reticulated nucleoli is considered to correspond to the typical active rRNA transcription units observed upon chromosome-spreading (FRANKE *et al.* 1979). Additional evidence for the location of rRNA transcription units within the fibrillar component was obtained from electron microscopic studies that revealed that Ag-NOR proteins are associated with the fibrillar component of different animal cells (BOURGEOIS *et al.* 1979, HERNANDEZ-VERDUN *et al.* 1980, 1982, DASKAL *et al.* 1980, DIMOVA *et al.* 1982). In chromosome-spreads the Ag-NOR proteins were found to be located mainly in the rDNA axes of active transcription units (ANGELIER *et al.* 1982).

Summing up the available evidence it may be concluded that the fibrillar component of nucleoli corresponds mainly to the *transcribed part of r-chromatin.* It remains to be elucidated how the different states of transcriptional activity of r-chromatin are reflected in the appearance of the fibrillar component as seen in ultrathin sections.

The nucleolar location of primary preribosomes (containing 45 S pre-rRNA) is still uncertain. Some authors accept the view that these particles are located within the fibrillar component (see FAKAN and PUVION 1980, BOUTEILLE *et al.* 1982). On the other hand, based on results of the morphological changes in nucleoli upon inhibition of transcription, DIMOVA *et al.* (1979) proposed that the fibrillar component is composed mainly of active rRNA transcription units and contains no primary preribosomes. The following observations support the latter concept:

a) As seen in chromosome-spread preparations the growing pre-rRNA chain is coated with proteins and already packed in granules (in particular towards the 3′-end of the L-rRNA gene). It is likely that these granules preserve their structure upon release of the primary preribosome from the r-chromatin template.

b) The bulk of isolated "80 S" primary preribosomes show a spherical or rod-like structure, although as a result of looser packing some of them may show "filamentous" or "fibrous" protrusions (MATSUURA *et al.* 1974, JOHNSON and KUMAR 1977). However, similar "fibrous" structures are seen also in "55 S" preribosomes (NARAYAN and BIRNSTIEL 1969, JOHNSON and KUMAR 1977).

c) The observation that the fibrillar component is labeled before the "80 S" primary preribosome (ROYAL and SIAMRD 1975; see above) indicates that the short-term labeling of the fibrillar component (detected by autoradiography) reflects mainly the label in growing pre-rRNA chains rather than in completed preribosomes.

d) A blockage of transcription without a concomitant blockage of processing does not result in disappearance of the fibrillar component although 45 S pre-rRNA is fully processed (DIMOVA *et al.* 1979, GAJDARDJIEVA *et al.* 1982).

The considerations presented above show that the exact location of primary preribosomes within the nucleolus is difficult to determine given the fact that these particles constitute only a relatively small portion of the nucleolus (see IV.3.). Further evidence is needed to decide whether they are confined to the fibrillar or granular component.

V.3.2. The Granular Component

In ultrathin sections of nucleoli the granular component is more easily discerned than the fibrillar component. The granules in these nucleolar areas are ribonucleoproteins of about 20 nm in diameter. The granular component in compact nucleoli occupies about 70–80% of the nucleolar volume. The granular component in reticulated nucleoli is less conspicuous. Here, the ribonucleoprotein granules are apparently associated with the nucleolonemal network, while the interstitial spaces contain euchromatin or heterochromatin, as well as light, probably proteinaceous, areas commonly designated as vacuoles (see SMETANA and BUSCH 1974, STAHL 1982, BOUTEILLE *et al.* 1982).

The granular component contains preribosomes at different stages of maturation, as well as nascent large and small ribosomal subparticles. The quantitative characteristics of the pre-rRNA and rRNA constituents of the granular component in nucleoli from most plant and animal cells are still largely unknown. Biochemical analyses show that in mammalian cells the bulk of the granular component is comprised of "55 S" preribosomes (containing 32 S pre-rRNA and 28 S rRNA) (see IV.3.). The labeling kinetics of the granular component *in situ* and in isolated "55 S" preribosomes (and constituent 32 S pre-rRNA) of CHO cells are identical (ROYAL and SIMARD 1975), thereby demonstrating that the granular component includes mainly preribosomes that contain 32 S pre-rRNA. The granular component also contains particles with L-rRNA (28 S rRNA) and most likely these particles include most, but not all, L-proteins found in cytoplasmic large ribosomal subparticles (see IV.3.). Under some conditions the particles containing L-rRNA may become the predominant constituent of the nucleolar granular component as shown by a combined biochemical and ultrastructural study of rat liver nucleoli following thioacetamide-induced hypertrophy (KOSHIBA *et al.* 1971). The granular component also contains "40 S" preribosomes (precursors to the small ribosomal subparticle) (TODOROV *et al.* 1983), although their pool is markedly smaller than that of the particles containing 32 S pre-rRNA and L-rRNA.

In summary, the granular component of the nucleolus is comprised of a heterogeneous pool of preribosomes and ribosomes containing variable amounts of different intermediate pre-rRNA, as well L-rRNA and S-rRNA. In animal cells, the bulk of this material is in the form of ribonucleoprotein particles containing 28 S rRNA and its immediate precursor, 32 S pre-rRNA. An accumulation in nucleoli of analogous particles (containing L-rRNA and its precursor) is likely to also be typical for plants and lower eukaryotes. The suggested presence in the nucleolar granular component of primary preribosomes (containing primary pre-rRNA) remains to be proven.

V.4. The Nucleolus and Other Nuclear Structures

V.4.1. Nucleolus-Associated Chromatin

The nucleolus, in particular of animal cells, is usually seen in close association with a layer of heterochromatin. This heterochromatin is normally recovered in fractions of isolated nucleoli (BACHELLERIE *et al.* 1977 a, b, LOENING and BAKER 1976).The structure of nucleolus-associated chromatin seen in the electron microscope has been described in detail (see BERNHARD and GRANBOULAN 1968, BUSCH and SMETANA 1970) and will not be considered here; it appears to be identical to chromatin structures found at other nuclear sites. It is possible that the nucleolus-associated chromatin originates from "nucleolar" chromosomes. Whether some other genes involved in ribosome biogenesis (*e.g.* the genes for r-proteins or 5 S rRNA) are located in the vicinity of the nucleolus is not known, but it is noteworthy that at least one r-protein gene in *Drosophila* is located in the neighborhood of rDNA sequences (FABIJANSKI and PELLEGRINI 1982 b). High resolution *in situ* hybridization will certainly help to learn more about the genetic environment of the nucleolus.

V.4.2. The Junction with the Nuclear Envelope

A direct or indirect connection of the nucleolus with the nuclear envelope has been suspected for a long time and the evidence bearing on the matter has been expertly reviewed (BOUTEILLE *et al.* 1982). Briefly, in many cases attachment of nucleoli to the nuclear envelope is observed and this conclusion is extended to practically all animal cells by three-dimensional reconstruction electron microscopic studies; also, different lamellar or canalicular lipoprotein structures (Fig. 31) appear to constitute the nucleolus-nuclear envelope junction in cases when there is no direct contact between the two structures. Although the nucleolus-nuclear envelope

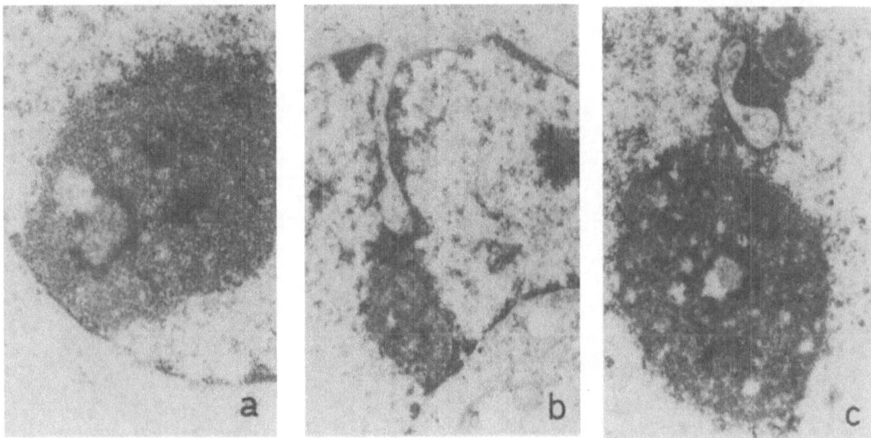

Fig. 31. Fine structure of the nucleolus-nuclear envelope junction in cultured TG cells, an established human cell line. *a* Direct attachment of the nucleolus to the nuclear envelope by an interposed layer of "dotted" chromatin (× 20,000). *b* A nucleolus attached both with the nuclear envelope and an invagination designated as a "nucleolar canal" (× 10,000). *c* A fully developed nucleolus associated with a "nucleolar canal" (× 20,000). (Courtesy by Drs. C. A. Bourgeois and M. Bouteille)

junction structures may play an important role in nucleolus-cytoplasm interactions (see below) little is known about the nature of this junction.

V.5. The Nucleolar Matrix

Early ultrastructural studies provided evidence for the existence in nuclei and nucleoli of a filamentous network of non-chromatin, proteinaceous structures designated as nuclear (or nucleolar) matrix (Bernhard 1966, Bernhard and Granboulan 1968). The development of appropriate isolation techniques gave an impetus to studies on the composition and interactions of the nuclear matrix. The rapidly expanding information has been reviewed (Berezney 1979, Shaper *et al.* 1979, Agutter and Richardson 1980) and may be summarized as follows. The isolated nuclear matrix contains 5–10% of the nuclear proteins, 0.1–0.5% of the DNA and

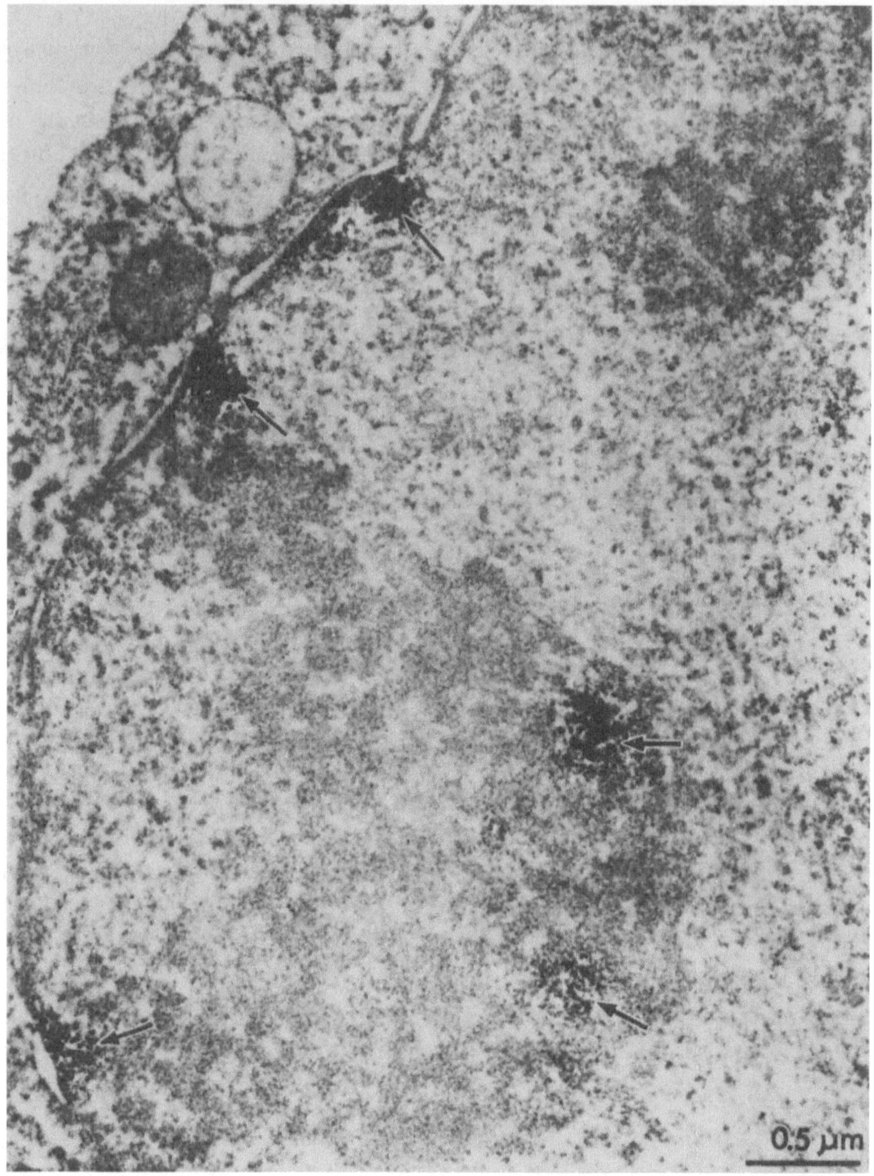

Fig. 32. Immune electron microscopy of an interphase CHO cell showing the localization (arrows) of centromere-associated antigens reacting with an autoimmune antibody from the sera of scleroderma patients. The centromere antigen is at the inner surface of the nuclear envelope or within the nucleolus. [From Moroi *et al.* (1981). Reproduced with permission by Dr. E. M. Tan and *The Journal of Cell Biology*]

1–2% of the RNA. Electron microscopy of the residual matrix structures show that they comprise three major components: (i) nuclear envelope with pore complexes; (ii) interchromatinic matrix structures and (iii) nucleoli (BEREZNEY and COFFEY 1977, BEREZNEY 1979). The gel electrophoresis pattern showed 3 prominent polypeptides in the molecular mass range of 60–70 kDa, which are further resolved into distinct fractions by 2D electrophoresis (COMINGS and PETERS 1981). Complex biochemical, immunochemical and ultrastructural studies revealed that some of these proteins are constituents of the nuclear envelope lamina-pore complex (see AGUTTER and RICHARDSON 1980). Gel electrophoresis analysis provided evidence that the major nuclear matrix proteins are identical to the proteins isolated from the scaffold of mitotic chromosomes (DETKE and KELLER 1982). Also, the elegant studies of MOROI *et al.* (1981) provided conclusive evidence that an antibody (from the serum of a scleroderma patient), specific for centromeric chromatin, interacts with antigenic structures confined to the inner surface of the nuclear envelope or the nucleolus (Fig. 32). These and other findings (see HANCOCK 1982, HANCOCK and BOULIKAS 1982) suggest that the nuclear matrix (in the nuclear envelope, the nucleoplasm or the nucleolus) may provide anchoring sites for chromatin in the interphase nucleus. The composition and organization of the nuclear matrix is still open to more precise characterization in particular concerning the components of the intranuclear matrix. Indeed, several authors observed a complete reorganization of the intranuclear matrix depending on the isolation conditions (KAUFMAN *et al.* 1981, GALCHEVA-GARGOVA *et al.* 1982). On the other hand, milder procedures of isolation revealed a prominent intranuclear filamentous network, whose protein composition was distinct from that of the cytoskeleton (CAPCO *et al.* 1982).

In most studies electron microscopy revealed the presence of residual nucleoli within the intranuclear matrix network. However, it was not clear whether the matrix structures of the nucleolus and the nucleoplasm are distinct or identical. Indications about the different composition of the nuclear and nucleolar matrix in liver was obtained when their protein constituents were compared (BEREZNEY and COFFEY 1977, TODOROV and HADJIOLOV 1979). Five distinct protein bands found in the nucleolar matrix were not present in its nuclear counterpart (TODOROV and HADJIOLOV 1979). Distinct nucleolar matrix proteins were found also upon 2D gel electrophoresis, and it was reported that the major protein specific for the nucleolar matrix has a M_r of 33 kDa (PETERS and COMINGS 1980, COMINGS and PETERS 1981). The characteristics of the minute amounts of residual DNA and RNA in the nucleolar matrix are not known. That RNA may play some role in the organization of the nucleolar matrix is suggested by experiments showing that treatment with ribonuclease under some conditions destroys residual nucleolar structures as seen in ultrathin sections (ADOLPH 1980, KAUFMAN *et al.* 1981, BOUVIER *et al.* 1982), but it should be noted that dissolution of residual nucleoli was observed also upon treatment with deoxycholate (KIRSCHNER *et al.* 1977) or EDTA (GALCHEVA-GARGOVA *et al.* 1982). An interesting observation showed that nuclear matrix DNA is about 5-fold enriched in rRNA genes (PARDOLL and VOGELSTEIN 1980). Whether this finding reflects a selective interaction of rDNA with the nuclear matrix remains to be clarified[5]. Generally, identification of an autonomous nucleolar matrix in eukaryotic cells is not easy to attain. A straitforward approach was used in studies on amplified extrachromosomal nucleoli of *Xenopus laevis* (FRANKE *et al.* 1981). The authors identified a nucleolar matrix structure constituted by protein filaments of a diameter of 4 nm, coiled into higher order fibrils (d = 30–40 nm), devoid of DNA and RNA. A single acidic protein was identified with a M_r of 145 kDa. Since amplified nucleoli are extrachromosomal these results provide direct proof for the existence of an autonomous nucleolar matrix. The observation that the nucleolar matrix contains actin and a 65 kDa protein present also in the nucleoplasmic matrix (see also CAPCO *et al.* 1982) indicates that a more complex structure for the nucleolar matrix could be envisaged. In this respect, the puzzling observations on anucleolate *X. laevis* mutants deserve special attention. It is known that O_{nu} mutants lack almost entirely the rRNA genes cluster (BIRNSTIEL *et al.* 1966). Yet, these mutants form "pseudonucleoli" constituted by a fibrillar network similar to nucleolar matrix structures (HAY and GURDON 1967). The possibility that "pseudonucleoli" might be nucleolar matrix structures visualized in the absence of rRNA gene activity remains to be ascertained.

V.6. Macromolecular Constituents

V.6.1. DNA and RNA

The characteristics of eukaryotic rDNA were reviewed earlier (see II.2.). In animal cells most of the rDNA is recovered in the nucleolus, whereas in plants a variable part may be associated with nucleoplasmic heterochromatin. Dissection of growing CHO cell nucleoli shows that about 75% of the rDNA is in intranucleolar chromatin (BACHELLERIE *et al.* 1977 a), but r-chromatin is only a small part of the total intranucleolar chromatin. The nature of the remaining intranucleolar DNA remains almost totally unknown.

The bulk of nucleolar RNA is pre-rRNA and rRNA (see IV.2.). Some small nuclear RNA (U 3 snRNA in particular) seem to accumulate in the nucleolus in association with pre-rRNA or rRNA (see REDDY and BUSCH

[5] Preferential association of rDNA with matrix structures was not observed in a recent study with *Dictyostelium* nucleoli (LABHART *et al.* 1984).

1981). Whether still other RNA species are confined to the nucleolus is not known, but in any case they would constitute a minor proportion of the total.

V.6.2. Nucleolar Proteins

V.6.2.1. General

Isolated nucleoli of animal cells have a protein/DNA ratio of about 8. Presumably most of the proteins are involved in the structure and operation of nucleolar chromatin, interphase NO, rDNA transcription complexes, assembly and processing of preribosomes. An estimated minimum of about 200 protein species is located in the nucleolus, including histone and non-histone proteins, ribosomal and non-ribosomal proteins in preribosomes (about 100), enzymes, etc. A census of nucleolar proteins has been attempted by several investigators (see BUSCH *et al.* 1978). Extraction with 0.075 M NaCl + 0.025 M EDTA, followed by 0.01 M Tris-HCl (pH 8.0) removes about 55% of the nucleolar proteins, mainly the proteins of ribonucleoprotein particles, the remaining 45% being mainly histone and non-histone proteins (ROTHBLUM *et al.* 1977). According to one study with whole nuclei of HeLa cells the non-histone proteins are composed of up to 450 components (PETERSON and MCCONKEY 1976). Almost all r-proteins in HeLa nucleoli are reported to be in preribosomes and ribosomes, with no significant nucleolar pool of free r-proteins (PHILLIPS and MCCONKEY 1976). Because the data on nucleolar r-proteins in preribosomes were discussed earlier (see IV.3.), I shall concentrate here on the proteins confined exclusively to the nucleolus, since they presumably play special roles in its structure and function.

One nucleolar protein expected to be exclusively nucleolar is RNA polymerase I. That both free and template-bound RNA polymerase I is indeed confined to the nucleolus is now firmly established (see III.1.1.). Several authors have shown that RNA polymerase I remains associated with nucleolar euchromatin during its isolation and fractionation (MARUSHIGE and BONNER 1971, TATA and BAKER 1974, DOENECKE and MCCARTHY 1975). Moreover, strong evidence was obtained that RNA polymerase I remains associated with chromosomes throughout the mitotic cycle (GARIGLIO *et al.* 1976, MATSUI *et al.* 1979, SCHEER *et al.* 1983). Although no rDNA transcription occurs during mitosis, the retention of inactive RNA polymerase I implies a stable and specific interaction with some NO structures. This finding raises the possibility that other proteins may also retain tight, specific nucleolar localization at all times.

V.6.2.2. Ag-NOR Protein(s)

Following introduction of a special Ag-staining reaction (GOODPASTURE and BLOOM 1975, HOWELL *et al.* 1975), evidence revealed that it stains selectively the NO in mitotic chromosomes and the fibrillar centers in interphase nucleoli (see BOUTEILLE and HERNANDEZ-VERDUN 1979, BOUTEILLE *et al.* 1982). After the technique was adapted for electron microscopy, the Ag-staining was found to be associated also with the fibrillar component of nucleoli in a broad variety of eukaryotic cells (BOURGEOIS *et al.* 1979, HERNANDEZ-VERDUN *et al.* 1978, 1980, 1982, DASKAL *et al.* 1980, PEBUSQUE and SEITE 1981, DIMOVA *et al.* 1982, GOESSENS and LEPOINT 1982). Further, it was shown that Ag-staining was associated with the rDNA axis in active transcription units, but not with ntS sequences (ANGELIER *et al.* 1982). All these observations raised the question of the chemical basis of Ag-staining, and it was recognized early that the staining is due to a non-histone, acidic protein, named Ag-NOR protein. Several authors attempted a detailed characterization of this apparently exclusively nucleolar protein. In one study, only a single "Ag-NOR" protein of those extracted by low-salt buffer was stained with silver on acrylamide gel electrophoregrams (HUBBELL *et al.* 1979), but in another study material extracted with higher ionic strength buffer from nucleoli of the same cells (Novikoff hepatoma) yielded two proteins stained by silver (LISCHWE *et al.* 1979). The latter were identified as the non-histone, phosphorylated proteins B 23 (M_r—37 kDa) and C 23 (M_r—110 kDa) present in nucleolar preribosomes (MAMRACK *et al.* 1979) and these two proteins were there considered to be responsible for the nucleolar Ag-staining reaction (LISCHWE *et al.* 1979, DASKAL *et al.* 1980). In the case of protein C 23, moreover, antibodies to this protein were shown to react selectively with antigenic sites in the fibrillar centers and in NOs of metaphase chromosomes (OLSON *et al.* 1981, LISCHWE *et al.* 1981), providing compelling evidence that C 23 is indeed a nucleolus-specific protein. It is not yet clear, however, whether this protein is associated with preribosomes, since such material is unlikely to persist in mitotic chromosomes. To what extent this protein is responsible for *in situ* Ag-staining is also unclear, since Ag-NOR proteins are associated with the rDNA axis, but not with the growing RNP fibrils (ANGELIER *et al.* 1982). Several argyrophilic proteins of *Xenopus laevis* oocytes that are stained on acrylamide gels have been identified, but they are not responsible for the *in situ* Ag-staining of nucleoli (WILLIAMS *et al.* 1982). The latter authors identified only two Ag-stained polypeptides (M_r 195 and 190 kDa) in purified nucleoli and suggested that they correspond to RNA polymerase I. This is a likely possibility and, if proven, Ag-staining may be considered as a highly selective technique for RNA polymerase I visualization.

V.6.2.3. Nucleolar Antigens

Due to their high specificity and the possibility of visualizing microscopically antigen-antibody complexes, immunochemical techniques are powerful tools in the search for specific nucleolar proteins. Structural r-proteins in nucleolar preribosomes and ribosomes can be easily identified by immunochemical methods (TODOROV and HADJIOLOV 1981, TODOROV *et al.* 1983) and the use of monoclonal antibodies against individual r-proteins (TOWBIN *et al.* 1982) will certainly increase the resolving power of these methods. In contrast, the proteins of nuclear hnRNP particles are exclusively nucleoplasmic and do not appear in the nucleolus (MARTIN and OKAMURA 1981). Here, I shall try to summarize the evidence on exclusively nucleolar antigens.

Numerous antinucleolar antibodies have been identified in the sera of patients with autoimmune diseases (TAN 1978), but only one has been shown to bind exclusively to nucleoli. This antibody is found in scleroderma patients and is directed against a nucleolar "4-6 S RNA", but the putative RNA antigen is poorly characterized. Whether it is one of the known nucleolar snRNA species remains to be established. Another antibody, identified in Sjögren syndrome patients, reacts with a poorly characterized acidic non-histone antigen named the SS-B antigen. The SS-B antigen is found in the nucleoplasm, but a distinctly nucleolar localization is observed during the G 1/early S phase of the mitotic cycle (DENG *et al.* 1981).

Another line of investigations aims at the immunochemical characterization of nucleolar proteins. Various antinucleolar antibodies have been obtained and some of them react predominantly with antigens restricted to the nucleolus (see BUSCH *et al.* 1978). One of these antigens is apparently the nucleolar protein C 23 described above, but much remains to be learned about the others.

In summary, it seems firmly established that many proteins are concentrated exclusively in the nucleolus, but their roles in the structure and function of the nucleolus still remain to be clarified. At present, RNA polymerase I appears to be the best characterized exclusively nucleolar protein. The basis for maintaining the stable association of this enzyme with NO structures (even in the absence of transcription) remains to be understood.

V.7. Outline

The experimental evidence accumulated during the last decade permits us to outline the molecular parameters of nucleolar structure, a structure that is essentially a ribosome-making machine. A general scheme, utilizing presently available knowledge, of the molecular architecture of the nucleolus is presented in Fig. 33. The scheme can be considered from two basic aspects: structural (static) and functional (dynamic).

The *structural aspects* of the scheme imply the following:

1. The nucleolus is formed at the nuclear site of active rRNA transcription

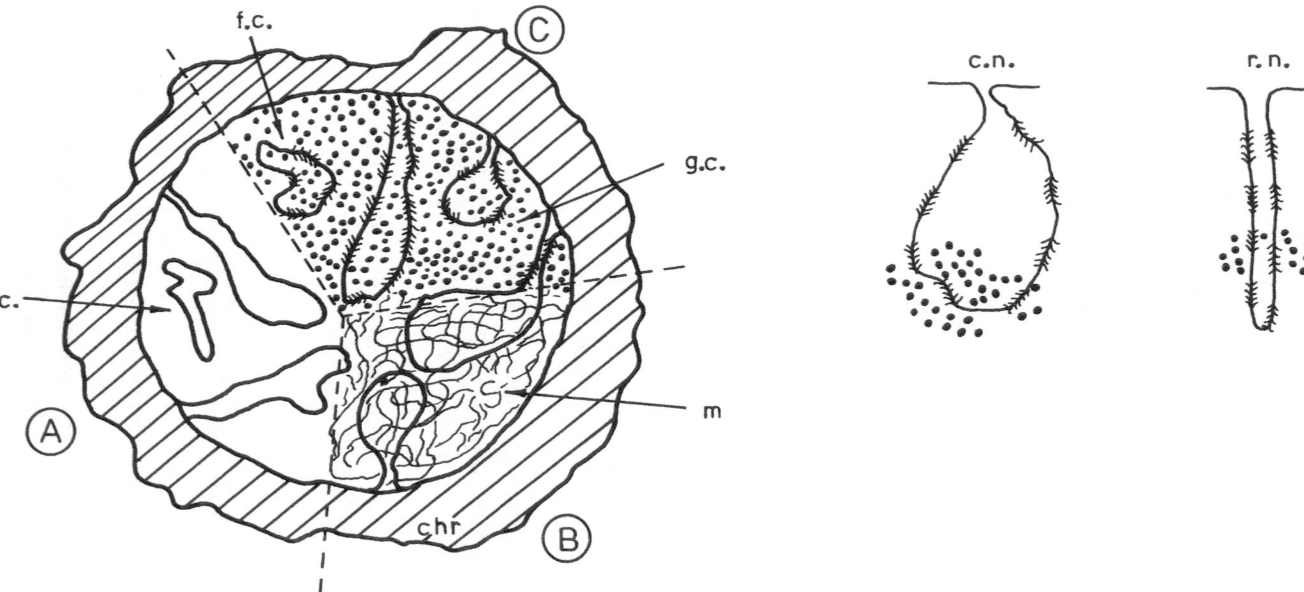

Fig. 33. A general model for the ultrastructure of the nucleolus. The nucleolus (left) is divided into 3 sectors to illustrate different aspects of its ultrastructure. *A* r-chromatin loops (*r.c.*), associated or not with the perinucleolar chromatin (*chr*). *B* A nucleolar matrix (*m*) meshwork envelops the r-chromatin loops. *C* The fibrillar (*f.c.*) and granular (*g.c.*) components in a fully active nucleolus. The fibrillar component is considered as the visualization of transcribed r-chromatin, while the granular component contains all types of preribosomes and nascent ribosomes. The three structures depicted in A, B, and C are superimposed in the active nucleolus. On the right two nucleolar structures representing compact (*c.n.*) or reticulated (with nucleolonema) (*r.n.*) nucleoli are depicted to display the possible appearance of fibrillar and granular components in these two types of nucleoli

units, which are seen as such in chromosome spreads or visualized as fibrillar components in ultrathin sections. The maximal number of rRNA transcription units is characteristic of the species. In the somatic mammalian cells about 100–200 rRNA transcription units per nucleolus operate simultaneously. The growing pre-rRNA chain is coated with proteins and packed in granular structures so that the transcription complex often acquires a fibrillogranular appearance.

2. Transcription termination release primary preribosomes, which have a more or less granular structure, and as a result contribute material to the nucleolar granular component. The nucleolus is the site where the maturation of preribosomes takes place. Thus, the nucleolar granular component comprises preribosomes and ribosomes at different stages of maturation and contains about 5×10^4 of these particles in the interphase animal cell nucleolus. What triggers the release of nascent ribosomal particles from the nucleolus is not known.

3. The fibrillar and granular components of active nucleoli exist in an environment of intra- and extra-nucleolar chromatin in varying states of condensation. The cluster of rDNA repeating units in transcribed r-chromatin is most likely linked to some elements of the intra- and extra-nucleolar chromatin. Non-transcribed r-chromatin, possibly associated with specific nucleolar proteins, may constitute structures seen as fibrillar centers in some cells.

4. The whole nucleolar structure is not floating free in the nucleus. It is anchored to skeletal structural elements of interphase chromosomes associated with the nuclear envelope. The system of transcribed and non-transcribed rDNA (r-chromatin) plus preribosomes and ribosomes in different stages of maturation appears to be held together by interactions with a fibrillar protein network—the nucleolar matrix, but it remains to be established whether the nucleolar matrix differs from the nuclear matrix.

The *dynamic aspects* of the scheme imply the following:

1. The morphology of the nucleolus during interphase reflects the existence of three major pools related by precursor-product relationships: (i) transcribed rRNA genes, (ii) preribosomes in different stages of maturation, and (iii) nascent large and small ribosomal subparticles. Under steady-state conditions the sizes of these three pools are relatively constant, resulting in an apparently stable structure for the nucleolus. In this sense the compact or reticulated types of nucleoli may represent differences in the steady-state sizes of the major nucleolar pools.

2. The nucleolus is an exceedingly dynamic structure basically reflecting the balance between the rate of production of primary preribosomes (input) and the rate of release of nascent, large and small, ribosomal subparticles (output). An idealized mammalian cell allows the following approximate estimates. If we assume the simulataneous activity of 100 rRNA

transcription units (approximate overall length 4 μm), each loaded with 160 molecules of RNA polymerase I and a transcription rate of 20–40 nucleotides per second per molecule of RNA polymerase, each second the nucleolus will make RNA chains containing a total of $3.2–6.4 \times 10^5$ nucleotides. Since the size of mammalian primary pre-rRNA is about 12.5×10^3 nucleotides, we shall have an input of 25.6–51.2 primary pre-rRNA molecules (and preribosomes) per second, or about 1,500–3,000 preribosomes per minute. To preserve the steady-state, the same number of nascent ribosomal subparticles should leave simultaneously the nucleolus. With the pool containing a total number of about 5×10^4 preribosomal and ribosomal particles, the pool will be replaced every 15–30 minutes. In other words if we block transcription and allow maturation and ribosome release to proceed at unchanged rates, we can expect the granular component of the nucleolus (75% of this organelle) to disappear within 15–30 minutes.

3. Modulation of the rates of transcription, maturation, and release is expected to have a rapid and profound repercussion on nucleolar morphology. Exactly such rapid changes are currently observed in studies on either nucleologenesis or disappearance of nucleoli under various physiological or pathological conditions. Again the major components of the nucleolus are expected to display pronounced quantitative alterations. In contrast, the nucleolar chromatin (including r-chromatin in the NO), as well as the nucleolar matrix, appear as relatively more stable structures in the interphase nucleolus.

The different facets of the model are not evenly supported by experimental evidence. Some are almost generally accepted, while others are controversial. The model is intended to provide a plausible framework for understanding the comportment of the nucleolus in the life cycle of eukaryotic cells and alterations induced by noxious agents.

VI. Regulation

VI.1. General Considerations

Ribosomes are needed for protein synthesis and the control of ribosome biogenesis has to respond to the requirements for protein production at any time in the life of a cell. Protein synthesis in prokaryotes involves the interaction of short-lived mRNAs with an excess of metabolically stable ribosomes, but such metabolic characteristics are clearly not valid for eukaryotes. In fact, the formation of ribosomes in growing eukaryotic cells occurs throughout most of the cell cycle, except for a short period during mitosis (see LLOYD *et al.* 1982). Ribosomes are in constant turnover even in non-dividing cells (LOEB *et al.* 1965, HADJIOLOV 1966), which—although unexpected—occurs in practically all resting cells of higher eukaryotes (see HADJIOLOV and NIKOLAEV 1976). Many cases are known wherein ribosome biogenesis displays a faster and deeper response to exogenous or endogenous stimuli than mRNA. For example, growth stimulation of resting fibroblasts results in a more than twofold increase in protein synthesis and formation of new ribosomes, while synthesis of mRNA remains unchanged (RUDLAND *et al.* 1975). Similarly, differentiation of myoblasts into myotubes results in a 10-fold decrease of ribosome formation, while synthesis of mRNA remains quantitatively unaltered (KRAUTER *et al.* 1979). These and other results indicate that the control of ribosome biogenesis involves mechanisms that permit rapid changes in its intensity finely tuned to the rate of protein synthesis.

The experimental evidence available reveals that ribosome biogenesis is a complex multistep process. The following major stages in ribosome biogenesis in eukaryotes may be outlined (Fig. 34):

1. Unlocking of r-chromatin to make rRNA genes potentially active.

2. Transcription of rRNA genes to produce primary pre-rRNA, containing S-rRNA, 5.8 S rRNA, and L-rRNA sequences; concurrent transcription of 5 S rRNA genes.

3. Synthesis of r-proteins following transcription of r-protein genes, processing of pre-mRNA, and translation of r-protein mRNA. Transport of r-proteins from the cytoplasm to the nucleolus.

4. Interaction of the rDNA transcript with ribosomal and non-ribosomal proteins to build the primary preribosome. Chemical modification of primary pre-rRNA. Association of 5 S rRNA with the preribosome.

5. Cleavage of the primary preribosomes into smaller precursors of the large and small ribosomal particles.

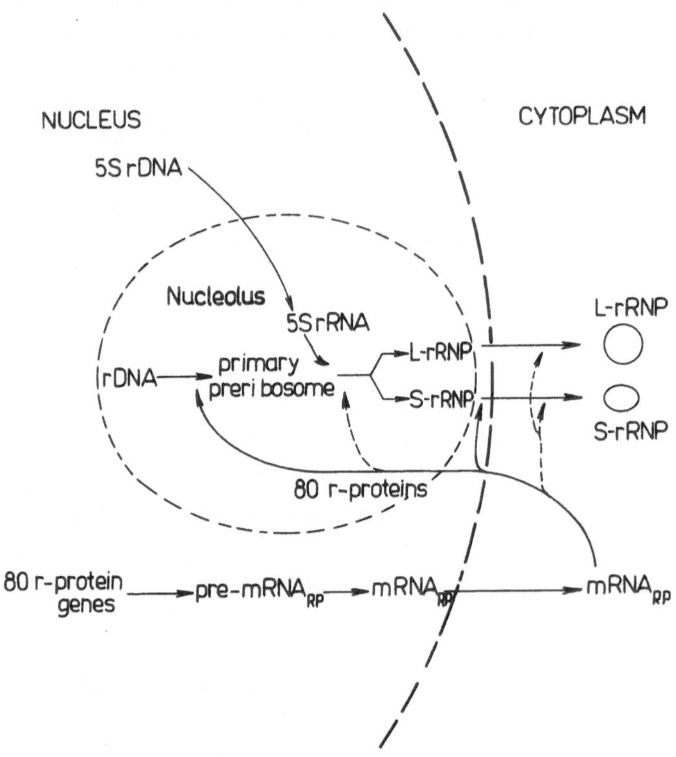

Fig. 34. A generalized scheme of ribosome biogenesis and its association with the nucleolus. Ribosome production is the result of the activity of three sets of genes. The rRNA genes are intranucleolar, while 5 S rRNA and r-protein genes are nucleoplasmic. The arrows indicate the sequence of stages in the expression of these three sets of genes and the sites of assembly of r-proteins and 5 S rRNA with pre-rRNA and preribosomes

6. Processing, including chemical modification (conversion of pre-rRNA into rRNA and generation of 5.8 S rRNA, "late" methylations of rRNA), of the two preribosomal particles to nascent large and small ribosomal particles.

7. Nucleolus → nucleoplasm → cytoplasm or nucleolus → cytoplasm transport of nascent large and small ribosomal particles, possibly including some additional maturation changes.

8. Engagement of the large and small ribosomal particles in the ribosome ⇄ polyribosome cycle and protein synthesis.

9. Degradation of rRNA and r-proteins of the large and small ribosomes.

It is evident that ribosome biogenesis involves not only numerous precursor-product relationships, but also many presumably independent sets of reactions (see Fig. 34). Regulation of ribosome biogenesis will thus include the control mechanisms operating at each stage. Several aspects of the regulation of ribosome biogenesis in different eukaryotes have been reviewed earlier (MADEN 1971, CRAIG 1974, WARNER 1974, PERRY 1976, HADJIOLOV and NIKOLAEV 1976, WARNER *et al.* 1979, HADJIOLOV 1980) and I shall concentrate only on some basic aspects, about which some precise molecular interactions are known.

VI.2. Transcriptional Control

The data on the total number of rRNA genes in eukaryotes shows (see Table 1) that broad variations exist among species. It is noteworthy that many species (including *Homo sapiens*) manage to survive with as few as 50–100 rRNA genes per haploid genome. Regulation of ribosome biogenesis by differential replication (*i.e.* amplification) of rRNA genes was considered earlier (see II.2.3.). This type of regulation operates mainly in oocytes to meet unusually high demands for ribosome production, but the amplified rDNA cannot replicate and is degraded during post-oocyte embryogenesis when these demands no longer exist (BUSBY and REEDER 1982). Normally, the number of rRNA genes is characteristic for the cell type and, at least in most somatic cells, remains constant throughout the cell cycle (see BIRNSTIEL *et al.* 1971, LONG and DAWID 1980). Transcriptional regulation of primary pre-rRNA synthesis possibly operates by one or both of two mechanisms: 1. regulation of the number of transcribed rRNA genes and 2. regulation of the rate of transcription of rRNA genes, to be considered in turn.

VI.2.1. Transitions in the State of Expression of rRNA Genes

Although the total number of rRNA genes in a broad variety of eukaryotes is now known, little is known of the number of actually transcribed rRNA genes in those organisms. As discussed earlier (see III.2.3.), even judging whether chromatin is "active" or "inactive" is somewhat difficult. The available evidence clearly shows that r-chromatin exists in three distinct states: locked (inactive), unlocked (potentially active) and transcribed (active). There is now converging evidence that the locked-unlocked state of total nuclear chromatin is stably programmed for a given cell and does not change during its interphase (see MATHIS and CHAMBON 1980, WEISBROD 1982, HANCOCK and BOULIKAS 1982). There are many cases of structural protein genes that are never expressed in a given cell type. Since rRNA genes are multiple the question about their proportion in locked or unlocked r-chromatin cannot be answered with a comparable clarity. Nevertheless, many examples of the stable state of expression of some rRNA genes are known.

VI.2.1.1. Inactive r-Chromatin

Convincing evidence for stable state of expression of rRNA genes was obtained by studies on nucleolar dominance, a phenomenon discovered in interspecies hybrids in the plant genus *Crepis* by NAVASHIN (1934) and observed later in numerous other plant and animal interspecies hybrids (see REEDER 1974, RIEGER *et al.* 1979, NICOLOFF *et al.* 1979). Briefly, hybrid cells and organisms express mainly or exclusively the rRNA genes of only one of the parent species. The phenomenon may be reversed during development and, therefore, cannot be due to the loss of chromosomes or rRNA genes. This conclusion was confirmed by determination of the number of rRNA genes in *Crepis* (DOERSCHUNG *et al.* 1976) and *Hordeum* (SUBRAHMANYAM and AZAD 1978) hybrids. In plant interspecies hybrids, r-chromatin inactivation seems to be rather stable since no secondary constriction is found at the site of the inactive nucleolus organizer. The mechanism of the nucleolar dominance phenomenon is not clear. Perhaps it is related to the segregation of some chromosomes and the loss of Ag-NOR proteins from the nonexpressed NO, as occurs in mouse-human hybrid cell lines that express only mouse rRNA genes (MILLER *et al.* 1976, DEV *et al.* 1979). *Xenopus laevis* × *X. borealis* hybrids also display an almost exclusive expression of one parent's (*X. laevis*) rRNA genes at the early stages of embryogenesis (HONJO and REEDER 1973) and evidence has been obtained that the expression of *X. laevis* genes is not due to hypomethylation of rDNA (MACLEOD and BIRD 1982), as seems to be the case with some other kinds of genes. MACLEOD and BIRD (1982) also found that only *X. laevis* rDNA was hypersensitive to DNase I digestion, thus showing that a particular structural state of r-chromatin may be involved in establishing the pattern of stable rRNA genes expression in interspecies hybrids.

Another case of stably locked rRNA genes is observed in the intron$^+$ rRNA genes of *Drosophila* (see II.2.4.3.). The intron$^+$ rRNA genes (with either type I or type II introns) are not expressed, even in flies carrying bobbed (*bb*) mutations, thus showing that such genes are stably switched-off (LONG *et al.* 1981 b). Since both active and inactive rRNA genes had the same promoter region sequence, the authors suggested that the intron DNA may fix the state of the rRNA genes by altering the phasing of nucleosomes along r-chromatin (LONG *et al.* 1981 a).

The presence of locked r-chromatin in some cells is also supported by *in situ* hybridization studies showing that in plant cells clusters of apparently inactive rDNA are confined to the nucleoplasmic heterochromatin (FLAVELL and MARTINI 1982). Yet, the molecular mechanisms causing the location of some rRNA genes within heterochromatin remain unknown.

It appears that, at least in some cases, confinement of part of the rRNA genes to locked r-chromatin (*i.e.* in heterochromatin) is reversible. For example, *X. laevis* × *X. borealis* hybrids do indeed express only *X. laevis* rRNA genes up to the tadpole stage, but in the liver of adult frogs *X. borealis* rRNA genes are also switched-on (HONJO and REEDER 1973, see REEDER 1974), suggesting that some rRNA genes may be reactivated during subsequent mitotic cycles. Recently, it was proposed that a stable state of expression of rRNA genes in a given species may be related to the structure of control sequences in the ntS of rDNA repeating units (BUSBY and REEDER 1983). The general validity of such a mechanism remains to be ascertained. In any case, the factors determining the distribution of rRNA genes in locked or unlocked r-chromatin of interphase nuclei must await further experimental evidence (see HANCOCK and BOULIKAS 1982).

The 5 S rRNA genes in *Xenopus laevis* oocytes and somatic cells may provide some clues to the mechanisms responsible for stabilizing the state of gene expression. Both oocytes and somatic cells of *Xenopus* possess 24,000 5 S rRNA genes per haploid complement that belong to three families distinguished by their sequences: *Xlo* (22,000 copies); *Xlt* (2,000 copies), and *Xls* (400 copies) (see II.3.1.). In oocytes, all three types of 5 S rDNA are expressed, whereas in somatic cells *Xlo* (and *Xlt*) 5 S rDNA are repressed and only *Xls* 5 S rDNA is expressed. The difference may be related to the action of protein factor TFIIIA (see III.3.). Accurate initiation of transcription of 5 S rDNA is known to be directed by the selective binding of TFIIIA to a specific region within the gene. Recently, the repression of *Xlo* and *Xlt* 5 S rDNA in somatic cells has been related to the absence of TFIIIA from the transcription control regions of these two genes, while it is simultaneously present in *Xls* 5 S rDNA (SAKONJU and BROWN 1982, GOTTESFELD and BLOOMER 1982). The preferential binding of a postulated limited amount of TFIIIA to *Xls* 5 S rDNA is explained either by the earlier replication of this gene set (GOTTESFELD and BLOOMER 1982) or by a higher affinity of TFIIIA for the somatic gene (SAKONJU and BROWN 1982). With either proposal, specific and stable inactivation of gene sets (*Xlo* and *Xlt* 5 S rDNA) is achieved by the absence of a critical control protein within the gene structures. The postulated mechanism provides an attractive perspective at the molecular level for the elucidation of selective and stable repression of genes. These experimental data provide evidence for the role of gene-specific non-histone proteins in the formation of inactive r-chromatin and may serve as a model for possibly more general mechanisms of chromatin inactivation.

In summary, the available evidence shows that the eukaryotic cell possesses mechanisms to stably repress part of its rRNA and 5 S rRNA genes for the duration of its life span. The assignment of some rRNA or 5 S rRNA genes to the inactive chromatin compartment most likely takes place during mitosis.

VI.2.1.2. Potentially Active and Transcribed rRNA Genes

The evidence discussed above shows that in some eukaryotic cells part of the rRNA and 5 S rRNA genes may be locked in stably inactive chromatin. The remaining part should be unlocked and present in either a potentially active or active (transcribed) state. The number of unlocked rRNA genes may vary considerably in the different eukaryotic species and also in different types of somatic or germ cells. Conceivably, the plant cells can easily spare part of their numerous rRNA genes (see Table 1) to be stored within inactive chromatin. In contrast, many animal species (with a total of only 100–200 rRNA genes per haploid genome) are likely to possess devices to keep all their complement of rRNA genes within the active r-chromatin compartment. The observed differences in the structure of NO in plant and animal mitotic chromosomes (see V.2.1.) may reflect such species-specific differences.

Are all or only some potentially active rRNA genes transcribed in any one cell? An answer may come from examination of the numbers of actively transcribed rRNA genes seen in chromosome spreads. In such a study of *Acetabularia mediterranea* (Spring *et al.* 1978) 3,500–4,800 transcribed rRNA genes per cell were observed in maximally growing cells and that figure corresponds to the estimate of 3,800 rRNA genes for that organism. Thus, at least in lower eukaryotes, switching on of all rRNA genes is attained in some stages of their life cycle. These results also demonstrate that all rRNA genes in lower eukaryotes are in a potentially active state.

Analysis of the transcription of rRNA genes under different states of cellular activity has been made by chromosome spreading studies of early embryonic stages in *Drosophila melanogaster* (McKnight and Miller 1976, McKnight *et al.* 1978) and *Oncopeltes fasciatus* (Foe *et al.* 1976, Foe 1978), of *Drosophila hydei* spermatocytes (Meyer and Hennig 1974) and of *Triturus alpestris* (Scheer *et al.* 1976, Franke *et al.* 1978), *Xenopus laevis* (Scheer *et al.* 1977) or *Rana pipiens* (Trendelenburg and McKinnell 1979) oocytes. All these studies provided conclusive evidence that on-off switching of rRNA genes is a basic mechanism in the regulation of ribosome biogenesis. These studies show that: 1. activation of rRNA genes is accompanied by changes in the conformation of r-chromatin that may *precede* the onset of transcription; 2. separate rRNA genes may be switched on and off independently, although simultaneous neighbor induction is observed in some regions, and 3. switched-on genes tend to be fully loaded with RNA polymerases and growing transcripts although lesser levels of loading also can be discerned. While activation of rRNA genes is clearly accompanied by changes in the conformation of r-chromatin (see also II.2.3.) the factors causing these changes remain unknown. It is noteworthy that activated rRNA genes are considered to be devoid of nucleosomes, but not yet loaded with RNA polymerase I molecules (FOE *et al.* 1976, FOE 1978, FRANKE *et al.* 1979), thus suggesting that the "opening" of rRNA genes is not directed by RNA polymerase I. The involvement of regulatory proteins exerting a negative or positive control on r-chromatin structure is a likely possibility (see WANG and KOSTRABA 1978), but their identification requires further studies. The state and number of potentially active and transcribed rRNA genes in adult somatic cells is more poorly understood. The activation or inactivation of these adult cell genes may follow the patterns described for embryonic cells or amplified nucleoli in germ cells, but are yet to be determined.

An interesting example of possibly wider significance is the TFIIIA dependent control of the transcription of 5 S rRNA genes in *Xenopus* oocytes (ENGELKE *et al.* 1980). In amphibian oocytes, intensive synthesis of 5 S rRNA takes place during oogenesis at a stage preceding the onset of L-rRNA and S-rRNA synthesis (see DENIS 1977). The 5 S rRNA associates with protein(s) to form 7 S ribonucleoprotein particles that are stored in the oocyte cytoplasm (PICARD and WEGNEZ 1979, KLOETZEL *et al.* 1981). The accumulation of 5 S rRNA triggers a control mechanism that inhibits its own synthesis. When the protein in the storage 7 S particles was found to be TFIIIA (PELHAM and BROWN 1980) it was recognized that 5 S rRNA and its gene compete for the same protein thereby regulating the level of 5 S rDNA transcription according to the cellular content of the gene product. Similar transcription control systems operate in the oocytes of other species of lower vertebrates (KLOETZEL *et al.* 1981) and in mammalian cells (GRUISSEM and SEIFART 1982, RINKE and STEITZ 1982, SHI *et al.* 1983). The nature of other 5 S rRNA binding proteins in storage particles remains to be determined, as do the mechanisms by which 5 S rRNA is assembled into ribosomes. The fact that 5 S rRNA is associated in ribosomes with r-proteins L 6, L 7, and L 19 (ULBRICH *et al.* 1980), which are distinct from TFIIIA and other storage proteins that bind 5 S rRNA (KLOETZEL *et al.* 1981), suggests that the control of 5 S rDNA transcription is intimately related to the mechanisms of 5 S rRNA storage and integration in the ribosome.

The development of efficient *in vitro* systems for RNA polymerase I driven transcription will certainly provide the tools needed for the understanding of control mechanisms in the transcription of rRNA genes and the identification of putative gene-specific positive or negative control factors.

VI.2.2. Control of Transcription Rate

Active rRNA genes may be transcribed at different rates, depending on regulatory factors affecting initiation, elongation, or termination of primary pre-rRNA chains. The following can be envisioned as playing roles in regulation:

1. Amount and activity of RNA polymerase I.
2. Concentration of nucleoside-5′-triphosphates.
3. Conformation of the growing pre-rRNA chain and its assembly with ribosomal and non-ribosomal proteins.
4. Participation of specific protein or other factors at any stage.
5. Steric hindrance in the nucleolus as a result of the accumulation of preribosomes or nascent ribosomes.

While the involvement of some of the above is obvious, the existence or role of others is clearly not and, moreover, their relative contributions may vary under different physiological or experimental conditions.

The results obtained by the chromosome-spreading technique reveal a general pattern in the transcription of rRNA genes; common observation in fully active nucleoli is that rRNA genes are fully loaded with RNA polymerases and growing RNP fibrils. In partly active nucleoli, one can find fully loaded genes adjacent to partly or totally fibril-free transcription units on the same rDNA axis (Scheer *et al.* 1976, Franke *et al.* 1978, Trendelenburg and McKinnell 1979). These observations suggest that, usually, transcription of rRNA genes is not limited by the supply of RNA polymerase I or by initiation control factors. In some cases a decrease of overall transcription rates is accompanied by a lower density of growing fibrils along the rDNA axis (Scheer *et al.* 1975, 1976, Franke *et al.* 1979) indicating a decreased frequency of transcription initiation. Of course, observations of chromosome spreads, while suggestive, cannot provide direct evidence on transcription rates.

VI.2.2.1. Role of RNA Polymerase I

It is evident that limitations in the supply of RNA polymerase I molecules could restrict overall transcription rates. If that does occur we would be interested in determining what factors control the synthesis and transport to the nucleolus of RNA polymerase I. But, indirect evidence indicates that normally the number of RNA polymerase I molecules is not rate limiting. A rough estimate shows that mammalian cells contain about 2.5×10^4 RNA polymerase I molecules per diploid genome (see III.1.1.). Assuming a polynucleotide chain elongation rate *in vivo* of 40 nucleotides/sec (see Maden 1971) and a chain length for primary pre-rRNA of 12.5×10^3 nucleotides, 2.5×10^4 RNA polymerase I would be able to produce about 4.8×10^3 pre-rRNA molecules/minute and presumably the same number of

Table 4. *Rates of ribosome synthesis in animal cells*[a]

Type of cell or tissue	Ribosomes/ min/nucleus[b]	Reference
Erythroid cells	220	HUNT (1976)
Fibroblasts		
Resting	500	EMERSON (1971)
Growing	1,200	EMERSON (1971)
Myoblasts (quail)	1,200	BOWMAN and EMERSON (1977)
Myotubes (quail)	630	BOWMAN and EMERSON (1977)
L cells	4,500	BRANDHORST and MCCONKEY (1974)
HeLa cells	3,100	WOLF and SCHLESSINGER (1977)
Rat liver		
Normal	1,100–1,400	DUDOV *et al.* (1978)
Regenerating (12 h)	3,700	DABEVA and DUDOV (1982)
Rat brain[c]	220–260	STOYKOVA *et al.* (1983)

[a] Data obtained in experiments measuring the rate of synthesis of ribosomes (or primary pre-rRNA) are included. It is assumed that degradation of newly synthesized pre-rRNA or rRNA is not taking place. The calculations are based on a molecular mass for 28 S + 18 S rRNA = 2.4×10^6.

[b] In some cases (*e.g.*, myotubes, rat liver) multinucleated cells are present and the number of ribosomes per cell is higher than the figures shown in the table.

[c] Data for a mixed population of neuronal and glial nuclei.

ribosomes, a rate that is at least sufficient for that which is actually observed in mammalian cells (Table 4). That the supply of RNA polymerase I is unlikely to be limiting the transcription rates, is supported by studies of amphibian oogenesis that show that the practically constant level of enzyme activity does not correlate with drastic changes in transcription rate (ROEDER 1974, HOLLINGER and SMITH 1976). A similar situation is found in encysted gastrulae of *Artemia salina* upon reactivation of their development and ribosome formation (HENTSCHEL and TATA 1974).

The metabolic stability of RNA polymerase I is also an important factor to consider as a possible control of transcription. A rapid turnover of nucleolar RNA polymerase I was deduced to occur from studies on the effect of cycloheximide inhibition of protein synthesis in rat liver (YU and FEIGELSON 1972), but on the other hand a block of protein synthesis by cycloheximide (BENECKE *et al.* 1973, SCHMID and SEKERIS 1973, ONISHI *et al.* 1977) or pactamycin (HAIM *et al.* 1983) does not affect RNA polymerase I activity in whole nuclei. Furthermore, complete block of transcription of nucleoplasmic genes by α-amanitin does not alter nuclear RNA polymerase I levels for several hours (TATA *et al.* 1972, HADJIOLOV *et al.* 1974). These and similar studies with different eukaryotes (*i.e.* SHIELDS and TATA 1976, HILDEBRANDT and SAUER 1976) permit the conclusion that RNA polymerase I is metabolically stable, making unlikely alterations in enzyme stability as a means for regulating transcription (see also BEEBEE and BUTTERWORTH 1980).

While alterations in the number of RNA polymerase I molecules does not seem to be a factor in the control of transcription rates, modification of its catalytical activity may be. The complexity of the RNA polymerase reaction renders difficult assessments of changes in its catalytic activity (see BEEBEE and BUTTERWORTH 1980), but, nevertheless, studies of possible controls have included investigations of: a) the binding of RNA polymerase I to template r-chromatin; b) chemical modifications of the enzyme and c) initiation, elongation, and termination control factors.

With the finding that RNA polymerase I exists in two forms: free and template-bound (LAMPERT and FEIGELSON 1974, YU 1974, 1975, CHESTERTON *et al.* 1975, MATSUI *et al.* 1976), YU (1976) proposed that regulation of transcriptional activity might operate through control of the association of RNA polymerase I with r-chromatin and an early response to inhibition of protein synthesis by cycloheximide (CHESTERTON *et al.* 1975, ONISHI *et al.* 1977) or aminoacid deprivation (GRUMMT *et al.* 1976) was reported to be a reduced binding of the enzyme to chromatin. Rat liver RNA polymerase I was shown to exist in two forms, I_A and I_B (COUPAR and CHESTERTON 1975), with mainly form I_B being template-bound (MATSUI *et al.* 1976). The two forms differ by the presence (making I_B) or absence (making I_A) of the 65 kDa subunit of RNA polymerase I. Similar observations of the role of RNA polymerase I subunits in the template binding of the enzyme were also reported for the yeast enzyme (COOPER and QUINCEY 1979).

Chemical modification of the RNA polymerase I molecule may also affect its activity, but at present only phosphorylation of the enzyme is documented. Inhibition of phosphorylation was proposed as the basis for the inhibition of RNA polymerase by thuringiensin (SMUCKLER and HADJIOLOV 1972), which was shown later to also inhibit adenylate cyclase (GRAHAME-SMITH *et al.* 1975). Although phosphorylation of RNA polymerase I *in vivo* has been carefully documented (HIRSCH and MARTELO 1976, BELL *et al.* 1976, BUHLER *et al.* 1976) and in the case of yeast RNA polymerase I, *e.g.*, 15 phosphate groups are identified on five of its subunits (BREANT *et al.* 1983), no evidence for the *in vitro* or *in vivo* modulation of RNA polymerase I activity by phosphorylation has been obtained.

VI.2.2.2. Supply of Nucleoside-5'-Triphosphates

No evidence exists that the local concentration of nucleoside-5'-triphosphates may alter rDNA transcription rates and we know nothing of mechanisms that would affect the supply of substrate nucleotides to the nucleus and the nucleolus. That this supply may involve more complex phenomena than simple diffusion is suggested by results showing the compartmentation of pyrimidine nucleotides precursor pools in the cell (WIEGERS *et al.* 1976, GENCHEV *et al.* 1980). The necessity of a continuous supply of free nucleotides for RNA synthesis is dramatically illustrated under some experimental conditions. A clear-cut example is provided by the action of D-galactosamine on rat liver. Given *in vivo* this drug reduces within 30 minutes the free uridine nucleotides to background levels (KEPPLER *et al.* 1970, 1974, GAJDARDJIEVA *et al.* 1977) causing an almost simultaneous and complete inhibition of RNA synthesis (GAJDARDJIEVA *et al.* 1980).

VI.2.2.3. The Role of Protein Synthesis

The effects of protein synthesis inhibition on rDNA transcription has been studied in a broad variety of eukaryotic cells and organisms (see HADJIOLOV and NIKOLAEV 1976, WARNER *et al.* 1979). Here, I discuss what the results of those studies tell us about: a) transcription control factors and b) the possible regulatory role of r-proteins.

Lower eukaryotes are more amenable to such investigations because combined biochemical and genetic analyses are possible. Because of the occurrence of a "stringent response" in *Saccharomyces* and some other unicellular eukaryotes, the inhibition of protein synthesis (by aminoacid deprivation, cycloheximide, temperature shifts, etc.) has been shown to result in a relatively rapid decrease in rRNA synthesis, while mRNA and tRNA are markedly less affected (see WARNER *et al.* 1979, WARNER 1982). It is noteworthy that the effect is not instantaneous; the synthesis of pre-rRNA is halted only 1–2 h after the block of protein synthesis (KULKER and POGO 1980). The mechanisms of the "stringent response" inhibition of pre-rRNA synthesis are unknown, but appear not to be related to the shortage of r-proteins, since cells unable to synthesize r-proteins continue to synthesize pre-rRNA, although the pre-rRNA is unstable (SHULMAN and WARNER 1978). Accordingly, the participation of some specific transcription control factor, dependent on protein synthesis, may be envisaged. However, unlike the case of bacteria, ppGpp does not seem to be involved and a distinct "stringent control" phosphorylated compound has been suggested (LUSBY and MCLAUGHLIN 1980). A "relaxed control" *S. cerevisiae* mutant continues to synthesize and process accurately pre-rRNA when protein synthesis is blocked (WALTSCHEWA *et al.* 1983). Hopefully, comparative studies using *Saccharomyces* cells with "stringent" or "relaxed" responses to protein synthesis inhibition will help locate and understand the still hypothetical rDNA transcription control factors.

As for the lower eukaryotes a block of protein synthesis results in a preferential inhibition of rRNA synthesis in a broad variety of animal and plant cells (see HADJIOLOV and NIKOLAEV 1976). In many cases the effect on rRNA synthesis appeared also to be faster and more pronounced than that on mRNA and tRNA. However, studies on the labeling of total rRNA or ribosomes do not discriminate between transcriptional and post-transcriptional control. Attempts at the dissection of the basis of rRNA synthesis inhibition have yielded controversial results. While most authors observed a rapid and pronounced alteration of primary pre-rRNA processing (see VI.3.1.), the effect on pre-rRNA synthesis was variable. In some animal cell studies inhibition of protein synthesis with cycloheximide (MANDAL 1969) or puromycin (SOEIRO *et al.* 1968) did not alter the labeling of pre-rRNA, whereas an early inhibition of primary pre-rRNA labeling was reported for HeLa cells (WILLIAMS *et al.* 1969) and rat liver (MURAMATSU *et al.* 1970) after cycloheximide treatment. However, the doses used in the latter cases were several-fold higher than needed to cause a complete inhibition of protein synthesis (ENNIS 1966, STOYANOVA and HADJIOLOV 1979). In fact, high doses of cycloheximide have been shown to cause a drastic inhibition in the labeling of free nucleotides and an artifactual decrease in the *in vivo* labeling of pre-rRNA (STOYANOVA and DABEVA 1980). Low doses of cycloheximide (1–5 mg per kg body weight) causing over 90% inhibition of protein synthesis in rat liver did not initially alter the synthesis of primary pre-rRNA (FARBER and FARMAR 1973, GOLDBLATT *et al.* 1975, STOYANOVA and HADJIOLOV 1979, STOYANOVA and DABEVA 1980, KARAGYOZOV *et al.* 1980). It was concluded that the effect of protein synthesis inhibition on rDNA transcription is slow and therefore the involvement of short-lived transcription control protein factors is unlikely (STOYANOVA and HADJIOLOV 1979, STOYANOVA and DABEVA 1980). This conclusion is supported by studies with CHO-tsH 1 cells, a mutant line with a temperature-sensitive leucyl-tRNA synthetase. It was found (Fig. 35) that even after 3 hours incubation at the non-permissive temperature and block of protein synthesis, the labeling of primary pre-rRNA was reduced only to about half of both control and 60 minutes values (ROYAL and

---

Fig. 35. Short-term incorporation of [^{3}H]-uridine into total cell RNA of wild type and mutant tsH 1 CHO cells at the permissive (34 °C) and restrictive (39.5 °C) temperatures. The mutant tsH 1 cell line has a thermosensitive leucyl-tRNA synthetase. Incubation at the restrictive temperature results in immediate block of protein synthesis. Labeling with [^{3}H]-uridine is for 15 min. Incorporation of [^{3}H]-uridine into RNA of wild type (*A*) and tsH 1 mutant (*B*) cells incubated at the permissive or the restrictive temperatures for the indicated times. Fractionation of RNA is by electrophoresis in agarose-acrylamide gels. [From ROYAL and SIMARD (1980). Reproduced with permission by the authors and *Biologie cellulaire*]

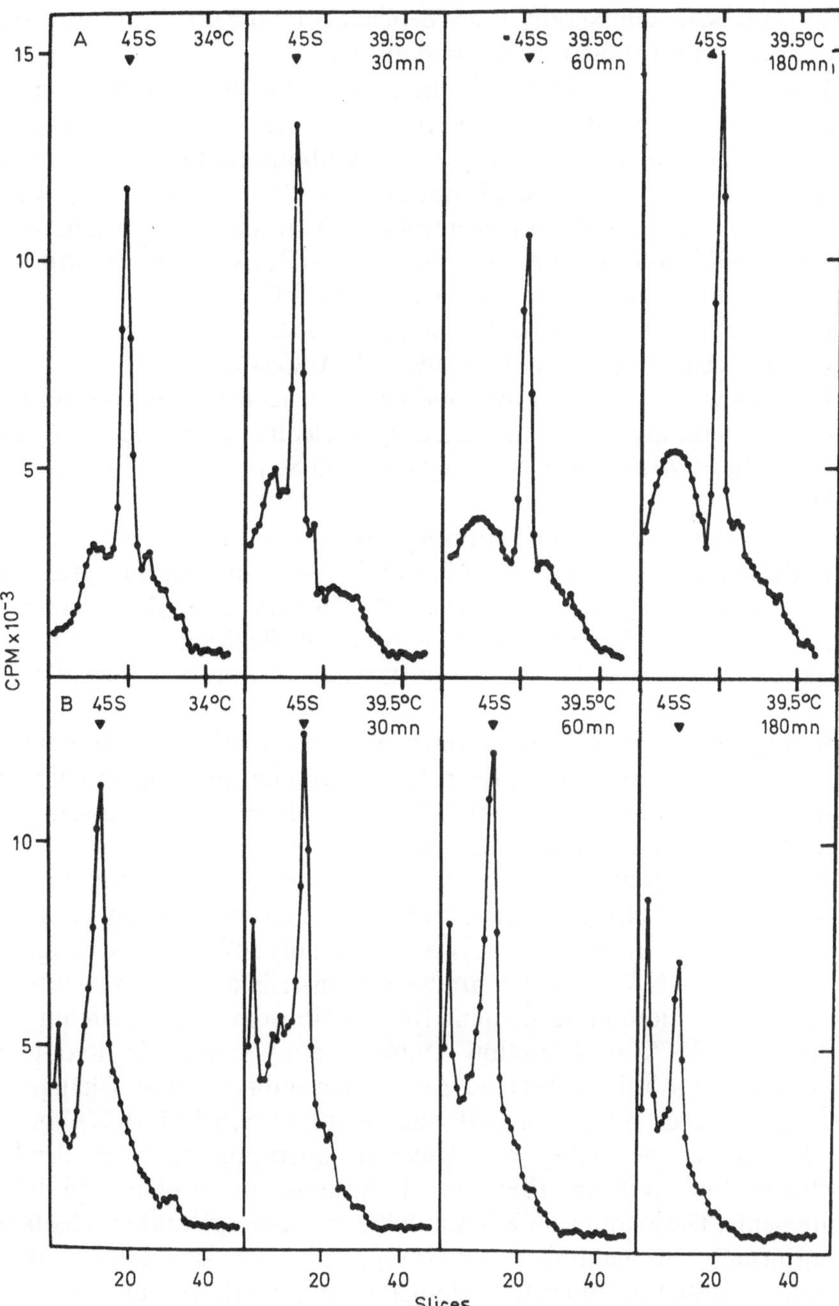
A
45S
34°C
45S
39.5°C
30mn
45S
39.5°C
60mn
45S
39.5°C
180mn
B
45S
34°C
45S
39.5°C
30mn
45S
39.5°C
60mn
45S
39.5°C
180mn
15
10
5
10
5
CPM x 10^-3
20
40
20
40
20
40
20
40
Slices

SIMARD 1980). The above observations show that *in vivo* transcription of rDNA, at least in animal cells, is not dependent on the action of short-lived transcription control factors of low pool size.

The possible role of r-proteins in the control of rDNA transcription rate deserves further comment. As observed in several studies (see above) a longer duration of the block in protein synthesis results in a decreased synthesis of primary pre-rRNA. It was proposed that this effect may be due to a shortage in the supply of r-proteins and the resulting alterations in the assembly and conformation of growing pre-rRNA chains (HADJIOLOV 1980). This interpretation is supported by the following:

a) The pool size of practically all free r-proteins in the nucleolus of mammalian cells is very low (PHILLIPS and MCCONKEY 1976).

b) Studies with isolated rat liver and HeLa nuclei provided evidence that added r-proteins are rapidly taken up by nucleoli (ROTH *et al.* 1976) and stimulate the *in vitro* synthesis of a putative "45 S" pre-rRNA (BOLLA *et al.* 1977).

c) Changes in rDNA transcription rates in the liver of starved and starved/refed rats (COUPAR *et al.* 1978) or in phytohaemagglutinin-stimulated lymphocytes (DAUPHINAIS 1981) do not alter the number of template-bound RNA polymerase I molecules, while the changes of *in vitro* elongation rates match the changes in the rates of ribosome formation *in vivo*.

Although the above evidence is primarily circumstantial and does not rule out alternative interpretations, the role of r-proteins and conformation of the growing, protein-coated, pre-rRNA chain in the control of transcription rate certainly deserves closer attention.

Our current understanding of the mechanisms controlling the transcription of rRNA genes can be stated as follows. In many cells and organisms the bulk of r-chromatin is in a potentially active state. Changes in the number of rRNA genes that are being transcribed seems to be a basic response to variations in the demand for new ribosomes. The switching on and off of rRNA transcription units certainly occurs in long-term adjustments of the cell, as observed, *e.g.*, during embryogenesis. Changes in transcription rates for a constant number of transcribed rRNA genes provide another possibility of regulation operating in more flexible adjustments of cellular ribosome biogenesis to protein synthesis requirements. The amounts of RNA polymerase I and substrate nucleotides are not usually rate limiting and their regulatory role is restrained. Both frequency of initiation and rate of elongation of pre-rRNA may contribute to changes in the overall rate of transcription. Switched-on rRNA transcription units tend to be maximally loaded with RNA polymerase I molecules and growing pre-rRNA chains thus indicating that variations in the rate of pre-rRNA elongation may be predominant as a short-term transcription control mechanism. In this respect, the role of the conformational state of the growing, protein-coated, pre-rRNA chain deserves attention as a factor modulating transcription rate. The role of r-proteins or of initiation, elongation or termination factors remains to be substantiated.

The transcription of 5 S rRNA genes is not directly dependent on rDNA transcription. There is strong evidence that 5 S rRNA synthesis is directed by specific initiation factor(s) exerting positive control and having the capacity to bind to both 5 S rDNA and 5 S rRNA. Thus, the cellular concentration of 5 S rRNA serves a major feedback control in regulating the transcription of 5 S rRNA genes. The supply of 5 S rRNA does not seem to normally limit the transcription of rRNA genes.

VI.3. Posttranscriptional Control

The large number of studies showing the importance of posttranscriptional control mechanisms in ribosome production has been reviewed and different models discussed in detail (PERRY 1973, WARNER *et al.* 1973, 1979, HADJIOLOV and NIKOLAEV 1976, HADJIOLOV 1980). Here I shall consider only some aspects, emphasis being placed on the elucidation of critical control sites in ribosome biogenesis.

VI.3.1. Synthesis and Supply of r-Proteins

The formation of mature ribosomes is dependent on continuous protein synthesis. This fact, first established with cultured HeLa (WARNER *et al.* 1966) and L (ENNIS 1966) cells, has been confirmed for all eukaryotes tested (see HADJIOLOV and NIKOLAEV 1976). In the consideration of the "stringent control" of ribosome biogenesis discussed above we stated that in most cases transcription of rRNA genes unexpectedly remains initially unaltered, whereas posttranscriptional mechanisms display a faster and deeper response to the limited protein supply. For example, ribosome formation in *S. cerevisiae* is blocked immediately after inhibition of protein synthesis, while synthesis of primary pre-rRNA continues for some time (DE KLOET 1966, UDEM and WARNER 1972, KULKER and POGO 1980). A similar response is observed with *Tetrahymena* (ECKERT *et al.* 1975) and other lower eukaryotes (see HADJIOLOV and NIKOLAEV 1976), as well as with various animal cells upon inhibition of protein synthesis with puromycin (SOEIRO *et al.* 1968), low doses of cycloheximide (MANDAL 1969, RIZZO and WEBB 1972, FARBER and FARMAR 1973, STOYANOVA and HADJIOLOV 1979), amino acid deprivation (MADEN *et al.* 1969, VAUGHAN 1972, CHESTERTON *et al.*

1975) and in a mutant cell line temperature-sensitive in protein synthesis (ROYAL and SIMARD 1980). What proteins are involved in the stringent posttranscriptional control? Ribosomal proteins seem to be the obvious candidates and therefore deserve first consideration.

Early experiments with HeLa and L cells showed that r-proteins are produced by cytoplasmic polyribosomes (HEADY and MCCONKEY 1971, CRAIG and PERRY 1971) that are free and not membrane-bound (NABESHIMA *et al.* 1975). Studies with yeast (MAGER and PLANTA 1976, BOLLEN *et al.* 1980, GORENSTEIN and WARNER 1979, FRIED *et al.* 1981), *Xenopus laevis* oocytes (PIERANDREI-AMALDI and BECCARI 1980), mouse cells (MEYUHAS and PERRY 1980) EHRLICH ascites tumor cells (HACKETT *et al.* 1978), and rat liver (NABESHIMA *et al.* 1979) showed conclusively that the mRNAs coding for r-proteins are monocistronic and are translated on small polyribosomes. The r-protein mRNAs are polyadenylated and about 30% of their sequence is non-coding (MEYUHAS and PERRY 1980).

Two basic questions regarding the regulation of r-protein synthesis may be asked: a) Is r-protein synthesis dependent on the continuous formation of their mRNA? b) Is the synthesis of different r-proteins coordinately regulated? Early studies showed that rp-mRNA (mRNA for r-proteins) are relatively stable (CRAIG and PERRY 1971, CRAIG *et al.* 1971, MAISEL and MCCONKEY 1971), a finding supported by more recent work. Thus, in HeLa cells, the synthesis of at least 34 r-proteins continues after 75% of hnRNA synthesis is inhibited by actinomycin D (WARNER 1977) and the half-life of mouse fibroblast rp-mRNA is in the range of 8 to 11 hours, identical to the half-life of total cellular mRNA (GEYER *et al.* 1982). In addition, when the rate of r-protein synthesis undergoes a pronounced increase in serum stimulated mouse fibroblasts (TUSHINSKI and WARNER 1982), the amount of rp-mRNA remains practically unchanged (GEYER *et al.* 1982). These results suggest that r-protein synthesis is regulated primarily by modulation of translation rates of comparatively stable rp-mRNAs, but the observed stability of rp-mRNA has not yet been demonstrated for each individual r-protein mRNA, making any generalization not quite definitive.

The second question also has been answered by measurement of individual r-protein synthesis in the lower eukaryotes *S. cerevisiae* (GORENSTEIN and WARNER 1976, PEARSON *et al.* 1982) and *Neurospora crassa* (STURANI and SACCO 1982) under conditions of drastic experimental modification of growth rates. Two general observations emerged: a) The synthesis of the individual r-proteins is coordinately regulated and b) the synthesis of r-proteins may not be related to total protein synthesis (see WARNER 1982).

These studies also demonstrated that the synthesis of r-proteins is adapted to the cell's need for new ribosomes and we can ask further: what regulatory mechanism triggers the change in the rate of r-protein synthesis?

In *E. coli* a simple control mechanism seems to operate via rRNA and rp-mRNA competition for the binding of r-proteins (NOMURA *et al.* 1980, 1982). In the absence of rRNA synthesis, r-proteins will bind to their respective mRNAs and thereby block their translation. Such a mechanism might operate in eukaryotes as well, although the compartmentalization of rRNA (nucleus) and r-protein (cytoplasm) synthesis may create a problem. It is also possible that r-proteins are produced in excess and, if pre-rRNA and rRNAs are not available, are rapidly degraded. This seems to be the case of mammalian cells. Actinomycin D blockage of rRNA synthesis in HeLa cells does not stop r-protein synthesis, but newly formed r-proteins are rapidly degraded (WARNER 1977). Differentiation of myoblasts into myotubes drastically reduces rRNA formation and the r-proteins, which apparently are made at an unchanged rate, are also rapidly degraded (KRAUTER *et al.* 1979). Similar results are obtained with regenerating rat liver, for which system it is known that the degradation of excess r-proteins takes place in the nucleus under the action of possibly specific proteases (TSURUGI and OGATA 1979, TSURUGI *et al.* 1983). Thus, in mammals at least, if r-protein synthesis is not regulated, post-translational stability is.

In view of the above, the transport of r-proteins to the nucleus may be of particular importance. The pool of free r-proteins in the cytoplasm is very low (WOOL and STÖFFLER 1976) and newly made r-proteins are rapidly transferred to the nucleus (WU and WARNER 1971, MAISEL and MCCONKEY 1971, TSURUGI and OGATA 1979) where they, either associate with pre-rRNA during transcription and processing, or, are degraded. In fact, r-proteins are taken up and concentrated 10- to 50-fold in the nucleolus (WARNER 1979) and it has been suggested that not only pre-rRNA, but perhaps other nuclear or nucleolar structures, may be involved in r-protein binding (TODOROV and HADJIOLOV 1979). The mechanism controlling the rapid transport of r-proteins to the nucleus is unknown and given that they belong to the class of *karyophilic proteins* (JELINEK and GOLDSTEIN 1973, DABAUVALLE and FRANKE 1982, DINGWALL *et al.* 1982, see BONNER 1978, GOLDSTEIN and KO 1981), we can expect that we may soon learn more about the basis for the affinity of these proteins for the nucleus.

Summarizing all the above findings, it may be safely concluded that the continuous synthesis of r-proteins and their supply to the nucleolus is a major factor involved in the stringent posttransriptional control of ribosome biogenesis. The synthesis of r-proteins seems relatively independent of pre-rRNA transcription and processing. Transcriptional and, principally, translational control mechanisms regulate r-protein synthesis, but nuclear degradation of excess r-proteins also play an important role. The low pool size of r-proteins in the cell is a major factor responsible for a cell's rapid and flexible response of ribosome biogenesis to changing conditions. How all this is achieved is only partially understood.

VI.3.2. The Role of Pre-rRNA Structure

The structure of pre-rRNA provides recognition sites for the accurate assembly of proteins that takes place during transcription and maturation and raises the question about the role of pre-rRNA structures *per se* in the assembly and processing of preribosomes. A basic question to ask is: are the modifications (methylation and pseudouridylation) of some nucleotides in pre-rRNA critical for its processing? There is no evidence about the role of pseudouridylation but a substantial amount about the possible role of pre-rRNA methylation. Undermethylation of primary pre-rRNA in cultured animal cells that results from methionine (VAUGHAN *et al.* 1967, WOLF and SCHLESSINGER 1977) or histidine (GRUMMT 1977) deprivation is accompanied by a drastic alteration in pre-rRNA processing, but metabolic effects other than undermethylation of pre-rRNA may be responsible for altered processing. Methylation of pre-rRNA itself may be dependent on protein synthesis, as shown by studies with CHO-tsH 1 cells whose protein synthesis is temperature sensitive (ROYAL and SIMARD 1980). When protein synthesis is inhibited a sharp decrease of pre-rRNA methylation and a disturbance of pre-rRNA processing occur. In contrast, inhibition of protein synthesis with puromycin (TAMAOKI and LANE 1968) or cycloheximide (SHULMAN *et al.* 1977) does not alter the methylation pattern of pre-rRNA, showing that the dependence of methylation on pre-rRNA synthesis is not a simple matter. In a different kind of investigation, cycloleucine, an inhibitor of S-adenosyl-L-methionine synthesis, abolished almost completely (95%) the methylation of 45 S pre-rRNA in animal cells but did not alter appreciably its synthesis (CABOCHE and BACHELLERIE 1977). Moreover, undermethylation of pre-rRNA did not prevent the formation of mature ribosomes and their involvement in polyribosomes, but the *rates* of pre-rRNA processing (particularly the stage of nucleo-cytoplasmic transport of 28 S rRNA) were drastically reduced. Altogether the above findings strongly suggest that methylation of primary pre-rRNA plays an important role in its processing, but the basis of that is unknown. It may be as simple as methylation being required for processing enzyme recognition.

The effect on pre-rRNA processing of the incorporation of various nucleoside analogues also has been investigated. Incorporation of analogues like toyocamycin (TAVITIAN *et al.* 1968, 1969) or 5-fluoroorotate (CIHAK and PITOT 1970, WILKINSON *et al.* 1971, HADJIOLOVA *et al.* 1973, HADJIOLOV *et al.* 1974b, ALAM and SHIRES 1977) does not inhibit the synthesis of pre-mRNA and its processing to mRNA, nor do they affect the formation of tRNA and 5 S rRNA, but ribosome formation is rapidly halted. Studies with 5-fluoroorotate in mice liver (HADJIOLOVA *et al.* 1973) or 5-fluorouridine in Novikoff hepatoma (WILKINSON *et al.* 1975)

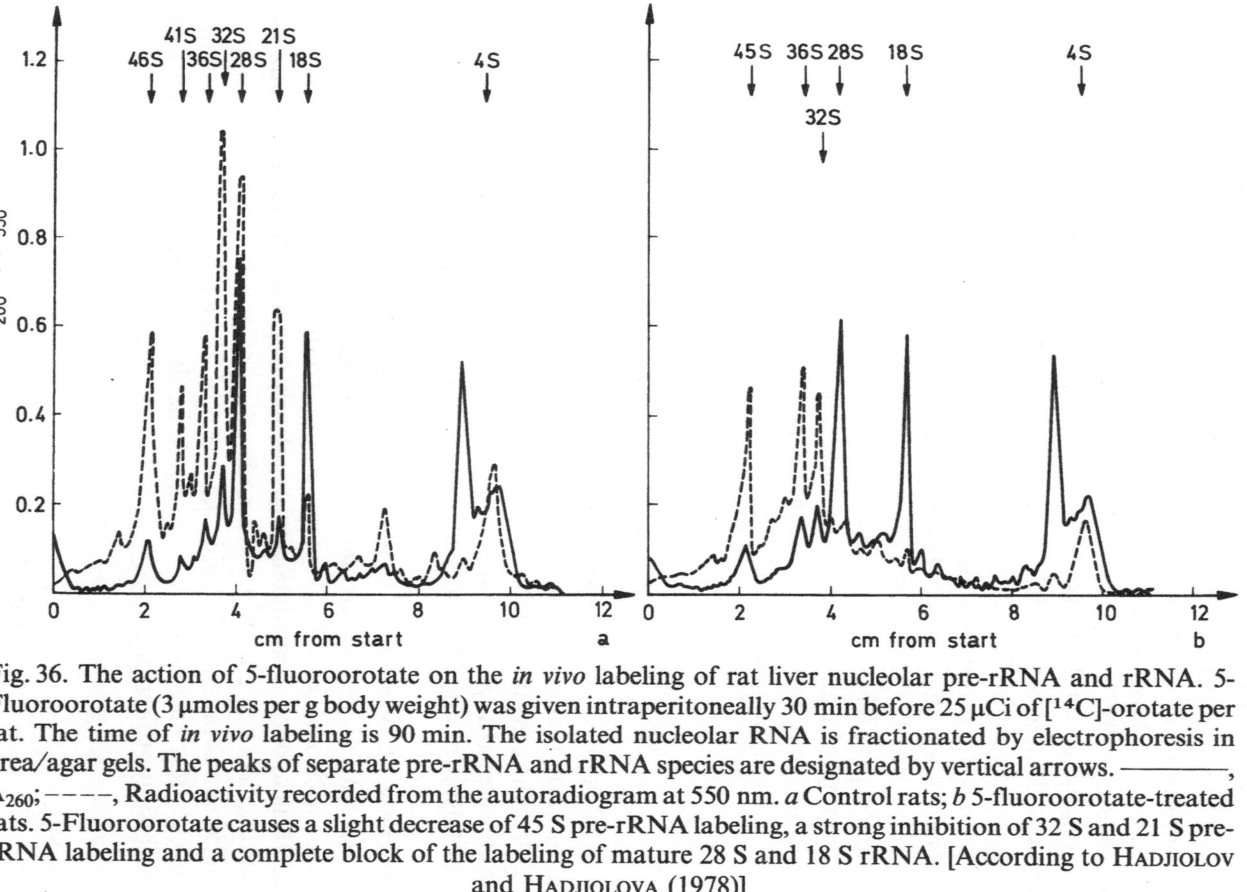

Fig. 36. The action of 5-fluoroorotate on the *in vivo* labeling of rat liver nucleolar pre-rRNA and rRNA. 5-Fluoroorotate (3 µmoles per g body weight) was given intraperitoneally 30 min before 25 µCi of [^{14}C]-orotate per rat. The time of *in vivo* labeling is 90 min. The isolated nucleolar RNA is fractionated by electrophoresis in urea/agar gels. The peaks of separate pre-rRNA and rRNA species are designated by vertical arrows. ———, A_{260}; – – – –, Radioactivity recorded from the autoradiogram at 550 nm. *a* Control rats; *b* 5-fluoroorotate-treated rats. 5-Fluoroorotate causes a slight decrease of 45 S pre-rRNA labeling, a strong inhibition of 32 S and 21 S pre-rRNA labeling and a complete block of the labeling of mature 28 S and 18 S rRNA. [According to HADJIOLOV and HADJIOLOVA (1978)]

demonstrated that ribosome formation is blocked preferentially at the last nucleolar stages of pre-rRNA maturation (Fig. 36). Toyocamycin has a similar effect on ribosome formation in *Saccharomyces* (VENKOV *et al.* 1977), Novikoff hepatoma cells (WEISS and PITOT 1974), mouse leukemia cells (AUGER-BUENDIA *et al.* 1978), and Friend erythroleukemia cells (HADJIOLOVA *et al.* 1981). The synthesis of proteins and the assembly of primary preribosomes is not appreciably altered either gy 5-fluorouridine (E. BERGER 1977) or by toyocamycin (AUGER-BUENDIA *et al.* 1978) but the formation of 18 S rRNA is blocked and 36 S pre-rRNA accumulates (AUGER-BUENDIA *et al.* 1978, HADJIOLOV and HADJIOLOVA 1979), suggesting that alterations in the structure of preribosomes may induce the chanelling of pre-rRNA processing along alternative pathways.

These and other similar findings (see HADJIOLOV and NIKOLAEV 1976, SUHADOLNIK 1979) strongly suggest that a correct structure of the whole primary pre-rRNA molecule is of critical importance for the accurate processing of preribosomes and therefore accurate pre-rRNA secondary structure and interactions with proteins are probably crucial in the posttranscriptional control of ribosome biogenesis.

VI.3.3. The Role of 5 S rRNA

Because 5 S rRNA joins the preribosome at an early stage of its maturation and at least one of the r-proteins (L 6) that interacts with 5 S rRNA is already present in the "80 S" primary preribosome, 5 S rRNA may be considered as possibly important component for the correct assembly of preribosomes and their processing. Little is presently known, about a possible role of 5 S rRNA in the regulation of posttranscriptional events. We know, of course, from experiments with bobbed *Drosophila* (WEINMANN 1972) and O_{nu} *Xenopus laevis* (L. MILLER 1973) mutants that the transcription of 5 S rRNA genes is not directly coordinated with that of rRNA genes and is regulated independently (see REEDER 1974). Whether a selective block of 5 S rRNA synthesis will alter preribosome maturation is not known at present. Plausibly, an excess of 5 S rRNA is maintained in the nucleus thus ensuring that it is never in shortage during assembly and processing of preribosomes.

VI.3.4. Critical Control Sites

Here, I shall try to discriminate those among different control sites that seem to be usually *critical* in adapting the overall process of ribosome biogenesis to the needs of the cell.

VI.3.4.1. Alternative Processing Pathways and Intranuclear Degradation of Preribosomes and Ribosomes

The continuous supply of r-proteins is a major factor adapting ribosome production to the efficiency of protein synthesis. This control is very fast

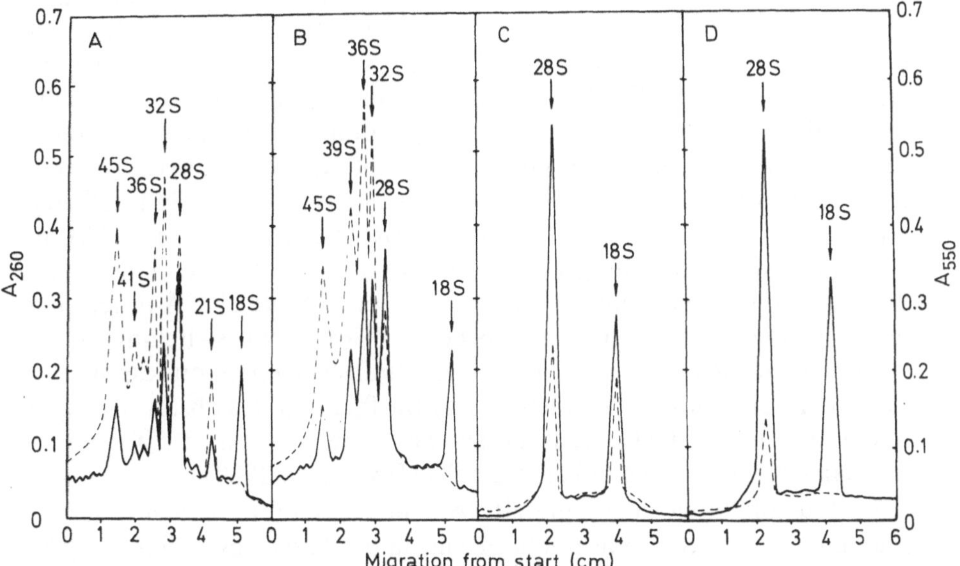

Fig. 37. Alterations in rat liver pre-rRNA processing induced by cycloheximide inhibition of protein synthesis. The rats from the experimental group are treated with 5 mg cycloheximide per kg body weight for 2 h. The RNA of control (*A, C*) and experimental (*B, D*) rats is labeled *in vivo* with [^{14}C]-orotate for 90 min. Nucleolar (A, B) and cytoplasmic (C, D) RNA is isolated and analyzed by urea/agar gel electrophoresis. The position of pre-rRNA and rRNA species is indicated by vertical arrows. ———, A_{260}; ----, Radioactivity, recorded from the autoradiogram at 550 nm. Cycloheximide inhibition of protein synthesis does not alter appreciably the labeling of 45 S pre-rRNA. However, the labeling of intermediate pre-rRNA and rRNA species is decreased (32 S pre-rRNA, nucleolar and cytoplasmic 28 S rRNA) or abolished (41 S and 21 S pre-rRNA, cytoplasmic 18 S rRNA). [Reproduced from STOYANOVA and HADJIOLOV (1979) with permission by *European Journal of Biochemistry*]

and versatile and operates largely at the posttranscriptional level, although rooted in the assembly of the primary preribosome during transcription. The strongly decreased rates of formation of mature ribosomes, while synthesis of primary pre-rRNA remains unchanged may be caused by: a) decreased rates of preribosome processing and b) degradation of incompletely or spuriously assembled preribosomes or ribosomes.

Channeling of preribosomes along alternative pre-rRNA processing pathways seems to be an early response to protein shortage and is illustrated by the rapid changes in pre-rRNA processing in rat liver exposed to low doses of cycloheximide (STOYANOVA and HADJIOLOV 1979). At early stages after inhibition of protein synthesis the production of 18 S rRNA is abolished and that of 28 S rRNA is reduced to one-half the level seen in controls (Fig. 37). This divergence in the production of the two ribosomal subparticles is correlated with a block in the formation of 41 S and 21 S pre-rRNA. Generation of 32 S and 36 S pre-rRNA is still possible, but their conversion into 28 S rRNA proceeds at a decreased rate. These results indicate that protein shortage blocks some endonuclease cuts, while others are less affected. Increased amounts of some intermediate pre-rRNAs were observed also at the restrictive temperature in the CHO-tsH 1 cell line temperature-sensitive in protein synthesis (ROYAL and SIMARD 1980), in resting human lymphocytes (PURTELL and ANTHONY 1975) and in the BHK-422 ts mutant cell line temperature-sensitive for 32 S pre-rRNA processing (WINICOV 1976, TONIOLO and BASILICO 1976). That switching-on of alternative pre-rRNA processing pathways reflects a common response to block in protein synthesis is illustrated by the fact that it is observed also in poliovirus infected HeLa cells (WEINBERG and PENMAN 1970) and in vesicular stomatitis virus infected L cells (JAYE *et al.* 1980).

The fate of "abnormal" pre-rRNA species is of particular interest, although in most cases it has not been elucidated. Three possibilities may be envisioned: a) processing to normal rRNA at reduced rates; b) accumulation in the nucleus and c) degradation to non-functional products. All three possibilities are apparently exploited by cells although the fates of "abnormal" pre-rRNA could vary in different cells and organisms.

Degradation of incompletely assembled preribosomes may be an important control mechanism in eukaryotes. In a series of studies with resting and phytohaemagglutinin-stimulated lymphocytes evidence was obtained that upon limited protein synthesis, "wastage" of excess rRNA (largely 18 S rRNA) may take place in the nucleus (COOPER 1969, 1970, 1973, COOPER and GIBSON 1971). To preserve the equimolar 18 S/28 S rRNA ratio, "wastage" of excess 28 S rRNA in the cytoplasm was postulated to take place at a later stage. Posttranscriptional degradation of excess rRNA was proposed as a regulatory mechanism operating also in resting fibroblasts (ABELSON *et al.* 1974) and differentiated myotubes (CLISSOLD and COLE 1973, BOWMAN and EMERSON 1977). Further evidence bearing on the extent and intracellular location of excess rRNA "wastage" is needed before its role in ribosome biogenesis can be evaluated. Thus, unequal labeling of 18 S and 28 S rRNA may be due to the known polarity in the labeling of primary pre-rRNA during transcription (HADJIOLOV and MILCHEV 1974) and at the very low transcription rates in some resting cells a lower labeling of 18 S rRNA can be expected for rather long periods of labeling (EMERSON 1971). Also, the pools of large and small ribosomes in the cytoplasm of cultured mammalian cells may not be equimolar (NISSEN-MEYER and EIKHOM 1976a, b) and the recorded excess of small ribosomes could explain the lower labeling of 18 S rRNA. In any case, it seems that sizeable degradation of excess or defective pre-rRNA and/or rRNA is taking place only when there is an extreme decrease in protein synthesis. No evidence for "wastage" of pre-rRNA or rRNA was found in growing HeLa cells (WOLF and SCHLESSINGER 1977), in normal and regenerating rat liver (DUDOV and DABEVA 1983) or in adult rat brain (STOYKOVA *et al.* 1983). Rather, precise adaptation of both transcription and processing to changes in growth (or protein synthesis) rates seems to occur in most cases. A limited contribution of "wastage" to the control of pre-rRNA or rRNA processing seems to be valid also for lower eukaryotes. For example, in *Tetrahymena* a 15–20-fold reduction in the rate of ribosome biogenesis did not result in the "wastage" of pre-rRNA (SUTTON *et al.* 1979).

VI.3.4.2. Release from the Nucleolus and Nucleo-Cytoplasmic Transport of Ribosomes

Since the early studies on ribosome biogenesis three basic facts about the nucleo-cytoplasmic transport of ribosomes have been established (see ATTARDI and AMALDI 1970, MADEN 1971): a) Nucleoli contain both 28 S and 18 S rRNA, the pool of 28 S rRNA being larger; b) the rates of formation and of nucleo-cytoplasmic transport of 18 S rRNA are about twofold faster than those for 28 S rRNA and c) the difference is largely accounted for by the time needed for the intranucleolar conversion of 32 S pre-rRNA into 28 S rRNA. These observations have since been confirmed and extended to a variety of eukaryotic cells and organisms and we now realize that the mechanisms controlling the flow of ribosomes along the pathway nucleolus → nucleoplasm → cytoplasm may constitute an important critical site in the posttranscriptional regulation of ribosome biogenesis.

Tracer kinetics studies with *Saccharomyces* revealed that immediately after the endonuclease cut of primary pre-rRNA, the 20 S precursor to 18 S rRNA is transferred to the cytoplasm, where the last maturation step takes place. Maturation of a 29 S precursor to 25 S rRNA takes place in the nucleolus simultaneously with the cytoplasmic conversions of 20 S pre-rRNA (UDEM and WARNER 1973, TRAPMAN and PLANTA 1976). The nucleo-cytoplasmic transport of 25 S rRNA was correlated with the formation of 5.8 S rRNA from its 7 S precursor (TRAPMAN *et al.* 1976). Thus, the delayed release of the large ribosomal subparticle appears to be dependent on the completion of the processing steps leading to 25 S rRNA. It is noteworthy that many r-proteins seem to join the ribosomes at a late stage of their formation, *i.e.*, after the formation of their 18 S and 25 S rRNA components. In the case of the large ribosomal subparticle some of the late-added L-proteins join the nascent particle (containing 25 S rRNA) in the nucleus, while others are found only in cytoplasmic ribosomes (KRUISWIJK *et al.* 1978). These observations suggest that addition of some r-proteins to nascent ribosomes in the nucleus may trigger their release and nucleo-cytoplasmic transport.

The maturation of preribosomes in *Tetrahymena* was studied in detail and nascent ribosomal subparticles (containing mature S-rRNA and L-rRNA) were found in the nucleus, but rapidly released to the cytoplasm (RODRIGUES-POUSADA *et al.* 1979). That the nucleo-cytoplasmic transport of nascent ribosomes is dependent on continuous transcription and protein synthesis comes from studying the effects of actinomycin D (LEICK 1969, ECKERT 1975) and cycloheximide or puromycin (ECKERT *et al.* 1975). Following evidence that preribosomes and nascent ribosomes are associated with a nuclear matrix structure (HERLAN *et al.* 1979) these associations were thought to be involved in the control of the maturation and nucleo-cytoplasmic transport of nascent ribosomes (see WUNDERLICH 1981).

The nucleo-cytoplasmic transport of ribosomal subparticles in higher eukaryotes may be more complex. Numerous studies have shown that the nucleo-cytoplasmic transport of nascent ribosomes is tightly coupled with protein synthesis (see HADJIOLOV 1980). In particular, experiments with different animal cells showed that particles containing L-rRNA, but defective in their protein complement, are retained in the nucleolus (SOEIRO *et al.* 1968, WILLEMS *et al.* 1969, LÖNN and EDSTRÖM 1977a, b). These findings are easily understood when we consider the fact that many L- and S-proteins join their respective particles at a late stage of their formation and are not found in isolated nucleoli and nucleolar "55 S" or "40 S" preribosomes (see IV.3.). Since most of these r-proteins do not exchange with ribosomes in the cytoplasm, the exact site at which they join the nascent ribosomal subparticles remains unknown. The possibility that the addition of some of these proteins may trigger the release from the nucleolus and the ensuing nucleo-cytoplasmic transport of nascent ribosomes offers an attractive perspective for future studies. The recent finding of excess 5′-terminal sequences in rat liver nucleolar 28 S rRNA (HADJIOLOVA *et al.* 1984a) suggests that 5′-terminal processing of this rRNA may be also involved in the control of the nucleo-cytoplasmic transport of large ribosomal subparticles.

A dramatic increase of nascent large ribosomal subparticles (containing 28 S rRNA) is found in liver nucleoli of thioacetamide-treated rats (KOSHIBA *et al.* 1971), suggesting that the drug causes a delay in the nucleo-cytoplasmic transport of the particle. However, the mechanism of this effect is still obscure (see BUSCH and SMETANA 1970), but is worth exploring to provide an understanding of factors controlling the release of ribosomal particles from the nucleolus.

Another important question is: Is nucleo-cytoplasmic transport

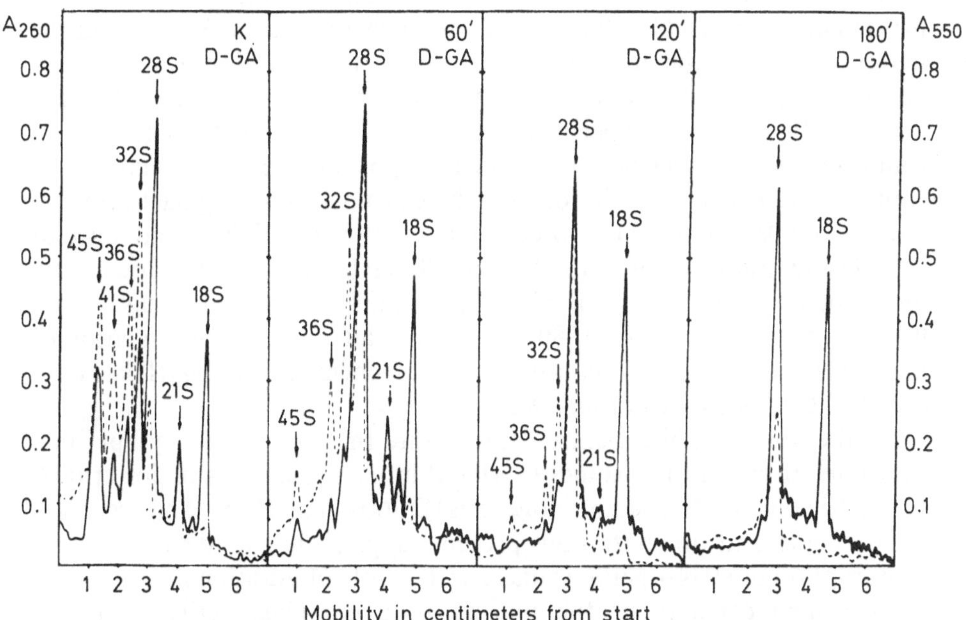

Fig. 38. Labeling of nucleolar pre-rRNA and rRNA after D-galactosamine inhibition of transcription. Nucleolar RNA is isolated and analyzed by urea/agar gel electrophoresis after 30 min *in vivo* labeling with [^{14}C]-orotate (K) and 60, 120, and 180 min chase with D-galactosamine (D-GA). The experimental rats treated with 250 mg of D-galactosamine/kg body weight show block in transcription within 15–30 min. The position of pre-rRNA and rRNA peaks is indicated by vertical arrows. ———, A_{260}; ----, Radioactivity, recorded from the autoradiogram at 550 nm. The label that disappears from nucleoli is recovered quantitatively in cytoplasmic 28 S and 18 S rRNA (not shown). [According to GAJDARDJIEVA *et al.* (1980). Reproduced with permission by *The European Journal of Biochemistry*]

dependent on continuous transcription of rRNA genes? Studies with animal cells showed that after an actinomycin D block of rDNA transcription, processing of presynthesized 45 S pre-rRNA is possible and leads to the formation of cytoplasmic 18 S-rRNA. However, both the conversion 32 S pre-rRNA → 28 S rRNA and the release of 28 S rRNA from the nucleolus are strongly inhibited (see HADJIOLOV and NIKOLAEV 1976). Since the effect of actinomycin D on pre-rRNA processing may not be related to its effect on transcription (WALTSCHEWA *et al.* 1976), these results, although suggestive, are open to alternative interpretations. More direct proof for a coupling of rDNA transcription and preribosome maturation was obtained in studies using camptothecin, a potent, but reversible, inhibitor of transcription that does not alter protein synthesis (WU *et al.* 1971). Inhibition of transcription in HeLa cells with camptothecin did not alter the processing of pre-rRNA, but blocked the release of 28 S rRNA from the nucleolus. When the inhibitor is removed, transcription resumes and accumulated 28 S rRNA is rapidly transferred to the cytoplasm (KUMAR and WU 1973). Other experiments, however, show that the coupling of transcription with nucleo-cytoplasmic transport of 28 S rRNA may not be compulsory. When transcription in rat liver is blocked by D-galactosamine-induced depletion of UTP, processing of 45 S pre-rRNA and nucleo-cytoplasmic transport of both 18 S and 28 S rRNA proceeded apparently unhampered (Fig. 38), at least at the initial stages of drug action (GAJDARDJIEVA *et al.* 1977, 1980). Moreover, a complete degranulation of nucleoli was observed, thus confirming the biochemical findings (DIMOVA *et al.* 1979, GAJDARDJIEVA *et al.* 1982).

All the above studies clearly demonstrate that the nucleo-cytoplasmic transport of nascent ribosomes (L-rRNP in particular) is a major critical site in the control of ribosome biogenesis (DUDOV *et al.* 1978). The process is clearly dependent on continuous protein supply, while its dependence on transcription of rRNA genes is unclear. The role of additional factors like flow-through capacity of the nuclear pore complex, active transport mechanisms and cytoplasmic compartmentalization of ribosomes may be rather complex (see WUNDERLICH 1981) and their regulatory role deserves closer attention. The observed restricted mobility of the large ribosomal subparticle in the cytoplasm (EDSTRÖM and LÖNN 1976, LÖNN and EDSTRÖM 1976, 1977 a) is a good example of possible posttranscriptional controls related to compartmentalization.

VI.3.4.3. Turnover of Ribosomes

The last stage in ribosome biogenesis, *i.e.* ribosome turnover, may be a major point in the regulation of ribosome abundance. The turnover of ribosomes was first observed in adult rat liver (LOEB *et al.* 1965, HADJIOLOV 1966, HIRSCH and HIATT 1966), but it is characteristic of all non-growing animal cells and tissues (see HADJIOLOV 1980). In the case of liver under steady-state conditions, both ribosomal particles turn over at essentially the same rate (HADJIOLOV 1966, TSURUGI *et al.* 1974, ELICEIRI 1976). The regulatory role of the turnover of ribosomes is suggested by observations showing that in protein-deprived animals the turnover rate of ribosomes is markedly increased, with half-life values ranging from 50 to 120 hours being reported under different feeding conditions (HIRSCH and HIATT 1966,

NORDGREN and STENRAM 1972, GAETANI *et al.* 1977). A similar situation exists in cultured animal cells. Intensive turnover of ribosomes is switched on when cells cease growing (ABELSON *et al.* 1974, BOWMAN and EMERSON 1977, LIEBHABER *et al.* 1978). Enhanced cytoplasmic degradation of ribosomes may provide signals for changes in the transcriptional (COUPAR *et al.* 1978) or posttranscriptional regulation of ribosome biogenesis. It is well known that decreased protein synthesis results in the accumulation of single ribosomes. Accordingly, several authors have proposed that the pool of free single ribosomes supplies signals for both their degradation and for deceleration of ribosome biogenesis (RIZZO and WEBB 1968, HENSHAW *et al.* 1973, PERRY 1973). Whether these changes are driven by modulation of pre-rRNA transcription and processing or of translation efficiency for rp-mRNA (GEYER *et al.* 1982) remains to be clarified.

VI.4. Autogenous Regulation of Ribosome Biogenesis in Eukaryotes: A Model

Analysis of the limited information presently available clearly demonstrates that there are many control sites which could be modulated to regulate the overall process of ribosome biogenesis. Both transcriptional and posttranscriptional regulatory mechanisms have been envisioned and the experimental evidence offered in support has been evaluated. Ultimately, acceleration or deceleration of ribosome biogenesis seems to be a rather flexible means by which every eukaryotic cell and organism responds to the temporal needs determining the level of protein synthesis. The relative contribution of the different control mechanisms varies from cell to cell, thus making precarious any attempts to generalize about the regulation of ribosome biogenesis in eukaryotes. However, some features seem to emerge from the wealth of accumulated evidence. Here, I shall try to define and discuss briefly a model for the autogenous regulation of ribosome biogenesis. Strong suggestive evidence for an autogenous regulation concept is given by the solidly documented examples of competitive interactions of r-proteins with both rRNA and mRNA in *E. coli* (NOMURA *et al.* 1980, 1982) or of the TFIIIA protein with both 5 S rRNA and 5 S rDNA in *Xenopus* oocytes (BROWN 1982). Similar interactions of regulatory significance may be involved in the basic control mechanisms of ribosome biogenesis in eukaryotes.

The production of a ribosome is the result of the concerted action of three sets of genes coding for: rRNAs, 5 S rRNA and r-proteins. Accordingly, the postulated autogenous regulation mechanisms reside in the interaction of these genes and their products outlined in Fig. 39. The basic postulates may be formulated as follows:

a) Ribosome biogenesis is a continuous process operating throughout the life cycle of the cell. Therefore, all three sets of genes should be *potentially active* at any time of the life cycle. If some of these genes are to be locked in inactive chromatin (examples: excess rRNA genes in plants or oocyte 5 S rDNA in somatic *Xenopus* cells), this is decided during mitosis and the accompanying reorganization of chromatin structure. However, at

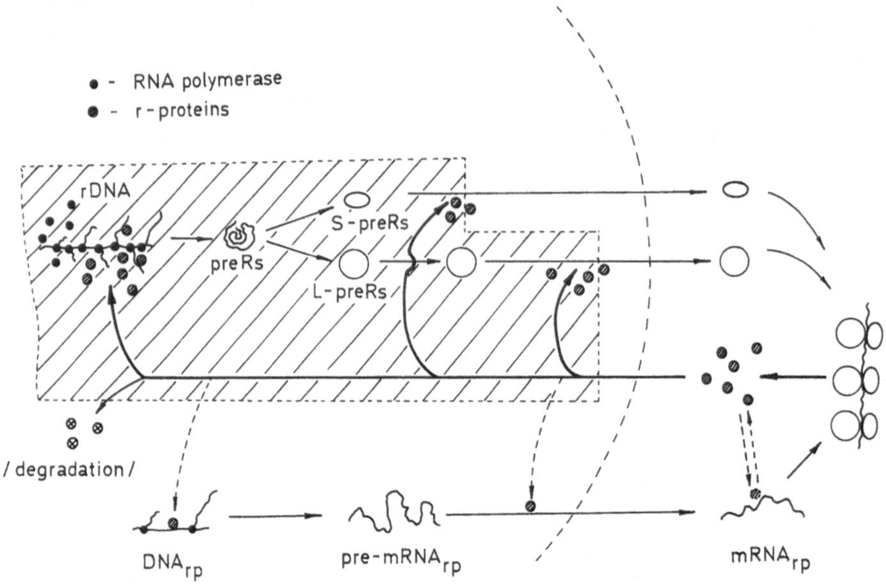

Fig. 39. Model for the main control factors in ribosome biogenesis. According to the model r-proteins are produced and their level maintained in slight excess over the needs for optimal rates of ribosome formation. Assembly of r-proteins with growing pre-rRNA chains during transcription or with maturing preribosomes is considered to modulate ribosome biogenesis at both its transcriptional and posttranscriptional stages. The regulatory action of r-proteins on the transcription of r-protein genes, processing of r-protein pre-mRNA or translation of r-protein mRNA is hypothetical. The low pool size of r-proteins in the nucleus is possibly maintained by degradation of excess r-proteins. See text for a more detailed discussion

telophase every cell enters its interphase life span with a stable number of potentially active ribosomal genes. That special devices exist to ensure that rRNA genes are always potentially active is suggested by the particular structure of the NO in mitotic chromosomes and during interphase. How the required constantly active state of 5 S rRNA and r-protein genes is maintained remains to be elucidated. Stable association of RNA polymerase I (even during mitosis) with r-chromatin is another outstanding feature of rRNA genes. Similar mechanisms may exist for the interaction of RNA polymerase II and III with the respective r-protein and 5 S rRNA genes. Thus, according to this model all ribosomal genes are potentially active (DNase I sensitivity expected) and characterized by a high affinity of interaction with RNA polymerases.

b) Normally, 5 S rRNA and r-protein genes produce their products in excess over the products of rRNA genes. Equal numbers of RNA polymerase I and III molecules seem to operate in the animal cell (YU 1980). Assuming equal rates of transcription the size difference of rRNA and 5 S rRNA transcription units would result in an 80-fold excess of 5 S rRNA molecules. Control of 5 S rRNA levels could be achieved by its interaction with TFIIIA-type proteins and the switching off of 5 S rRNA gene transcription. Excess production of r-proteins is also plausible. The animal cell produces 500–3,000 ribosomes/minute (see Table 4). A complete r-protein set is comprised of about 17,000 amino acids (WOOL 1979). If 6×10^6 ribosomes/cell are engaged in polyribosomes and assuming a polypeptide elongation rate of 5–10 amino acids/second (LACROUTE 1973), then 1% of the polysomal ribosomes will be sufficient to produce the required r-protein complements. Yet, markedly higher percentages of r-protein polyribosomes were reported for animal cells (GEYER *et al.* 1982). Excess r-proteins may: a) slow-down translation of their individual mRNAs; b) be rapidly degraded in the nucleus, or c) modulate the transcription of r-protein genes or the processing of rp-pre-mRNA. These assumptions are still largely hypothetical. It seems that possibility b) is the best documented at least for some animal cells, while the evidence for a) is indirect (PEARSON *et al.* 1982, GEYER *et al.* 1982). Mechanism c) is included, despite the absence of supporting evidence, with the sole intent of keeping an open mind in devising and interpreting future experiments. In any case, there is now general agreement that the pool of free r-proteins in the cell is kept low. Hence changes in r-protein level are likely to play a prompt regulatory role.

c) Ribosome biogenesis is centered on the activity of rRNA genes. The average mammalian cell operates with 100–200 such genes. The maximum output of primary pre-rRNA (and ribosomes) is limited by both the number of transcribed rRNA genes and transcriptionally competent (form I_B) RNA polymerase I molecules. Rough estimates show that for mammalian cells this maximum corresponds to about 5,000 primary pre-rRNA molecules/minute (see VI.2.2.1.), which is close to the experimental findings for rapidly growing cells (see Table 4). It seems that normally there is no critical shortage of RNA polymerase I (or even RNA polymerase I_B) molecules.

d) The model implies that the major regulatory factor is the supply of r-proteins to the nucleolus. All r-proteins should be present: the shortage of any single one may become rate limiting. It is proposed that r-protein control operates at two sites, corresponding to the major sites of addition of r-proteins to pre-RNA or rRNA. Excess r-proteins at the transcription site (site 1) enhances maximal elongation rates, which drives also maximal initiation rates. Note that if no factors limit the recycling of RNA polymerase I, the efficiency of the whole transcription process will reflect the rate of elongation. The above situation corresponds to the commonly observed pattern of fully loaded rRNA transcription units. A shortage of r-proteins at site 1 entails one major, early consequence: the production of faultily assembled primary preribosomes. This in turn results in their slower processing or degradation. Subsequent deceleration of transcription due to altered conformation of growing RNP fibrils or to other transcription control factors (*e.g.*, pre-rRNA degradation products) is a likely and ultimate consequence. Shortage of r-proteins at the release site (site 2) slows down the release of ribosomal subparticles (the large one being markedly more sensitive) and limits their supply to the cytoplasm. Since the production of r-proteins is coordinately regulated it seems that usually both sites 1 and 2 are simultaneously involved. This could be the reason why transcription and processing appear to be tightly coupled in most cases.

e) The whole process takes place within the nucleolus. It is envisioned that this structure is held together by specific interactions of rRNA genes, preribosomes, and nascent ribosomes with the nucleolar matrix. Apparently, the matrix structures are elaborated at telophase and fix spatial limits to the nucleolus and its molecular constituents. A conveyor mechanism of the transcription of rRNA genes and processing of preribosomes can be envisaged. In this case input (transcription) and output (release) rates may influence the efficiency of the whole nucleolar complex and provide a tool for the control of rRNA genes transcription rates by the products of transcription, *i.e.*, pre-rRNA and preribosomes. Again a leading role for r-proteins may be to ensure proper interactions of preribosomes and nascent ribosomes with nucleolar structures.

The proposed model seems to explain reasonably well the majority of the experimental findings on the regulation of ribosome biogenesis in eukaryotes. It places emphasis on r-protein supply as a major and very flexible regulatory factor. Unfortunately, how this supply is itself regulated is still largely unknown. Recent evidence points to translation control mechanisms changing the balance between free and polyribosome-engaged rp-mRNAs (PEARSON *et al.* 1982, GEYER *et al.* 1982). However, at least in regenerating liver, evidence for changes also in the amount of cytoplasmic rp-mRNA has been obtained (NABESHIMA and OGATA 1980, FALIKS and MEYUHAS 1982).

If the proposed model reflects truly the regulatory mechanisms operating in the eukaryotic cell, then the balance of cytoplasmic and nuclear r-proteins could be decisive for the flexible and remarkably fast adaptation of ribosome biogenesis to the continuously changing needs of the cell during its life cycle.

VI.5. Synopsis

Ribosome biogenesis results from the expression of three separate sets of genes coding for rRNAs, 5 S rRNA and r-proteins. The products of these genes interact during transcription or maturation to achieve the gradual formation of mature ribosomes. The features of the regulation of ribosome biogenesis in eukaryotes may be outlined as follows.

a) Ribosome biogenesis starts with the transcription of rRNA genes. There is evidence that in lower eukaryotes and animals, the bulk of r-chromatin is in a potentially active state. This may be related to the specific structure of r-chromatin in chromosomes and interphase nucleoli. However nucleolar dominance and similar phenomena show that in some cases r-chromatin (or 5 S rDNA chromatin) may be in a fully inactive state. The factors restricting part of the r-chromatin into inactive structures are unknown. r-Chromatin undergoes the apparently reversible transition: potentially active ⇄ transcribed. There is evidence that the number of rRNA genes being transcribed can vary, each rRNA transcription unit being independently switched on and off. The factors switching transcription of rRNA genes are not yet elucidated. At least a minimal number of nucleoplasmic r-protein and 5 S rRNA genes also should be in a potentially active state, but it is not known how this is achieved.

b) At a constant number of transcribed rRNA genes the rate of transcription for individual and total genes may vary depending on: a) the structure of r-chromatin; b) the activity of RNA polymerase I; c) the local supply of substrate nucleotides, and d) the conformation of growing pre-rRNP fibrils. There is a tendency for transcribed rRNA genes to be fully loaded with transcription complexes, suggesting that RNA polymerase I and putative initiation factors are not normally limiting. This correlates with findings showing that RNA polymerase I is metabolically stable and usually present in excess. In some cases increased overall transcription rates *in vivo* were correlated with an increase in the rate of elongation at a constant number of growing pre-rRNP chains. The factors controlling elongation rates and transcription rates in general are still hypothetical. Nevertheless, changes in overall transcription rates are clearly observed upon activation or inactivation of nucleolar activity and ribosome biogenesis.

c) The process of maturation of preribosomes is under stringent control, *i.e.*, dependent on the continuous supply of proteins. Since the cellular pool size of r-proteins is very low, their synthesis and supply to the nucleolus plays a leading regulatory role. The r-protein mRNAs are relatively stable and translational control factors seem to define the synthesis of r-proteins taking place on small polyribosomes. The r-proteins are rapidly taken up in the nucleus and their supply determines the efficiency of ribosome production. It is shown in some cells that in the absence of pre-rRNA and rRNA synthesis, excess r-proteins are rapidly degraded in the nucleus. The correct primary and secondary structure of pre-rRNA and conformation of preribosomes is also critical for their maturation.

d) The synthesis of 5 S rRNA is regulated independently of rRNA gene transcription. Positive initiation control factors interacting with both 5 S rRNA and 5 S rDNA regulate the transcription of 5 S rRNA genes. Thus, an excess of 5 S rRNA may decelerate transcription by binding the specific initiation factor. Whether the integration of 5 S rRNA into preribosomes has an effect on 5 S rDNA transcription is not known although it is a likely possibility. It seems that an excess of 5 S rRNA is normally synthesized in the nucleus so that ribosome biogenesis is not limited by the supply of 5 S rRNA.

e) Changes in the rate of ribosome formation seem to result primarily from the action of regulatory factors at critical sites in the sequential process of ribosome biogenesis. These include: a) switching on of alternative processing pathways and intranucleolar degradation of preribosomes; b) release of ribosomal subparticles from the nucleolus and their nucleo-cytoplasmic transport; and c) changes in the rates of cytoplasmic degradation (turnover) of ribosomes. The mechanisms of the regulatory processes involved in determining the overall efficiency of ribosome formation and the accumulation or ribosomes in the cytoplasm remain to be elucidated. There is little doubt that the continuous supply of r-proteins is a major factor regulating ribosome biogenesis at both transcriptional and posttranscriptional levels.

VII. Ribosome Biogenesis in the Life Cycle of Normal and Cancer Cells

Since the pioneer work of CASPERSSON (1950) and BRACHET (1957) we have known that the activity of the nucleolus is directly related to RNA and protein synthesis. A correlation between ribosome biogenesis and normal and malignant cell growth has since been supported by numerous studies (see BUSCH and SMETANA 1970). Here I shall try to outline presently available evidence on the role of ribosome biogenesis in the life cycle of eukaryotic cells and the changes observed under different physiological and pathological conditions. Emphasis will be placed on the correlation between the biochemical studies on ribosome biogenesis and the cytological observations on nucleolar structure in higher eukaryotes. Hopefully, elucidation of the molecular texture of cellular structures will help to understand better the comportment of the cell in health and disease.

VII.1. Nucleologenesis and Nucleololysis

VII.1.1. Nucleolar and Ribosome Biogenesis During the Mitotic Cycle

Nucleoli of varying number, size, and shape are observed in all active eukaryotic cells. Generally, the nucleoli are formed in early telophase and are functionally active throughout interphase. The activity of nucleoli ceases normally in late G_2 when r-chromatin is condensed at the NO sites of "nucleolar" chromosomes. Thus, it appears that NO structures display continuity throughout the whole mitotic cycle (see GOESSENS and LEPOINT 1979, BOUTEILLE *et al.* 1982). That rRNA genes in mitotic chromosomes are transcriptionally inactive is generally accepted (PRESCOTT 1964). However, several studies provided converging evidence that active RNA polymerase I is present in metaphase chromosomes (JOHNSON and HOLLAND 1965, WRAY and STUBBLEFIELD 1970, GARIGLIO *et al.* 1974, MATSUI *et al.* 1979), probably associated with NO structures. Thus, RNA polymerase I apparently is also a stable component of r-chromatin. The fate of the fibrillar and granular components of nucleoli during mitosis is a heavily debated question and will not be considered here. It seems that in most cells a complete dissolution of nucleoli takes place during mitosis. However, evidence for the persistence of some nucleolar material and its utilization in the reformation of nucleoli at telophase has been obtained, for example, in *Physarum polycephalum* (LORD *et al.* 1977) and plants (see GIMENEZ-MARTIN *et al.* 1977). At least in one case biochemical evidence for the persistence of preribosomes throughout mitosis was reported (FAN and PENMAN 1971).

Formation of nucleoli at telophase is centered at NO sites. Accordingly, the number of nucleoli corresponds generally to the number of NOs characteristic for a given cell type. However, this correlation is observed only in early G_1 phase. There is a marked tendency of NOs to coalesce later in interphase, resulting in the formation of a reduced number of larger nucleoli (see ANASTASSOVA-KRISTEVA 1977, SIGMUND *et al.* 1979, WACHTLER *et al.* 1982). From the point of view of ribosome biogenesis a more interesting question is: how does the total nuclear volume vary during interphase? A few studies of growing *Allium cepa* cells (SACRISTIAN-GARATE *et al.* 1974), chinese hamster cells (NOEL *et al.* 1971) and Ehrlich ascites cells (LEPOINT and GOESSENS 1982) showed that the nucleolar volume doubles during S phase and remains unchanged during G_2. Both fibrillar and granular components seem to expand, but the exact situation is difficult to assess. As expected, the changes in nucleolar volume are correlated with increased rates of ribosome formation (KURATA *et al.* 1978), as well as of protein synthesis during S and G_2 phases, but only a few direct measurements have been reported (see LLOYD *et al.* 1982). Most resting cells in adult tissues are known to be arrested at early G_1 phase (designated also as G_0), but populations blocked at late G_1 or G_2 are also known (see GELFANT 1981). These cells are studied routinely in biochemical investigations, but the heterogeneity of the populations used does not allow conclusions on the relations between their rates of ribosome biogenesis and position in the cell cycle. A common observation is that a highly active interphase nucleolus more closely reflects the metabolic state of the cell than its position in the cell cycle (see below).

VII.1.2. Nucleologenesis

The formation of nucleoli at telophase has been studied in great detail in a broad variety of plant and animal cells. The basic characteristics may be summarized as follows (see STAHL 1982, BOUTEILLE *et al.* 1982, DE LA TORRE and GIMENEZ-MARTIN 1982): a) formation of nucleoli takes place in close association with fibrillar centers, the counterpart of NO in interphase nucleoli; b) initially the fibrillar component of nucleoli is formed, corresponding to active rRNA transcription units as shown by gene-spreading and autoradiography; c) formation of the granular component is the second stage of nucleologenesis and correlates with the formation of preribosomes and ribosomes. This simple pattern of nucleologenesis corresponds well to the molecular process of rRNA gene transcription and preribosome maturation and provides a sound basis for the evaluation of biochemical and morphological findings.

An interesting and informative case of nucleologenesis is encountered in micronucleated cells that are generated by disruption of the mitotic spindle. The numerous micronuclei in such cells were found to be capable of forming nucleoli if they contain NO elements. Interestingly, some micronuclei that did not form nucleoli contained fibrillar nucleolar "blobs" capable of reacting with antinucleolar antibodies (PHILIPS and PHILIPS 1969, 1973, 1979, HERNANDEZ-VERDUN *et al.* 1979). These results demonstrate that the presence of fibrillar center(s) in micronuclei is sufficient to generate nucleoli with fibrillar and granular components, but leave unresolved the molecular characteristics of the micronuclear material that forms nucleolar "blobs".

The latter phenomenon may be related to instances of *abortive nucleologenesis* that result in the formation of fibrillar *pseudonucleoli.* Such pseudonucleoli were observed in anucleolate *Xenopus laevis* mutants (JONES 1965, HAY and GURDON 1967) and in cultured animal cells upon inactivation of chromosomal NOs by ultraviolet or laser microbeam irradiation (BURNS *et al.* 1970, SAKHAROV *et al.* 1972). These intriguing fibrillar structures give a positive histochemical reaction for RNA, but are inactive in RNA synthesis. Since O_{nu} mutants do not contain rRNA genes, a closer examination of such material could provide useful information regarding still unknown macromolecular components involved in nucleologenesis.

VII.1.3. Nucleololysis

The designation "nucleololysis" is not precise because different cases are encountered, depending on cell type and experimental conditions. In most cells the nucleolar activity is reduced in late G_2 phase and disappears entirely during mitosis at the time of nuclear envelope breakdown. During late G_2 or at the onset of mitosis the fibrillar component gradually disappears in parallel with extensive degranulation of nucleoli. The fate of the material from fibrillar and granular components during mitosis is not yet elucidated and it seems that, depending on cell type, different situations may be encountered (see above). Whether there is causal relationship between nucleololysis and mitosis is not known. In numerous interphase cells, cessation of rRNA synthesis and ribosome formation does not result in complete disappearance of nucleoli. Indeed, some nucleolar structures, designated as *remnant nucleoli* or *micronucleoli* (see BOUTEILLE *et al.* 1982), persist. That is seen, *e.g.*, in chick erythrocyte nuclei at the terminal stages of differentiation, when replication and transcription are practically halted. Here, the nucleolus is reduced to dense fibrillar remnants (LIKOVSKY and SMETANA 1978). The macromolecular components of these remnants are still unknown, but they retain the Ag-NOR proteins characteristic of the fibrillar center (HERNANDEZ-VERDUN *et al.* 1980). Yet, the activity of RNA polymerase I in remnant nucleoli is totally blocked and there is no detectable rRNA synthesis (GAZARYAN *et al.* 1973, ZENTGRAF *et al.* 1975, LAVAL *et al.* 1981). Nevertheless, these structures are capable of nucleologenesis, as observed after hybridization with other types of cultured cells (DUPUY-COIN *et al.* 1976, HERNANDEZ-VERDUN and BOUTEILLE 1979). Another example of nucleololysis is provided by liver cells treated with D-galactosamine to block transcription. In this case the nucleoli are reduced to dense fibrillar remnants within 4 hours after transcription is blocked (SHINOZUKA *et al.* 1973, DIMOVA *et al.* 1979). These nucleolar remnants have been shown to contain Ag-NOR proteins (DIMOVA *et al.* 1982) and all nucleolar rRNA genes and matrix proteins (GAJDARDJIEVA *et al.* 1982). Thus, the block of nucleolar activity in interphase nucleoli seems not to result in complete disaggregation of the nucleolus. Instead, residual nucleolar remnants are formed, apparently retaining the structures needed for a new cycle of nucleologenesis upon reactivation.

VII.2. Inhibition of Ribosome Biogenesis

Inhibitors and activators have been widely used to analyze the mechanisms of ribosome biogenesis. By causing drastic alterations at specific steps of this complex process, these agents helped elucidate the sequence of events of ribosome formation and the roles played by defined molecular transformations and interactions. Such studies are fully informative if: a) we know the precise molecular mechanisms of action of a given agent; b) a single (or a few) well defined site of action is the primary target, with a minimum of side effects; and c) the consequences for the structure and function of the cell and its organelles are neatly outlined. These requirements are seldom met and only a small number of agents have been sufficiently studied. Therefore, I shall discuss only the data obtained from inhibitor studies that permit more precise correlations between the observed molecular and cellular events.

VII.2.1. Inhibitors Interacting with DNA and Chromatin

This group is by far the largest studied and includes various antibiotics (actinomycin D, daunomycin, nogalamycin, camptothecin, hedamycin, etc.) and intercalating agents (ethidium bromide, proflavin, etc.). All these inhibitors block the synthesis of RNA by altering the structure of the DNA or chromatin template. Most of them cause an inhibition of both nucleoplasmic and nucleolar transcription, but some show a preference for rRNA genes and/or r-chromatin.

Among the inhibitors mentioned above and many others studied (see

SIMARD *et al.* 1974, HADJIOLOV and NIKOLAEV 1976, SUHADOLNIK 1979, SARIN and GALLO 1980), actinomycin D is by far the most widely used. It is known that actinomycin D blocks RNA synthesis in all organisms tested. An important early observation showed that low doses of actinomycin D inhibit preferentially rRNA synthesis in cultured animal cells (PERRY 1963). This effect was studied in detail and primary pre-rRNA synthesis in mouse L cells was found to be 50 to 100 times more sensitive to actinomycin D than is the synthesis of hnRNA (PERRY and KELLEY 1970). There is evidence that the action of actinomycin D on transcription is due to its binding with DNA (see GOLDBERG and FRIEDMANN 1971). However, autoradiographic studies with ^{3}H-actinomycin D failed to detect a preferential binding to nucleolar chromatin (BERNIER *et al.* 1972). Moreover, it was reported that ^{3}H-actinomycin D at low concentrations binds with a higher affinity to isolated non-ribosomal DNA than to rDNA (KHAN and LINDELL 1980). Thus, the mechanism of the preferential inhibition of rDNA transcription *in vivo* remains unknown. It should be noted, however, that a preferential inhibition of rDNA transcription is observed with many agents interacting with DNA and chromatin (see HADJIOLOV and NIKOLAEV 1976). Therefore, a simple explanation of the preferential action of actinomycin D may reside in the specific structure of r-chromatin or the thight packing of RNA polymerase I transcription complexes. This explanation is in line with the results of chromosome-spreading studies with *Triturus alpestris* oocytes (SCHEER *et al.* 1975) and CHO cells (PUVION-DUTILLEUL and BACHELLERIE 1979) showing that inhibition of rRNA synthesis by low doses of actinomycin D causes a detachment from the rDNA axis of growing pre-rRNP fibrils and condensation of nucleolar chromatin. The involvement of short-lived mRNA or protein(s) controlling rDNA transcription was also proposed (LINDELL 1976, LINDELL 1978), but remains to be substantiated.

What are the effects of actinomycin D on nucleolar ultrastructure? These effects have been studied in different plant and animal cells, all of which follow a characteristic pattern of sequential changes (see SIMARD *et al.* 1974, DASKAL 1979). The initial changes involve a condensation of fibrillar components and their migration toward the nucleolar periphery. The next stage is the *segregation* of the nucleolus into its constituents including (Fig. 40): a) more or less condensed fibrillar component; b) granular component; c) proteinaceous mass designated earlier as "amorphous part"; d) electron dense and coarsely granular masses designated as "spherical bodies", "caps" etc., usually associated with the granular component (REYNOLDS *et al.* 1964). After longer exposure to actinomycin D the granular component is dispersed and at the final stages a dense fibrillar remnant is seen. It has been shown that actinomycin D does not abolish the migration of prelabeled material from the fibrillar into the granular component (GEUSKENS and BERNHARD 1966). Most authors consider the "amorphous part" in segregated nucleoli as equivalent to the fibrillar center, the main evidence being: a) its continuity with both the fibrillar component and the nucleolus-associated chromatin (RECHER *et al.* 1976, VAGNER-CAPODANO and STAHL 1980) and b) the presence of a small amount

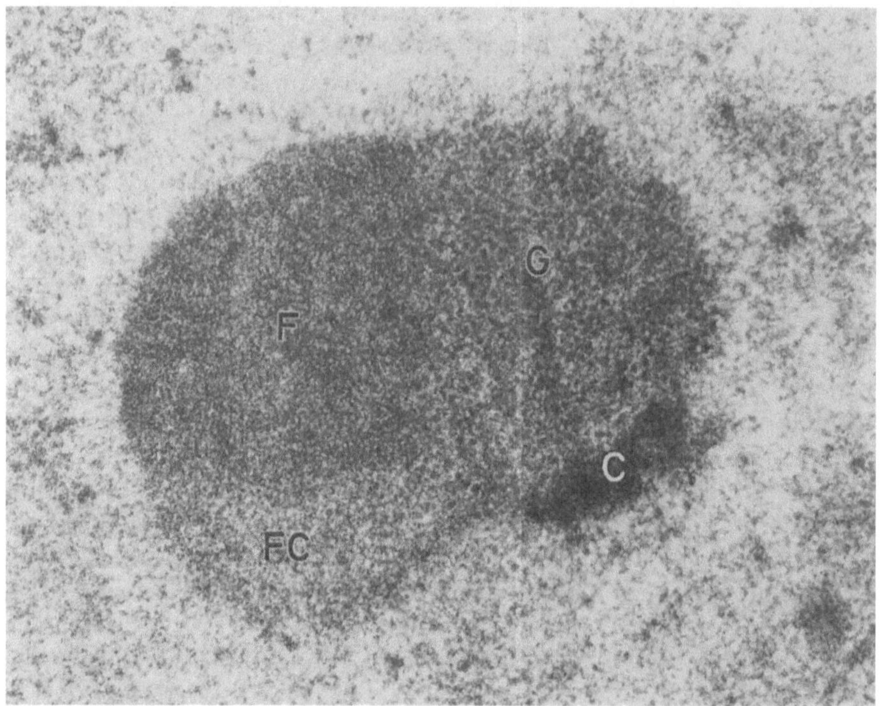

Fig. 40. Segregation of components of a hepatocyte nucleolus induced by actinomycin D. The experimental rats are treated for 4 h with actinomycin D (1 mg per kg body weight). *F* fibrillar component; *G* granular component; *FC* fibrillar center; *C* spherical body (synonyms: "contrasted zone", "cap" etc.). Fixation in OsO_4, × 66,000. [Courtesy by Dr. R. N. DIMOVA]

of DNA (rDNA?) and Ag-NOR proteins (seen also in the segregated fibrillar component) (HERNANDEZ-VERDUN *et al.* 1980, 1982).

The structural effects induced by actinomycin D may be considered typical of the effects of all transcription inhibitors that interact with DNA or chromatin although temporal or quantitative differences can be discerned in some cases. For example, camptothecin causes a rapid segregation of nucleolar components (RECHER *et al.* 1976), but it seems that in this case the formation of spherical bodies and nucleolar perichromatin-like granules is more pronounced (GAJKOWSKA *et al.* 1977).

The effect of aflatoxin B_1 deserves further comment since analysis of its action may provide further insight into the mechanism of nucleolar lesions induced by transcription inhibitors. Aflatoxin B_1, a potent carcinogen, has a varied action on both the cytoplasm and the nucleus of the target liver cell. An early action of aflatoxin B_1 (or its metabolic products) is the inhibition of all RNA synthesis, that of rRNA being more deeply affected than other RNAs (LAFARGE and FRAYSSINET 1970, NEAL 1972, YU 1977). Both RNA polymerases I and II are inhibited and, in particular, a strong inhibition of elongation rates of template-bound RNA polymerase I is recorded (YU 1981). Processing of primary pre-rRNA is also inhibited with a pronounced block of the 32 S pre-rRNA → 28 S rRNA conversion (SARASIN and MOULE 1975). The action of aflatoxin B_1 on pre-rRNA synthesis and processing may be related to its binding to DNA and/or histones (GROOPMAN *et al.* 1980, BAILEY *et al.* 1980, CROY and WOGAN 1981). Aflatoxin B_1 causes also profound cytoplasmic alterations, including an early inhibition of protein synthesis (BUSBY and WOGAN 1979). Investigation of the effect of aflatoxin B_1 on nucleolar structure showed the expected induced segregation of nucleolar components (BERNHARD *et al.* 1965, PONG and WOGAN 1970). However, the observed changes appear to be less pronounced and to occur more slowly than with drugs, such as actinomycin D, that act more selectively on transcription. Thus, it appears that the simultaneous inhibition of transcription, processing and protein synthesis prevents the full development of the alterations in nucleolar structure induced by more selective transcription inhibitors.

VII.2.2. Inhibitors That Act on RNA Polymerases

Among the limited number of inhibitors interacting with RNA polymerases, α-amanitin is the most thoroughly investigated one. It is known that in most eukaryotes low doses of α-amanitin preferentially inhibit RNA polymerase II and at higher doses also RNA polymerase III, while RNA polymerase I is fully resistant (see III.1.1.). When given to rats or mice *in vivo* α-amanitin causes a rapid (within 30–60 min) inhibition of RNA polymerase II, while RNA polymerase I remains unaffected for at least 2 hours (Fig. 41) (TATA *et al.* 1972, SMUCKLER and HADJIOLOV 1972, SEKERIS and SCHMID 1972, HADJIOLOV *et al.* 1974 a). These results show that both *in vitro* and *in vivo* α-amanitin acts as a selective inhibitor of RNA polymerase II and the transcription of nucleoplasmic genes. Surprisingly, administration of α-amanitin *in vivo* causes a strong inhibition of both nucleoplasmic and nucleolar RNA synthesis (JACOB *et al.* 1970, NIESSING *et al.* 1970). Further studies revealed that α-amanitin initially causes a strong inhibition of pre-rRNA processing followed by a decrease in pre-rRNA synthesis, while the labeling of 5 S rRNA is only slightly affected (HADJIOLOV *et al.* 1974 a). The *in vivo* effect of α-amanitin on pre-rRNA synthesis and processing is not limited to rat and mouse liver. It was observed with a variety of eukaryotic cells and organisms, including *Chirnomus* salivary glands (SERFLING *et al.* 1972), *Calliphora* larvae (SHAAYA and CLEVER 1973), cultured mouse AKR-2 B cells (WELLS *et al.* 1979), and germinating wheat embryos (JENDRISAK 1980).

The effect on pre-rRNA synthesis is not observed with other eukaryotic

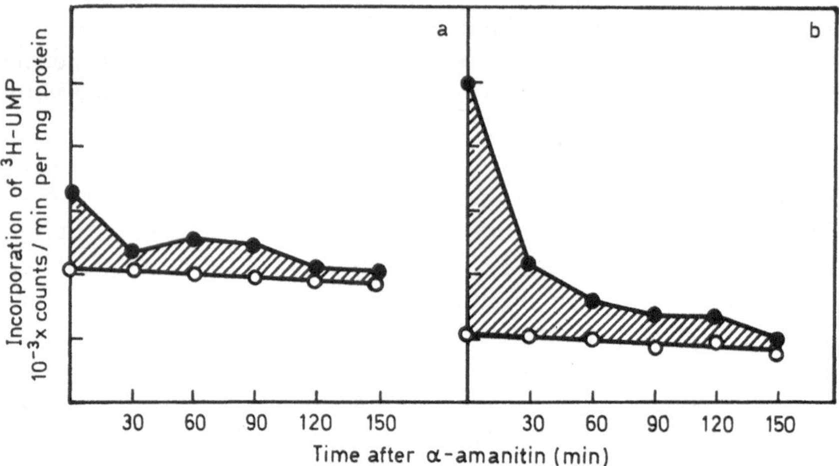

Fig. 41. Action of α-amanitin *in vivo* on the RNA polymerase activity of isolated mouse liver nuclei. α-Amanitin (0.2 μg/g body weight) is given intraperitoneally at different times before sacrifycing the animals. RNA polymerases are assayed under ionic conditions favorable for RNA polymerase I (*a*) or II (*b*) with (○———○) or without (●———●) the addition of 2.5 μg α-amanitin per ml incubation medium. The shaded area shows the decrease of RNA polymerase activity after α-amanitin administration *in vivo*. [From HADJIOLOV *et al.* (1974). Reproduced with permission by *Biochem. J.*]

cells. For example, in *Triturus* oocytes (BUCCI *et al.* 1971), *Aedes aegypti* adults (FONG and FUCHS 1976), chick embryo or rat fibroblasts (HASTIE and MAHY 1973, KUWANO and IKEHARA 1973), and CHO cells (KEDINGER and SIMARD 1974) nucleoplasmic RNA synthesis is preferentially inhibited by α-amanitin, but pre-rRNA synthesis and processing are not appreciably altered. In the case of CHO cells it was observed that hnRNA synthesis is still fully expressed for 8 hours after α-amanitin addition and pre-rRNA remains unaltered until at least the 12th hour (KEDINGER and SIMARD 1974). These negative results are difficult to interpret. They could reflect peculiarities in ribosome biogenesis or merely the effect of permeability barriers on the intracellular concentration and timing of action of α-amanitin. Indeed, it has been shown that cultured hepatoma cells are impermeable, hence resistant, to α-amanitin (Boctor and Grossman 1973) and permeability potentiation by polyene antibiotics was shown to facilitate the effect of α-amanitin (Kuwano and Ikehara 1973).

Given the selective interaction of α-amanitin with RNA polymerase II, its effect on pre-rRNA synthesis is unexpected and the elucidation of its

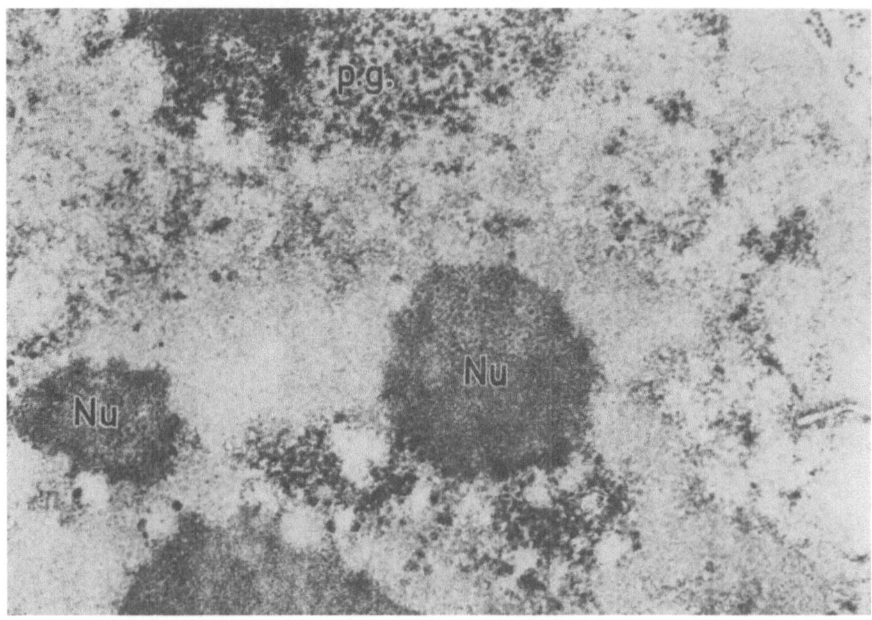

Fig. 42. Fragmentation and microsegregation of hepatocyte nucleoli induced by α-amanitin. The animals are sacrificed 60 min after the intraperitoneal injection of mice with 10 μg α-amanitin per g body weight. *Nu* fragmented and microsegregated nucleoli; *p.g.* clumps of perichromatin granules. Fixation in glutaraldehyde and preferential bleaching of heterochromatin with EDTA, × 60,000. [Courtesy by Dr. P. Petrov]

mechanisms may provide clues to the understanding of the regulation of ribosome biogenesis.

Ultrastructural studies revealed the induction by α-amanitin of rapid and drastic changes in nuclear and nucleolar structures (Marinozzi and Fiume 1971, Petrov and Sekeris 1971, Fiume and Laschi 1965, Sinclair and Brasch 1978, Romen *et al.* 1977). An early and pronounced condensation of nuclear euchromatin is accompanied by nucleolar segregation. However, the segregation stage is of short duration and is followed by *fragmentation* of nucleoli (Fig. 42). The nucleolar fragments retain their segregated fibrillar and granular components, a phenomenon described as *microsegregation.* At the final stage nucleoli are reduced to dense fibrillar remnants. Interestingly, the studies with CHO cells indicate that fragmented nucleoli may continue to synthesize pre-rRNA (KEDINGER and SIMARD 1974), but this is not the case with mouse liver (HADJIOLOV *et al.* 1974a, EMANUILOV *et al.* 1974). The difference may be related to the rapid onset of fragmentation of practically all liver nucleoli (Fig. 43). Another interesting

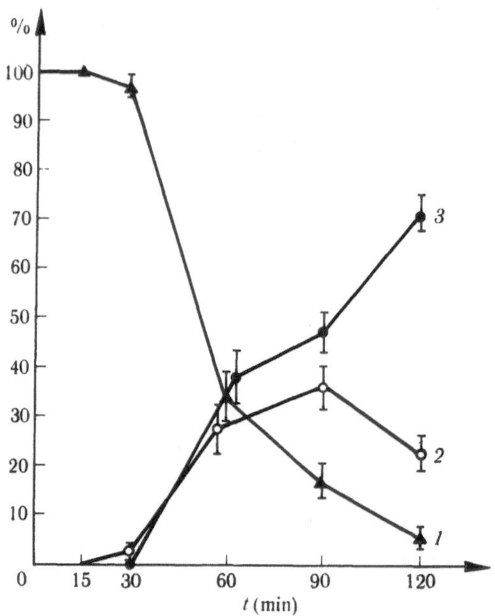

Fig. 43. Quantitative analysis of the process of mouse liver nucleolar fragmentation induced by α-amanitin. The animals are treated with 0.2 μg α-amanitin per g body weight for the indicated times. Analysis by electron microscopic inspection of 300–700 nuclei and classification of nucleoli into three types: *1* (▲)—intact nucleoli; *2* (○)—nucleoli showing segregation and partial fragmentation, and *3* (●)—complete fragmentation of nucleoli. Vertical bars indicate the confidence intervals. [From EMANUILOV *et al.* (1974) with permission by *Experimental Cell Research*]

finding is that simultaneous block of protein synthesis with cycloheximide prevents the α-amanitin-induced fragmentation of nucleoli (BARSOTTI *et al.* 1980).

The structural alterations of the nucleolus induced by thuringiensin are also noteworthy. As shown by biochemical studies, this ATP analogue acting *in vivo* preferentially inhibits primary pre-rRNA synthesis (MACKEDONSKI *et al.* 1972), while *in vitro* it suppresses both RNA polymerase I and II (SMUCKLER and HADJIOLOV 1972, BEEBEE *et al.* 1972). A whole spectrum of striking nucleolar structural changes is induced by thuringiensin, including formation of ring-shaped nucleoli and segregation

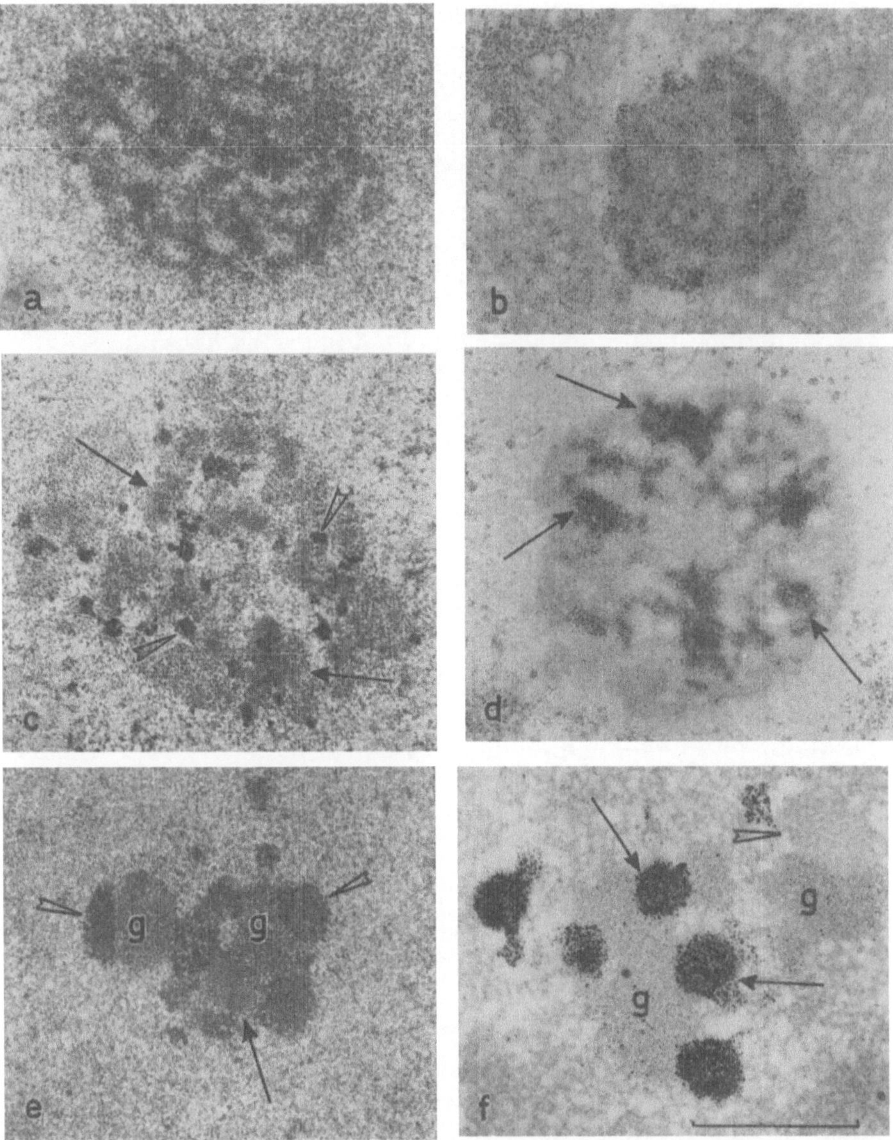

Fig. 44. Ultrastructure and localization of Ag-NOR proteins in rat liver nucleoli at different times after D-galactosamine block of transcription. Standard OsO_4 fixation (a, c, e, g, j); Ag-staining (b, d, f, h, i, k). *a* and *b*—Normal rat liver. Rat liver nucleoli at 30 (c, d), 60 (e, f), 120 (g, h and i), and 240 (j, k) min after administration of 250 mg D-galactosamine per kg body weight. *g*—granular component. Nucleolar segregation starts at 30 min (c, d) with clumping of fibrillar components (positive in Ag-staining) and appearance of microspherules (arrowheads). At 60 min (e, f) fragmentation and segregation of nucleoli is fully developed: arrows—fibrillar component; arrowheads—spherical bodies. At 120 min (g–i) advanced degranulation of segregated nucleoli is observed. The Ag-NOR protein(s) are confined to the segregated fibrillar component (arrows); arrowheads—spherical bodies. At 240 min (j, k) nucleolar fibrillar remnants showing strongly positive Ag-staining. Bar—1 μm, a–j, × 25,000. [Reproduced from Dimova *et al.* (1982) with permission by *European Journal of Cell Biology*]

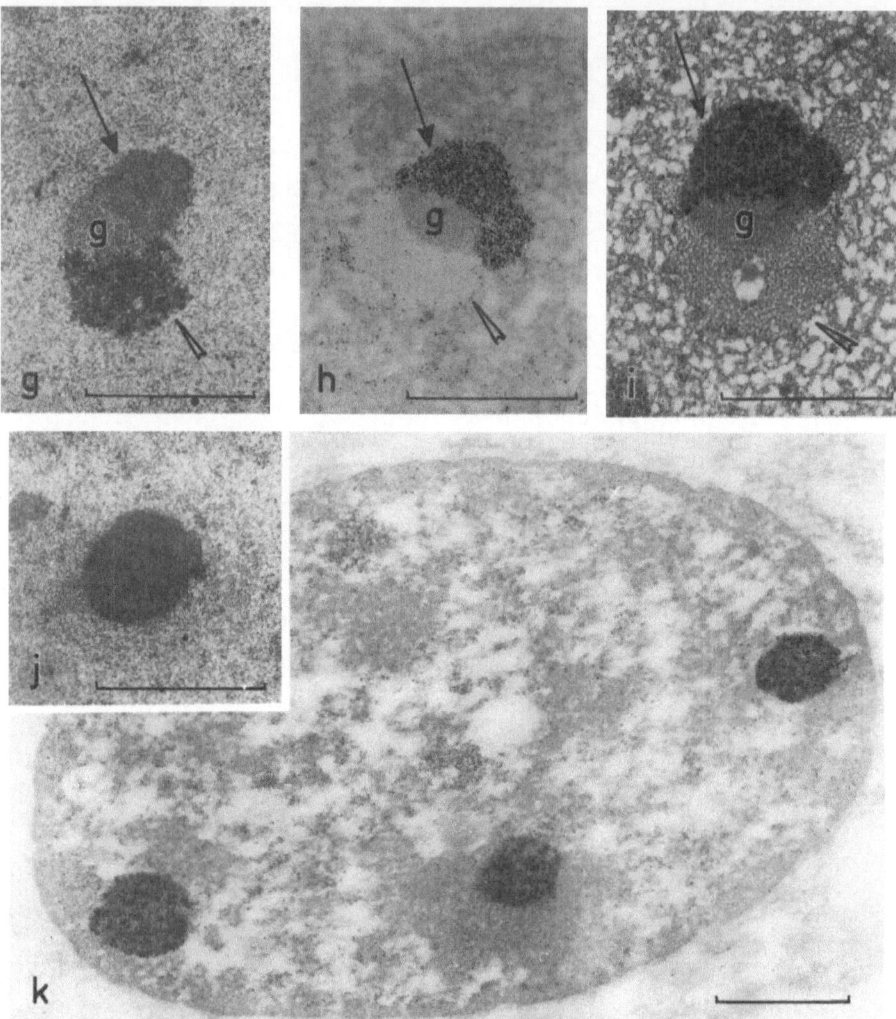

Fig. 44 g–k

of nucleolar components. Interestingly, in the segregated nucleolus two distinct forms of fibrillar components, as well as two forms of granular components, were discerned (SMETANA *et al.* 1974).

VII.2.3. Inhibitors of Nucleoside-5′-Triphosphate Formation

Several investigators have attempted to influence RNA synthesis by modulating the level of precursor nucleoside-5′-triphosphates. It is noteworthy that under these conditions too, rRNA synthesis is markedly more sensitive than hnRNA synthesis, even when such unspecific inhibitors as the uncouplers of oxidative phosphorylation, 2,4-dinitrophenol and sodium azide, are used (MACKEDONSKI and HADJIOLOV 1970, STELLETSKAYA *et al.* 1973). At present, the most clear-cut case of an induced decrease of a precursor nucleoside-5'-triphosphate *in vivo* is provided by the action of D-galactosamine on liver. D-Galactosamine causes a rapid and complete block of transcription, but does not alter processing of preribosomes, as well as the nucleo-cytoplasmic transport of nascent ribosomes (see VI.3.4.2.). The D-galactosamine block of transcription causes the following sequential changes of nucleolar structure (Fig. 44): a) early appearance of microspherules; b) extensive fragmentation and microsegregation (60 min); c) in segregated nucleoli numerous spherical bodies are formed adjacent to the segregated granular component, with formation of perichromatin-like granules at their rim: d) later (120 minutes) extensive degranulation takes place leaving mainly a condensed fibrillar remnant associated with spherical bodies; e) finally (240 minutes) dense fibrillar remnants remain in the nucleoplasm, while the spherical bodies gradually collapse (SHINOZUKA *et al.* 1973, DIMOVA *et al.* 1979). The fibrillar nucleolar remnants contain all rRNA genes and matrix proteins (GAJDARDJIEVA *et al.* 1982) and serve as centers of nucleologenesis at later (12–16 hours) recovery stages (SHINOZUKA *et al.* 1973). Analysis of the distribution of Ag-NOR proteins (see Fig. 44) reveals that they are associated with the fibrillar component and the derived nucleolar fibrillar remnants (DIMOVA *et al.* 1982). Biochemically, it is important to stress that the D-galactosamine block of transcription causes a complete depletion of nucleolar ribonucleoprotein structures which are either processed and transported to the cytoplasm or degraded (HERZOG and FARBER 1975, GAJDARDJIEVA *et al.* 1977, 1980, 1982).

VII.2.4. The Effects of Analogues Incorporated into Polyribonucleotide Chains

A large number of natural or synthetic analogues of the four major RNA nucleotides that have been identified can be phosphorylated to the respective nucleoside-5'-triphosphates and be incorporated more or less efficiently into RNA chains (LANGEN 1975, SUHADOLNIK 1970, 1979). The major consequences from such substitution of the normal nucleotides are: a) distortion of base-pairing interactions and of the secondary structure of the pre-rRNA chains and b) an altered interaction with proteins and assembly of conformationally defective preribosomes. Here, I shall consider only the action of *toyocamycin* and *5-fluoropyrimidines* as typical examples for the action of analogues on ribosome biogenesis.

Toyocamycin (7-CN-7-deaza-adenosine) is rapidly incorporated *in vivo* into RNA chains. At low doses it blocks the processing of primary pre-rRNA without affecting the synthesis of hnRNA, tRNA, and 5 S rRNA (TAVITIAN *et al.* 1968, 1969). Experiments with various cultured animal cells reveal that the processing site that is altered preferentially by toyocamycin depends on the concentration of the drug. At low doses, processing to 32 S pre-rRNA occurs, but its conversion into 28 S rRNA is blocked. At higher doses the processing of 45 S pre-rRNA is also inhibited. At still higher doses rDNA transcription is also inhibited, although to a markedly lesser extent than processing of pre-rRNA (WEISS and PITOT 1974, AUGER-BUENDIA *et al.* 1978, HADJIOLOVA *et al.* 1981). Thus, the primary consequence of toyocamycin action is a strong inhibition of pre-rRNA processing, the last stages of which are the most sensitive. Similar results were obtained in studies with *Saccharomyces cerevisiae* (VENKOV *et al.* 1977), thereby showing that the effect is probably similar in all eukaryotes. Toyocamycin inhibition of 45 S pre-rRNA processing in mouse L cells resulted in a 2.5-fold accumulation of "80 S" preribosomes in nucleoli (AUGER-BUENDIA *et al.* 1978), characterized by the same complement of L- and S-proteins as found in control cells (AUGER-BUENDIA and TAVITIAN 1979).

The morphological studies on the effect of toyocamycin are few. A detailed study of cultured BSC monkey kidney cells revealed that toyocamycin at doses up to 1 μg/ml produces some nucleolar hypertrophy and the accumulation of granular component structures, but the "normal" appearance of nucleoli is basically preserved (MONNERON *et al.* 1970). The changes in nucleolar structure, *i.e.* fragmentation of nucleoli, observed at higher concentrations of toyocamycin (5 μg/ml) are related to the inhibition of transcription that occurs at that concentration. Thus, the main conclusion is that the block in preribosome processing is not accompanied by an appreciable alteration of nucleolar structure.

The action of *5-fluoropyrimidines* is similar in many respects to that of toyocamycin. Experiments with yeast cells showed that 5-fluorouracil inhibits the formation of mature ribosomes, while pre-rRNA accumulates (DE KLOET 1968, MAYO *et al.* 1968). Administration of 5-fluorouracil to rats *in vivo* causes a delayed release of ribosomes into the cytoplasm and nucleolar hypertrophy (see STENRAM 1972). The use of 5-fluoroorotate (analogue of the main pyrimidine precursor in animals) results in a more pronounced inhibitory effect. Briefly, the drug does not alter rDNA transcription or the early steps of pre-rRNA processing, but the formation of both large and small ribosomal particles is blocked (see HADJIOLOV and HADJIOLOVA 1979 and VI.3.2.). As with the effects of 5-fluorouracil, no gross alterations in nucleolar morphology could be detected (HADJIOLOV *et al.* 1974 b). These observations reinforce the conclusion that even a complete block in pre-rRNA processing and ribosome formation does not change appreciably the ultrastructure of nucleoli.

A special case is presented by 3′-deoxynucleosides, the best studied being the adenosine analogue *cordycepin*. This nucleoside is phosphorylated *in vivo* and incorporated into RNA chains, but the absence of the 3′-OH group prevents elongation and results in premature transcription termination and formation of incomplete transcripts (see SUHADOLNIK 1979). Studies with HeLa cells have shown that the cordycepin-induced inhibition of transcription does not prevent processing of already synthesized primary pre-rRNA and the formation of 18 S rRNA. However, the formation of 28 S rRNA is apparently altered (SIEV *et al.* 1969). The morphological consequences of cordycepin action have been studied in detail with plant (GIMENEZ-MARTIN *et al.* 1973) and animal cells (PUVION *et al.* 1976) and, as with other inhibitors of transcription, it causes fragmentation and microsegregation of nucleoli. Numerous spherical bodies are formed, surrounded by a rim of perichromatin-like granules. It was also shown by radioautography that cordycepin does not prevent the labeling of the nucleolar fibrillar component, but totally blocks migration of the label to the granular component (PUVION *et al.* 1976).

VII.2.5. Inhibitors of Protein Synthesis

The effects of protein synthesis inhibitors were considered earlier (see VI.2.2.3. and VI.3.1.). The blockage of protein synthesis results in an early and profound alteration in the processing of primary pre-rRNA and preribosomes, while the effect on transcription is usually delayed and partial. What ultrastructural changes are induced by inhibition of protein synthesis? Most studies have been devoted to the action of cycloheximide. In rat liver, low doses of cycloheximide (1–5 mg/kg body weight), sufficient to cause a complete block of protein synthesis, did not produce any noticeable alterations in nuclear and nucleolar morphology (HWANG *et al.* 1974, GOLDBLATT *et al.* 1975, STOYANOVA *et al.* 1980). That a block in primary pre-rRNA (preribosome) processing induced by inhibition of protein synthesis does not alter nucleolar structure is clearly demonstrated by studies with CHO-tsH 1 cells that are temperature-sensitive for protein synthesis (ROYAL and SIMARD 1980). Higher doses of cycloheximide (20–100 mg/kg body weight) inhibit both transcription and processing of pre-rRNA, but nevertheless the changes in nucleolar ultrastructure are negligible and segregation or similar phenomena are not observed (UNUMA *et al.* 1973, DASKAL *et al.* 1975, DERENZINI and BONETTI 1975, STOYANOVA *et al.* 1980). These results suggest that nucleolar segregation is hampered when transcription and processing are blocked simultaneously.

VII.2.6. Interpretation of Nucleolar Alterations

I have outlined above several typical examples of alteration of nucleolar structure induced by different inhibitors of ribosome biogenesis. More detailed and extensive surveys have been published (SIMARD 1970, SIMARD *et al.* 1974). The cases considered here were selected to provide information on parallel biochemical and morphological studies that make possible interpretations of altered nucleolar structure in a variety of pathological conditions (see BOUTEILLE *et al.* 1982). The following interpretation of observed alterations in nucleolar ultrastructure is based largely on the concept of the molecular architecture of the nucleolus outlined earlier (see V.7.). As pointed out, the nucleolus may be considered to be composed of four major structural components: a) fibrillar component = transcribed rRNA genes; b) granular component = preribosomes and ribosomes at different stages of maturation; c) fibrillar center(s) = interphase NOR = non-transcribed r-chromatin plus nucleolar proteins, and d) nucleolar matrix and intranucleolar non-ribosomal chromatin. I also stressed that the nucleolus is a highly dynamic structure representing the balance between the rates of several continuous processes, the major ones being: a) transcription of rRNA genes resulting in the release of primary preribosomes; b) processing of primary preribosomes to form nascent small and large ribosomal particles; and c) transport of nascent ribosomes to the nucleoplasm and cytoplasm. With these basic notions in mind we can attempt to understand the molecular basis of experimentally induced alterations in nucleolar ultrastructure.

VII.2.6.1. Nucleolar Segregation

It is important to stress the point that segregation is only a transient stage of nucleolar inactivation, although its duration may vary (see SIMARD *et al.* 1974, BOUTEILLE *et al.* 1982). There is now general agreement that nucleolar segregation reflects the inhibition of rDNA transcription (SIMARD *et al.* 1974, ROMEN and ALTMANN 1977, DASKAL 1979, FAKAN and PUVION 1980, BOUTEILLE *et al.* 1982), a conclusion that is substantiated by all presently available data. Nevertheless, several features deserve closer attention.

An inhibition of rDNA transcription is necessary, but not sufficient, to induce nucleolar segregation. The transcription block should be accompanied by: a) retained capacity of processing primary preribosomes and b) a delayed release of nascent ribosomes from the nucleolus. Apparently, the combination of these three conditions determines both the onset and the duration of the segregation state. An inhibition of rDNA transcription with a concomittant block of preribosome processing (*i.e.* with cycloheximide) may hamper nucleolar segregation. On the other hand, an inhibition of transcription without a block of ribosome release shortens the duration of the segregated state as exemplified by D-galactosamine action. We could ask further: why is nücleolar segregation observed so often? It is plausible that the reason may be looked for in the fact that all agents inhibiting transcription by interaction with rDNA or r-chromatin (the majority of inhibitors tested) enhance also a delayed release of nascent ribosomes from the nucleolus (best studied examples being the effects of actinomycin D, camptothecin etc.). The mechanism of the delayed release of nascent ribosomes following transcription inhibition is not known, but could be related to the distortion of chromatin or of the nucleolar matrix.

Looking more closely at the consequences of rDNA transcription block, two molecular phenomena deserve attention: a) the gradual release of incomplete pre-rRNP fibrils from the rDNA template and b) the degradation of spuriously or incompletely assembled preribosomes and ribosomes. The release of pre-rRNP fibrils may be rather slow in some cases (SCHEER *et al.* 1975), but may be completed within 30–60 minutes in other cases (PUVION-DUTILLEUL *et al.* 1977). The timing of pre-rRNP fibrils release may be dependent on local nuclease activity or unknown factors. In any case, release of the growing pre-rRNP fibrils results in the condensation of the residual r-chromatin (FRANKE *et al.* 1979). Therefore, the fibrillar component in segregated nucleoli is gradually condensed so that its structure is no longer identical to that of its active counterpart. What is the fate of the released incomplete pre-rRNP fibrils and faultily assembled preribosomes?

VII.2.6.2. Nucleolar Spherical Bodies and Perichromatin Granules

These two structures are not observed in the nucleoli of normal cells. The spherical bodies (synonyms: "caps", "blebs", "contrasted zone", "peripheral dense substance", etc.) constitute electron dense, granular structures, usually adjacent to the slowly disappearing nucleolar granular component in cells treated with actinomycin D (REYNOLDS *et al.* 1964, SIMARD and BERNHARD 1966, ROMEN *et al.* 1977), α-amanitin (MARINOZZI and FIUME 1971, ROMEN *et al.* 1977), cordycepin (PUVION *et al.* 1976), D-galactosamine (DIMOVA *et al.* 1979), camptothecin (RECHER *et al.* 1972, GAJKOVSKA *et al.* 1977) or other drugs (Fig. 45). Thus, spherical bodies may be the images of some preexisting structures in nucleoli ordinarily masked by the abundance of fibrillar and granular components. Spherical bodies are usually associated with the ribonucleoprotein nucleolar perichromatin granules, also seen at the border of the fibrillar component or the nucleolus-associated chromatin (PUVION *et al.* 1981 and references therein). The origin of the nucleolar perichromatin granules has been sought in several studies. A plausible explanation is that perichromatin granules represent incomplete or faultily assembled pre-rRNPs in the process of slow degradation (PUVION *et al.* 1977, 1981, DIMOVA *et al.* 1979). This interpretation is in line with the biochemical findings showing that pre-rRNPs (and preribosomes) with abnormal pre-rRNA or protein constituents cannot be processed to ribosomes and are slowly degraded. The origin of the spherical bodies is still puzzling. It is known that they are: (i) largely proteinaceous and formed in association with the granular component (RECHER *et al.* 1976); (ii) devoid of r-chromatin, since they do not contain Ag-NOR proteins (DIMOVA *et al.*

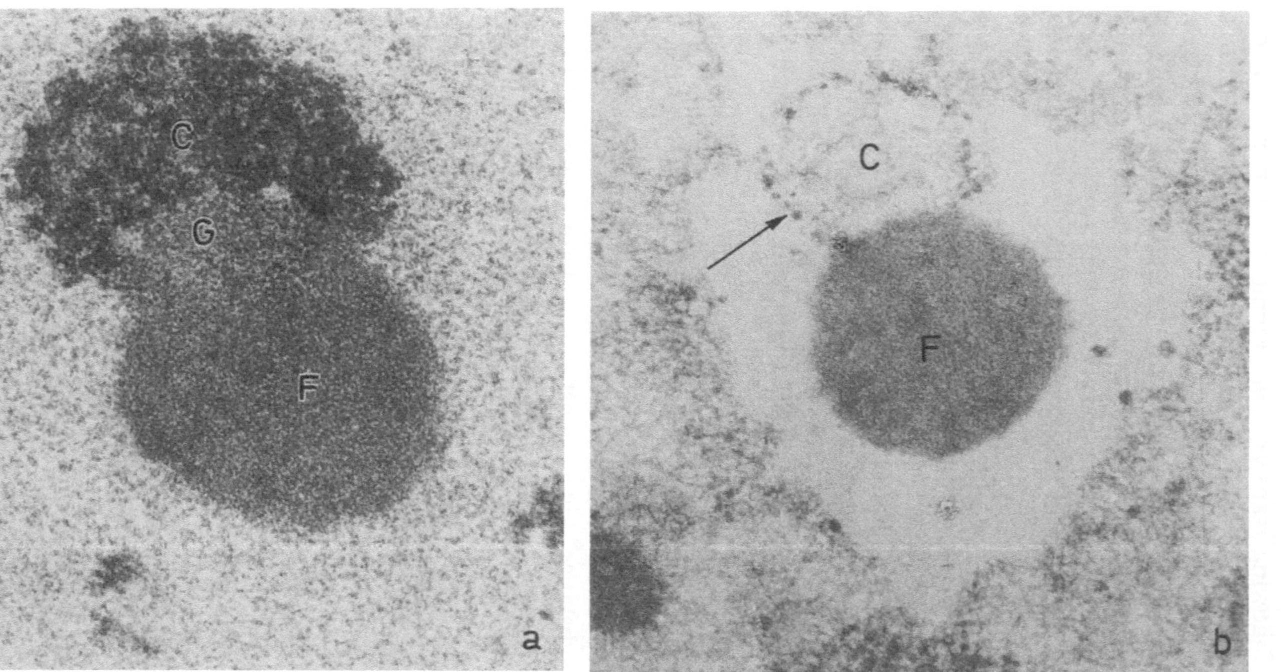

Fig. 45. Spherical bodies in rat liver cell nucleoli after D-galactosamine block of transcription. The rats are treated for 2 h with 250 mg D-galactosamine per kg body weight. *a* Electron dense spherical body (*C*) with coarse granular structure is formed at the periphery of the fibrillar nucleolar remnant (*F*) at the site where the granular component (*G*) is partly preserved. Standard fixation with OsO_4, × 60,000. *b* The same structure as in *a*, but after fixation in glutaraldehyde and bleaching with EDTA of nucleolus-associated heterochromatin. The nucleolar fibrillar remnant (*F*) is associated with a spherical body (*C*) with numerous perichromatin-like granules at its rim (arrow), × 62,000. [According to DIMOVA *et al.* (1979) (unpublished results)]

1982); (iii) relatively stable in the process of nucleolar dissolution (DIMOVA *et al.* 1979); and (iv) best seen in those cases (*e.g.*, after cordycepin, D-galactosamine or heat shock treatments) where extensive degranulation of nucleoli takes place. These features justify the proposal that spherical bodies represent the visualization of nucleolar matrix protein structures, normally covered by the preribosomes and ribosomes constituting the granular component of nucleoli. Whether this hypothesis is correct remains to be ascertained in future studies.

VII.2.6.3. Microspherules

A variety of drugs cause an early formation of numerous intranucleolar electron dense bodies designated as *microspherules.* Since microspherules are apparently constituted by proteins and RNA (UNUMA and BUSCH 1967), they are considered to represent an early stage in the formation of spherical bodies (see SIMARD *et al.* 1974). However, other possibilities also have been envisaged and the importance of a closer investigation of microspherules has been stressed (DASKAL 1979).

VII.2.6.4. Nucleolar Fragmentation

It is known that in active cells the separate NO and nucleoli tend to coalesce during interphase, thereby reducing the number of nucleoli. An apparently reverse phenomenon is observed as a result of the action of various transcription inhibitors as cordycepin (PUVION *et al.* 1976, 1977), D-galactosamine (SHINOZUKA *et al.* 1973, DIMOVA *et al.* 1979), ethionine (SHINOZUKA *et al.* 1968), α-amanitin (MARINOZZI and FIUME 1971), and other agents. Since an early nucleolar fragmentation and microsegregation is caused by α-amanitin (see VII.2.2.), it was proposed that nucleolar fragmentation is due to a block in extranucleolar transcription (ROMEN and ALTMANN 1977). While this is a likely explanation, the factors causing coalescence of nucleoli in interphase and fragmentation upon transcription block are still largely unknown.

VII.3. Growth Transitions

Ribosome biogenesis plays a central role in the life cycle of eukaryotic cells. The fact that the formation of new ribosomes is a continuous process in both cycling and non-cycling cells suggests that not only cell growth, but protein synthesis in general are closely related to the presence of a fixed level of cytoplasmic ribosomes. Moreover, numerous observations indicate that the nucleolar apparatus for ribosome production displays an early response to a variety of exogenous (and endogenous?) stimuli. How the process of ribosome biogenesis is finely tuned to the metabolic needs of the cell during its life cycle remains one of the challenging problems of cell biology (see

LLOYD *et al.* 1982). In multicellular eukaryotes ribosome biogenesis has to be coordinated also with cell-to-cell contact and long-range (metabolic or hormonal) interactions between cells throughout development and adult life of the organism. The content of ribosomes in a cell can be regulated by changes in: a) rates of rDNA transcription; b) rates and efficiency of preribosome processing and release of nascent ribosomes, and c) rates of ribosome degradation. Of the many experimental models that have been investigated, I shall restrict myself to those in which both molecular and cellular details have been elucidated.

VII.3.1. Modulation of Growth Rates in Yeasts

Growth rates in yeast are easily manipulated by changes of nutrients in the medium, temperature shifts and other factors. The consequences for ribosome biogenesis of those kinds of manipulations have been expertly reviewed (WARNER 1982) and only the basic findings and conclusions will be summarized here. The content of ribosomes in the yeast cell is proportional to the growth rate (WALDRON and LACROUTE 1975, KIEF and WARNER 1981). Thus, changes in growth rate result in rapid and marked changes in the rates of ribosome biogenesis. The yeast is a typical cycling cell and under normal conditions of growth there is not appreciable degradation of ribosomes. Numerous studies have established that ribosome biogenesis in yeast is adjusted to growth rates by mechanisms ensuring: a) coordinated rates of rRNA and r-protein synthesis and b) coordinated rates of synthesis of the individual r-proteins (see WARNER 1982). It seems that the rate of rRNA synthesis may be regulated at either the transcription or processing steps. The "stringent control" of rDNA transcription dependence upon protein synthesis seems now proven (see VI.2.2. and WARNER 1982). On the other hand, upon inhibition of r-protein synthesis in temperature-sensitive *S. cerevisiae* mutants, transcription of primary pre-rRNA continues, but processing to mature rRNA species is halted and accumulated pre-rRNA is degraded (SHULMAN and WARNER 1978). In both cases production of mature rRNA is inhibited. Whether r-protein synthesis is dependent on continuous formation of rRNA in yeast is not yet clarified because permeability barriers hamper the use of selective inhibitors of rRNA synthesis. Experiments using yeast mutants with increased drug-permeability showed that after a preferential block of rRNA synthesis with rifampin (VENKOV *et al.* 1975) or actinomycin D (WALTSCHEWA *et al.* 1976), protein synthesis continues for some time at an unaltered or slightly altered rate. However, it is not yet directly proven whether r-protein synthesis also remains unaltered. In higher eukaryotes, the level of r-proteins is not regulated primarily by inhibition of their syntheses, but rather by intranuclear degradation of excess r-proteins (see VI.3.1.). This may not be the case for yeast in which free r-proteins appear to be more stable, while the mRNAs coding for those proteins are unstable (GORENSTEIN and WARNER 1977, ROSBASH *et al.* 1981). In fact, the synthesis of individual r-proteins in yeast has been shown convincingly to be regulated at the level of translation by changes in the lifetime and efficiency of translation of their mRNAs (PEARSON *et al.* 1982). These studies provided also evidence for the coordinated synthesis of individual r-proteins. Also, by analogy with bacteria (NOMURA *et al.* 1982), it was suggested that r-proteins may alter the synthesis and turnover of their mRNA by autogenous regulation. How the yeast cell r-proteins "sense" the level of rDNA transcription remains to be clarified. WARNER has proposed that a hypothetical positive effector exists, dependent on rDNA transcription, that regulates the transcription of r-protein genes. When rDNA transcription is blocked, the effector presumably is degraded and the expression of r-protein genes is thereby halted (WARNER 1982). If the model of autogenous regulation of ribosome biogenesis proposed above (see VI.4.) is applied to yeasts, the role of such effectors could be played by r-proteins capable of interacting with either pre-rRNA or their respective gene. In the absence of pre-rRNA the excess of a particular r-protein would presumably bind to its gene and act as a negative effector thereby shutting off the synthesis of its own mRNA.

VII.3.2. Activation of Lymphocytes

Lymphocytes from peripherial blood can be maintained in culture as quiescent, G_0 arrested cells. The synthesis of RNA and proteins in such cells is exceedingly slow. If cells are incubated in the presence of radioactive RNA precursors it takes several hours before label appears in cytoplasmic rRNA (KAY 1968). Addition of phytohaemagglutinin causes initially a marked increase in protein and RNA synthesis, followed several hours later by the entry of lymphocytes into S-phase. Measurement of RNA polymerase activity in isolated nuclei shows that RNA polymerase I activity doubles within 4 hours after addition of phytohaemagglutinin, reaching a 5–6-fold increase by 20 hours, while the level of RNA polymerase II does not change for the first 12 hours and shows only 2-fold increase at 20 h (COOKE and KAY 1973, JAEHNING *et al.* 1975). These results reveal an early response of rRNA synthesis to mitogen stimulation. Further analysis provided evidence that stimulation of rDNA transcription is related exclusively to an increase of transcript elongation rates, while the number of growing pre-rRNA chains and the frequency of transcription initiation remained unchanged (DAUPHINAIS 1981). Posttranscriptional control is also operating in quiescent and phytohaemagglutinin-stimulated lymphocytes. In fact, evidence was obtained with resting lymphocytes that showed that continuous degradation ("wastage") of pre-rRNA and rRNA (up to 50%) is taking place and it was proposed that a decrease of rRNA degradation is largely responsible for the increase in rRNA accumulation upon growth stimulation (COOPER 1969, 1970, see also VI.3.4.1.). Further studies provided evidence that channeling of pre-rRNA along alternative processing pathways may also contribute to an increase in the production of cytoplasmic ribosomes (PURTELL and ANTHONY 1975). A quantitative evaluation of the contribution of transcriptional and posttranscriptional control mechanisms in the increased formation of ribosomes in growth-stimulated lymphocytes is not possible at present. It is plausible that the increased rate of rDNA transcription is enhanced by a more complete processing of preribosomes. Detailed cytological studies revealed several interesting changes in nucleolar structure proceeding in parallel with the increased rate of ribosome biogenesis during growth stimulation of lymphocytes (WACHTLER *et al.* 1980, 1982). These authors suggested that during lymphocyte stimulation additional NOs are switched on and that at later stages they fuse to produce one or two large nucleoli. Also, the obviously inactive ring-shaped nucleoli are gradually transformed into reticulated nucleoli with nucleolonema and later into compact nucleoli, thus showing that the three forms of nucleoli correspond to different intensities of transcription and processing (see SMETANA and BUSCH 1974).

VII.3.3. Growth Stimulation of Cultured Cells

The growth rate of cultured normal diploid cells decreases markedly when they form a continuous monolayer—a phenomenon described as contact inhibition (TODARO and GREEN 1963). Addition of serum releases the cells, arrested in G_0-phase, from contact inhibition to reenter the mitotic cycle and resume growth. The phenomenon is best studied with mouse (lines 3 T 3 and 3 T 6) and human (line WI 38) diploid fibroblasts. The amount of ribosomes in resting mouse fibroblasts is 2–3-fold lower than in growing cells (JOHNSON *et al.* 1974). When resting cells are stimulated to grow they react by an increase of rRNA synthesis (MAUCK and GREEN 1973, ZARDI and BASERGA 1974). The total amount of mRNA in growing cells is also higher (JOHNSON *et al.* 1974). Interestingly, the increase of the amount of new ribosomes in the cytoplasm upon growth stimulation precedes the appearance of new cytoplasmic mRNA, thus supporting an earlier proposal that synthesis of new rRNA is an early step in growth control (see TATA 1970). Further analysis showed that activation of pre-existing cytoplasmic mRNA and its engagement into polyribosomes is an early response to growth stimulation (RUDLAND 1974, RUDLAND *et al.* 1975, JOHNSON *et al.* 1975). This effect is correlated with a rapid recruitment of pre-existing free single ribosomes into both free and membrane-bound polyribosomes, a process completed within 1–2 hours after serum stimulation of resting fibroblasts (STANNERS and BECKER 1971). All these studies suggest the following sequence of events involved in the response of ribosome biogenesis to serum growth stimulation of resting cells: (i) formation of new polyribosomes by "activation" of pre-existing cytoplasmic mRNA and single ribosomes and an increase in protein synthesis rates; (ii) increased rate of rRNA synthesis and formation of new ribosomes; (iii) increased rates of mRNA formation; and (iv) a cessation of ribosome degradation and turnover (ABELSON *et al.* 1974).

The mechanisms leading to the early enhancement of protein synthesis have been intensively investigated, but unequivocal descriptions of those mechanisms are not yet available (see THOMAS and GORDON 1979). Some of the features that have been observed include the following. An early, within 5 minutes of growth stimulation, phosphorylation of protein S6 in cytoplasmic ribosomes has been observed, possibly related to the increased efficiency of translation in growth-stimulated cells (THOMAS *et al.* 1979b). Also, a recent study of the content and translation of r-protein mRNAs revealed important facets in the regulation of ribosome biogenesis in resting and growth-stimulated cells (GEYER *et al.* 1982). Although the synthesis of r-proteins is markedly increased soon after growth stimulation (TUSHINKI and WARNER 1982), GEYER *et al.* (1982) found that at these early times the amount, rate of synthesis, and turnover of rp-mRNA did not change appreciably, but the increase of r-protein synthesis can be accounted for by a shift of pre-existing free rp-mRNA into polyribosomes. We do not yet know whether the comportment of rp-mRNA is distinct from that of other cytoplasmic mRNA species (IGNOTZ *et al.* 1981, TOLSTOSHEV *et al.* 1981, GEYER *et al.* 1982), but, clearly, factors controlling translation of cytoplasmic mRNAs (rp-mRNAs in particular) play a leading role in the early response of resting cells to serum growth-stimulation. As with yeasts (see VII.3.1.) the putative link between enhanced synthesis of r-proteins and accelerated transcription of rRNA genes (and ribosome biogenesis) remains to be substantiated. Whether the increased rate of rDNA transcription is triggered by r-proteins (see VI.4.) or some other factor dependent on protein synthesis remains an intriguing puzzle.

The relationship between the observed changes in ribosome biogenesis in resting and growth-stimulated cells and the quantitative distribution of nucleolar components might be informative. That relationship was studied with strain MRC-5 of human diploid fibroblasts by JORDAN and MCGOVERN (1981), who noted a marked decrease in the volume of fibrillar centers in nucleoli of growing cells, confirming the view that this structure is more prominent when transcriptional activity is low. Unexpectedly, the absolute and relative volumes of the fibrillar and granular components (91 to 96% of the nucleolar volume) did not change appreciably. These results suggest that the higher *rates* of rDNA transcription and preribosome processing in growing cells do not alter the amount of the respective nucleolar compartments, namely, the fibrillar and the granular component.

VII.3.4. Differentiation of Myoblasts in Culture

Myoblasts isolated from cultured embryonic or skeletal muscles possess the capacity to differentiate *in vitro* by fusing into multinucleate myotubes. This experimental system provides an interesting model to study the changes in ribosome biogenesis related to cell differentiation (see PEARSON 1980). Growing myoblasts, as other growing cells, maintain a constant amount of rRNA per nucleus (WEBER 1972, LIEBHABER *et al.* 1978). However, after fusion into myotubes, which do not divide, a 5-10-fold decrease in the rate of rRNA accumulation takes place. This phenomenon has been studied in great detail using a subclone of the L 6 rat myoblast cell line (KRAUTER *et al.* 1979), and provided evidence that: (i) no degradation of cytoplasmic rRNA occurs in myotubes; (ii) the decreased rate of rRNA formation corresponds quantitatively to the reduction in primary pre-rRNA synthesis, and (iii) there is not "wastage" of pre-rRNA or rRNA during ribosome biogenesis. Thus, in the myoblast-myotube system rRNA formation is primarily regulated through the control of rDNA transcription rate. On the other hand, experiments with primary quail cells provided evidence that the decreased accumulation of rRNA in myotubes is largely accounted for by: (i) "wastage" during processing of pre-rRNA, and (ii) the switching-on of degradation of cytoplasmic ribosomes (BOWMAN and EMERSON 1977). Whether the observed differences reflect the operation of distinct control mechanisms in established cell lines and in primary cell cultures remains to be clarified.

VII.3.5. Regeneration of Rat Liver

One useful model for the study of control mechanisms of ribosome biogenesis operating in the whole animal is regenerating rat liver. It is known that following removal of up to 90% of the liver, the complete organ may be restored by regeneration (see BUCHER and MALT 1971). The sequence in the switching-on of synthetic processes in regenerating rat liver is represented in Fig. 46 (MARKOV *et al.* 1975).

Several studies demonstrate the increase in RNA content (as reflected in the ratio RNA/DNA) in the regenerating hepatocyte and its nucleus. This accumulation of RNA correlates with the increased rate of RNA synthesis (see BUCHER and MALT 1971). Both the synthesis of rRNA and hnRNA is increased, but that of rRNA is more pronounced (MARKOV *et al.* 1975, GLAZER 1977). Detailed studies showed that the number of nucleoli decreases, but their total volume is markedly increased, while the number of rRNA genes per nucleus remains unchanged (TAKATSUKA *et al.* 1976). The increased synthesis and accumulation of rRNA in the hepatocyte is an early response to hepatectomy that precedes the hepatocyte entry into S-phase. An approximately 3-fold increase in nuclear RNA polymerase I activity starts at the 6th hour after hepatectomy and peaks at about 20 hours (YU 1975, ORGANTINI *et al.* 1975, KRAWCZYK and CHORAZY 1978). Both total and template-bound RNA polymerase I, particularly the I_B-form, are increased (YU 1975, MATSUI *et al.* 1976). Further analysis provided evidence for an increase in both the number of transcribing enzyme molecules and the pre-rRNA elongation rate (PICCOLETTI *et al.* 1982). These findings are in good agreement with *in vivo* studies that demonstrated a 2.7-fold increase in

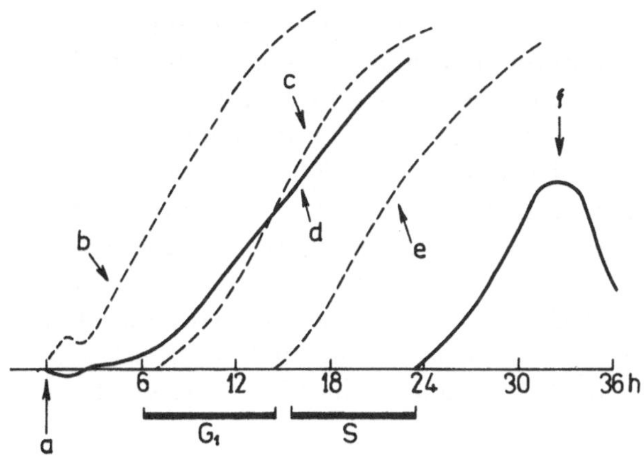

Fig. 46. Representation of the sequential changes in macromolecular synthesis during rat liver regeneration after partial hepatectomy. *a* Hepatectomy; *b* Incorporation of ^{32}P-phosphate into RNA; *c* Activity of enzymes of DNA replication; *d* Accumulation of RNA; *e* Synthesis of DNA—S phase of the mitotic cycle; *f* Mitoses. [According to MARKOV *et al.* (1975)]

the rate of 45 S pre-rRNA synthesis at 12 hours after hepatectomy (DABEVA and DUDOV 1982b and references therein). These authors established that the synthesis rate of 45 S pre-rRNA is equal to the rates of 28 S and 18 S rRNA formation and concluded that: (i) there is no "wastage" of pre-rRNA in either normal or regenerating liver and (ii) the 2.7-fold stimulation of rRNA formation is exclusively a result of the increase in transcriptional activity. It was shown further that the increased rate of pre-rRNA synthesis corresponds to a faster pre-rRNA processing (DUDOV and DABEVA 1983). The estimated rate of 3,700 ribosomes formed per minute per nucleus in regenerating rat liver corresponds to the rate observed in rapidly growing cultured cells (see Table 4). Whether the observed increase in transcription rate involves the switching-on of new rRNA genes or an enhancement of initiation and elongation on rRNA genes already being transcribed in normal liver is not yet known. The degradation of cytoplasmic ribosomes is slowed-down, thereby further increasing their accumulation in the regenerating hepatocyte (NIKOLOV *et al.* 1983). The ribosome biogenesis response to hepatectomy appears to be faster and markedly more pronounced than that of mRNA synthesis. Moreover, the sequence complexity of cytoplasmic mRNAs from normal and regenerating liver does not differ appreciably, thus indicating that the changes in mRNA production are mainly quantitative rather than qualitative (WILKES *et al.* 1979, SCHOLLA *et al.* 1980).

The increase of pre-rRNA transcription in regenerating liver, is quenched by administration of low doses of cycloheximide (SEIFART *et al.* 1978, PICCOLETTI *et al.* 1982) suggesting a need for continuous protein synthesis in ensuring the ribosome biogenesis response to hepatectomy. This finding may be correlated with the increased synthesis of r-proteins in regenerating rat liver (TSURUGI *et al.* 1972). Also, the template activity (amount?) of cytoplasmic rp-mRNA is already higher than its normal level at the 6th hour, reaching a peak at the 18th hour with a level 2.4-fold higher than in controls (NABESHIMA and OGATA 1980). It is remarkable that the level of the mRNAs for individual r-proteins increased in a coordinate manner (NABESHIMA and OGATA 1980, FALIKS and MEYUHAS 1982). Whether the increased rates of cytoplasmic rp-mRNA translation and r-protein synthesis are directly related to the increase in rDNA transcription remains to be clarified.

The cytological investigations of the hepatocyte of regenerating liver reveal a marked increase in the amount of nucleolar RNA (TAKATSUKA *et al.* 1976), but the relative proportions of the fibrillar and granular components of the nucleolus remain apparently unchanged (see BUSCH and SMETANA 1970). These results may be correlated with biochemical estimates that show that under balanced conditions of accelerated ribosome biogenesis in regenerating rat liver all nucleolar pre-rRNA and rRNA components are increased proportionally (DABEVA and DUDOV 1982 a). It is a common notion that enlarged compact nucleoli are indicative of an increased synthesis of preribosomes. However, the increased amount of the granular component indicates some bottleneck at the last nucleolar stage of ribosome formation and the release of nascent ribosomes from the nucleolus. This possibility is illustrated dramatically by the thioacetamide-induced nucleolar hypertrophy. In this case, there is a 3–4-fold increase in RNA polymerase I activity (LEONARD and JACOB 1977), but the release of nascent ribosomes from the nucleolus is altered, transforming the nucleus into a large spherical mass of densely packed ribosomes (containing 28 S rRNA) (KOSHIBA *et al.* 1971). It is plausible that the maximal size of the nucleolar fibrillar component is limited by the number and packing of RNA polymerase I molecules in active rRNA transcription units, but the factors determining the maximal size of the nucleolar granular component remain to be elucidated.

VII.4. Senescent and Cancer Cells

Consideration of ribosome biogenesis in aging and cancerogenesis under the same heading is apparently artificial, but possibly useful. From the point of view of growth capacity senescent and cancer cells seem to represent opposite poles in the cellular states reached compulsory or occasionally by the populations of cells in the multicellular organism. Nevertheless, it is usually among senescent populations that cancer cells are recruited. The involvement of ribosome biogenesis in aging and cancerogenesis may be considered under two aspects: a) as an alteration *accompanying* the key changes characteristic for the process, and b) as an alteration playing a *pathogenetic* role in the sequence of events resulting in a given state of the cell.

VII.4.1. Senescent Cells and Tissues

Although the aging phenomenon in different tissues has many facets, a common pattern of nucleolar changes has been discerned. Generally, the nucleoli of senescent cells are marked by structural changes reflecting reduced synthetic activity, including an increased number of ring-shaped and even of partly segregated nucleoli and a disorganization and dissociation of nucleoli from surrounding chromatin (ADAMSTONE and TAYLOR 1977, 1979). Unfortunately, more detailed studies on ribosome biogenesis in different tissues during aging are practically impossible.

A fruitful model of cellular aging arose from the observation that cultured diploid fibroblasts have a limited capacity for division cycles, 50 ± 10 in human diploid fibroblasts (HAYFLICK and MOOREHEAD 1961), and HAYFLICK (1965) proposed that this phenomenon reflects the *in vitro* aging of cells (see CRISTOFALO 1970). Such senescent cells are characterized by a reduced potential for cell division, with most cells in an extended G_1 phase and part of the cell population arrested in G_2. The delayed division is related to an increase in cell volume and amount of rRNA, mRNA, tRNA, and protein per cell (SCHNEIDER and SHORR 1975). These changes reflect, most likely, a prevailing deceleration of ribosome degradation, since the synthesis of RNA, rRNA in particular, is also reduced in senescent cells (MACIEIRA-COELHO and LORIA 1976). The decreased rates of rRNA synthesis correspond also to a lower activity of RNA polymerase I (BOWMAN *et al.* 1976, NOSE and OKAMOTO 1980). Nuclei from senescent fibroblasts have a limited capacity for reactivation of nucleolar transcriptional activity when included into heterokaryons by cell fusion (RENER and NARDONE 1980). Also, inactivation and even possible loss of rRNA genes was reported for aging human myocardium (STREHLER *et al.* 1979) and this finding was correlated with a marked decrease in the number of Ag-stained NORs in the cells of aged human individuals (DENTON *et al.* 1981).

The observed deceleration of ribosome biogenesis in senescent cells may be related to changes in protein synthesis and turnover. A drastic decrease in protein degradation and accumulation of cytoplasmic proteins was observed in senescent chick fibroblasts (KAFTORY *et al.* 1978) and this conclusion is supported by experiments showing a decreased degradation of labeled proteins microinjected into senescent human fibroblasts (DICE 1982). The accumulation of proteins with altered structure and function has been implicated as a plausible molecular mechanism of the aging phenomenon. How these changes are related to alterations in nuclear and nucleolar activity, if at all, remains to be clarified.

VII.4.2. Cancer Cells

I shall concentrate on the nucleolus and ribosome biogenesis as elements of the neoplastic cell phenotype. Pioneer studies stressed the prominence of nucleoli in cancer cells and several authors (*i.e.* MACCARTY 1936, CASPERSSON 1950, KOPAC and MATEYKO 1958) considered the enhanced nucleolar activity as a characteristic feature of cancer cells. With the accumulation of experimental evidence and the use of a wider spectrum of cytological and cytochemical methods, a more detailed view was elaborated. The extensive information has been expertly reviewed (BUSCH *et al.* 1963, BERNHARD and GRANBOULAN 1968, BUSCH and SMETANA 1970, 1974, BOLOGNARI and CONTINI 1979) and the basic conclusions may be summarized as follows.

a) No nucleolar structural change is typical solely of cancer cells. Many of the structural features of nucleoli in cancer cells seem to reflect the growth rate of the cells, rather than the malignant state *per se*. Nevertheless the structural features of cancer cell nucleoli are certainly an important constituent of the constellation of cytological markers for malignancy.

b) An increased nucleolar volume (and increased nucleolus/nucleus and nucleolus/cytoplasm ratios) is characteristic for most cancer cells, in particular when compared with the tissue of origin. Compact nucleoli are typical for many neoplasms, but in some cases (*e.g.*, breast cancer) numerous reticulated nucleoli with highly contorted nucleolonema are observed (PLOTON *et al.* 1983).

c) Irregular size and shape of nucleoli is consistently found in experimental and spontaneous neoplasms. This nucleolar *pleomorphism* is typical for many malignant cells and is clearly distinct from cases of experimentally induced (*e.g.*, by thioacetamide treatment) nucleolar hypertrophy (see SIMARD *et al.* 1974). Alterations in the skeletal structures of nuclei and nucleoli may be responsible for the pleomorphism of cancer cell nucleoli (see BUSCH and SMETANA 1974).

d) Some authors have drawn attention to the great variability in the size and shape of fibrillar centers in tumor cells. These observations may be correlated with reported alterations in the localization of Ag-NOR proteins (BUSCH *et al.* 1979, PLOTON *et al.* 1983), but a broader range of tumors has to be investigated before generalizations can be drawn, and the same can be said of the presence of microspherules, partial segregation and other structural anomalies of the nucleolus attributed to cancer cells.

Increased nucleolar activity at the onset of neoplastic transformation is consistently found in studies of experimental cancerogenesis. For example, several authors have reported marked increases in nucleolar size and cytoplasmic basophilia in primary neoplastic nodules during chemically induced liver cancerogenesis (STOWELL 1949, HADJIOLOV and TENCHEVA 1958, GRUNDMANN and FECHLER 1965, KARASAKI 1969). An old, and perhaps unfortunately forgotten, observation emphasizes that cytoplasmic basophilia persists throughout mitosis in some cancer cells (*e.g.*, BRACHET 1947, ROSKIN 1949). That these observations might reflect some alterations in the regulation of nucleolar activity and ribosome formation is indicated by findings that nucleoli in some malignant cells persist throughout mitosis, whereas this is not the case for embryonic cells, or other normal cells, characterized by rapid growth and cell division (HSU *et al.* 1967, SHELDON *et al.* 1981).

Generally, the molecular characteristics of the cancer cell phenotype do not seem to be basically distinct from those of the homologous normal cells (see BUSCH 1974, 1979). For example, practically all protein constituents (see, *e.g.*, TAKAMI and BUSCH 1979, WU *et al.* 1981) and their respective mRNAs (*e.g.* MOYZIS *et al.* 1980) are identical in normal and neoplastically transformed cells. These findings provide support to the commonly shared view that neoplastic transformation is related to altered intensity and timing in the expression of oncogenes, rather than to switching on of some new genes alien to the cell of origin or the species in general (see WEINBERG 1982, DULBECCO 1983). As pointed out *c-onc* genes are likely to have been expressed at some early stages of embryonic development (see BUSCH 1979). Studies on the products of *c-onc* genes and the target(s) of their action are still at an early stage (see WEINBERG 1982). Here, I shall consider some recent evidence on putative *c-onc* and *v-onc* gene proteins and their possible relationship with nucleoli and ribosome biogenesis.

Among the best studied cases of neoplastic transformation are those induced by the small oncogenic DNA viruses SV 40 and polyoma (see LEBOWITZ and WEISSMAN 1979, DULBECCO 1983). There is conclusive evidence that neoplastic transformation here is directly related to specific T-antigens (T, t, and "middle" T) encoded in the virus genome. The action of transforming T-antigens ("middle" T seems to play the leading role) is certainly pleiotropic and includes enhanced expression of several cellular genes. BASERGA *et al.* (1977) proposed that the SV 40 large T-antigen stimulates host DNA synthesis and cell division by first stimulating rDNA transcription and experiments with purified T-antigen indeed showed that it stimulates rDNA transcription in isolated nuclei and nucleoli (IDE *et al.* 1977, WHELLY *et al.* 1978). A selective binding of T-antigen to RNA polymerase I was implicated to explain its action (POTASHKIN and SCHLEGEL 1980).

Several studies provided evidence for the activation during neoplastic transformation of specific *onc* genes coding for proteins with protein kinase activity (see DULBECCO 1983). The action of *onc* gene(s) encoded protein kinase(s) is certainly targeted on a broad variety of substrates possibly including some nucleolar proteins (see BUSCH *et al.* 1979, 1982), and some r-proteins, particularly protein S 6 (see THOMAS *et al.* 1979 a, b). How phosphorylation of these proteins may participate in establishing, if at all, the neoplastic phenotype is not known. It is worth mentioning that a protein kinase associated with RNA polymerase I subunits and capable of phosphorylating them has been isolated from hepatoma cells, while apparently absent from normal liver (ROSE *et al.* 1981).

Specific nucleolar antigens have been shown to be present in many human tumors, but not in normal adult tissues (BUSCH *et al.* 1982). One of these antigens was purified and shown to be a protein with M_r of about 68 kDa associated with the fibrillar component of tumor cell nucleoli. How this antigen is related to the phenotype of human cancer cells and what its function is remains unknown. However, its localization in the nucleolus provides further support to the suggestion that alterations in the protein complement of nucleoli may be characteristic of the cancer cell phenotype and possibly involved in enhanced nucleolar activity.

That altered control of ribosome biogenesis is a characteristic feature of neoplastic cells is shown further by comparative studies on non-transformed (parent line) and transformed mouse and human fibroblasts. Stationary, non-transformed diploid fibroblasts have been shown to possess about half the amount of cytoplasmic ribosomes found in growing cells (BANDMAN and GURNEY 1975, MEEDEL and LEVINE 1978, TUSHINSKI and WARNER 1982). This condition is due mainly to a decrease in rDNA transcription and switching on of ribosome turnover (see VI.2. and VI.3.4.). In contrast, transformed cells maintain the synthesis and the amount of ribosomes at levels characteristic for growing cells, even upon deprivation of nutrients (MEEDEL and LEVINE 1978, LIEBHABER *et al.* 1978, STANNERS *et al.* 1979). STANNERS *et al.* (1979) proposed that a basic feature of neoplastically transformed cells is the loss of control over ribosome biogenesis. The molecular mechanisms of this phenomenon remain to be elucidated, but, if proven to be true for cancer cells in general, ribosome biogenesis will be implicated more directly in the mechanism of neoplastic transformation. The relationship between ribosome biogenesis, protein synthesis, and cell growth is straitforward and supported by numerous studies. However, the connections of these processes with DNA replication and cell division is still obscure. Recent experiments provide evidence that ribosome biogenesis is not a factor needed for DNA replication. Thus, experiments with the BHK 422 E cell line, that is temperature-sensitive for the conversion 32 S pre-rRNA → 28 S rRNA, showed that these cells enter S-phase at the restrictive temperature although cell division is blocked (Toniolo *et al.* 1973, Grummt *et al.* 1979, Mora *et al.* 1980). Further, microinjection of an antibody to RNA polymerase I into quiescent 3 T 3 cells, stimulated subsequently with serum, resulted in 50–70% decrease of nucleolar RNA synthesis and accumulation of total cellular RNA, but did not prevent the entry of these cells into S phase (Mercer *et al.* 1984). These studies show that the participation of ribosome biogenesis in cell growth and cell division is unlikely to be through an action on DNA replication.

All the above studies using different experimental models converge to show that alterations in ribosome biogenesis control mechanisms seem to be a characteristic trait of the cancer cell phenotype. Whether these alterations are more directly involved in the onset of neoplastic transformation remains to be elucidated. In this respect, as noted by A. Stahl (see Hernandez-Verdun 1983), it is perhaps significant that several consistant translocations, specifically associated with human leukaemia and neoplasia, involve "nucleolar" chromosomes, including a chromosomes 22/9 translocation in chronic myeloid leukemia, a 21/8 translocation in acute myeloblastic leukemia, a 15/17 translocation in acute promyelocytic leukemia and a 14/8 translocation in Burkitt lymphoma.

VII.5. Synopsis

Ribosome biogenesis is a continuous process operating throughout the interphase span of almost all living eukaryotic cells. Moreover, the intensity of ribosome biogenesis seems to be finely adapted to the temporal protein synthesis needs of the cell. The changes in ribosome biogenesis under different physiological or pathological conditions are reflected in changes in nucleolar ultrastructure.

1. Nucleologenesis and the onset of ribosome biogenesis start at telophase in association with NO structures in "nucleolar" chromosomes. The fibrillar component is formed initially, followed by formation of the granular component. Both nucleolar ribonucleoprotein structures gradually disappear upon nucleololysis, leaving the nucleolar remnants—residual structures containing rRNA genes. Ag-NOR proteins and possibly some nucleolar matrix proteins.

2. Several agents may inhibit, more or less selectively, ribosome biogenesis and cause parallel alterations in nucleolar structure. Inhibition of rDNA transcription causes the segregation of nucleolar constituents (with

or without fragmentation of nucleoli). Subsequent degranulation of nucleoli may result ultimately in their gradual transformation into dense fibrillar remnants. Inhibition of preribosome maturation and/or release of nascent ribosomes has a less pronounced effect on nucleolar structure although the relative proportions of the fibrillar and granular components may be altered in some cases. Generally, the structure of the nucleolus reflects the balance between transcription of rRNA genes (and preribosome formation) and the release of nascent ribosomes from the nucleolus.

3. The intensity of ribosome biogenesis changes rapidly in parallel with changes in cellular growth rates induced under different experimental conditions. Enhanced protein synthesis (including r-proteins) is an early response to growth stimulation of different cells. The increased rates of protein synthesis appear to trigger ribosome biogenesis by causing primarily an increase in the rates of rDNA transcription and preribosome formation.

4. Senescent cells are characterized by a deceleration of ribosome biogenesis possibly related to changes in protein synthesis and turnover. However, the characteristics of the senescent cell genotype and phenotype remain largely unknown. Cancer cells are generally characterized by an increase in the rates of ribosome biogenesis. A parallel increase of nucleolar volume is also characteristic of most cancer cells, but the size and shape of nucleoli may differ markedly from cell to cell. Increased rates of rDNA transcription and accumulation of cytoplasmic ribosomes seem to be characteristic of the cancer cell phenotype and are possibly indicative of alterations in ribosome biogenesis control mechanisms. Whether ribosome biogenesis is more directly involved in the neoplastic transformation of cells is not yet known.

References

AARSTAD, A. A., ØYEN, T. B., 1975: On the distribution of 5 S RNA cistrons on the genome of *Saccharomyces cerevisiae*. FEBS Lett. **51**, 227—231.

ABELSON, H. T., JOHNSON, L. F., PENMAN, S., GREEN, H., 1974: Changes in relation to growth of the fibroblast. II. The lifetime of mRNA, rRNA and tRNA in resting and growing cells. Cell **1**, 161—165.

ACKERMAN, S., FURTH, J. J., 1979: Selective *in vivo* transcription of the 5 S rRNA genes of a DNA template. Biochemistry **18**, 3243—3248.

ADAMSTONE, F. B., TAYLOR, A. B., 1977: Nucleolar reorganization in cells of the kidney of the rat and its relation to aging. J. Morphol. **154**, 459—477.

— — 1979: Nucleolar reorganization in liver cells of the aging rat. J. Morphol. **161**, 211—219.

ADOLPH, K., 1980: Organization of chromosomes in HeLa cells: isolation of histone-depleted nuclei and nuclear scaffolds. J. Cell Sci. **42**, 291—304.

AGUTTER, P. S., RICHARDSON, J. C., 1980: Nuclear non-chromatin proteinaceous structures: role in organization and function of the interphase nucleus. J. Cell Sci. **44**, 395—435.

AIELLO, L. O., GOLDENBERG, C. J., ELICEIRI, G. L., 1977: *In vivo* incorporation of ribosomal proteins into HeLa cell ribosomal particles. Biochim. Biophys. Acta **475**, 652—658.

ALAM, S. N., SHIRES, T. K., 1977: The labelling of polysomes and rough microsomal membranes by 5-fluoroorotic acid. Biochem. Biophys. Res. Commun. **74**, 1441—1449.

ALLFREY, V. G., JOHNSON, E. M., SUN, I., LITTAU, V. C., MATTHEWS, H. R., BRADBURY, E. M., 1978: Structural organization and control of the ribosomal genes in *Physarum*. Cold Spring Harbor Symp. Quant. Biol. **42**, 505—513.

ALLIS, C. D., ZIEGLER, V. S., GOROVSKY, M. A., OLMSTED, J. B., 1982: A conserved histone variant enriched in nucleoli of mammalian cells. Cell **31**, 131—136.

ALONSO, C., BERENDES, H. D., 1975: The location of 5 S rRNA genes in *Drosophila hydei*. Chromosoma **51**, 347—356.

AMALDI, F., BECCARI, E., BOZZONI, I., LUO, Z.-Y., PIERANDREI-AMALDI, P., 1982: Nucleotide sequences of cloned cDNA fragments specific for six *Xenopus laevis* ribosomal proteins. Gene **17**, 311—316.

ANASTASSOVA-KRISTEVA, M., 1977: The nucleolar cycle in man. J. Cell Sci. **25**, 103—110.

ANDREW, C., HOPPER, A. K., HALL, B. D., 1976: A yeast mutant defective in the processing of 27 S rRNA precursor. Mol. Gen. Genet. **144**, 29—37.

ANGELIER, N., LACROIX, J. C., 1975: Complexes de transcription d'origines nucléolaire et chromosomique d'ovocytes de *Pleurodeles waltlii* et *Pleurodeles poireti*. Chromosoma **51**, 323—335.

ANGELIER, N., HEMON, D., BOUTEILLE, 1979: Mechanisms of transcription in nucleoli of amphibian oocytes as visualized by high resolution autoradiography. J. Cell Biol. **80**, 277—290.

— HERNANDEZ-VERDUN, D., BOUTEILLE, M., 1982: Visualization of Ag-NOR proteins on nucleolar transcriptional units in molecular spreads. Chromosoma **86**, 661—672.

ANTEUNIS, A., POUCHELET, M., GANSMULLER, A., ROBINEAUX, R., 1979: Organization of nucleolar DNA in resting lymphocytes as revealed by the diaminobenzidine technique. J. Ultrastr. Res. **69**, 22—27.

ARNHEIM, N., 1979: Characterization of mouse ribosomal gene fragments purified by molecular cloning. Cell **7**, 83—96.

— SOUTHERN, E. M., 1977: Heterogeneity of the ribosomal genes in mice and men. Cell **11**, 363—370.

— KRYSTAL, M., SCHMICKEL, R., WILSON, G., RYDER, O., ZIMMER, E., 1980 a: Molecular evidence for genetic exchanges among ribosomal genes on nonhomologous chromosomes in man and apes. Proc. Nat. Acad. Sci. U.S.A. **77**, 7323—7327.

— SEPERACK, P., BANERJI, J., LANG, R. B., MIESFELD, R., MARCU, K. B., 1980 b: Mouse rDNA nontranscribed spacer sequences are found flanking immunoglobulin C_H genes and elsewhere throughout the genome. Cell **22**, 179—185.

ARROUA, M. L., HARTUNG, M., DEVICTOR, M., BERGE-LEFRANC, J. L., STAHL, A., 1982: Localization of ribosomal genes by *in situ* hybridization in the fibrillar centre of the nucleolus in the human spermatocyte. Biol. Cell **44**, 337—340.

ARTVANIS-TSAKONAS, S., SCHEDL, P., TSCHUDI, P., PIRROTTA, V., STEWARD, R., GEHRING, W. J., 1977: The 5 S genes of *Drosophila melanogaster*. Cell **12**, 1057—1067.

ATTARDI, G., AMALDI, F., 1970: Structure and synthesis of ribosomal RNA. Annu. Rev. Biochem. **39**, 183—226.

AUGER-BUENDIA, M.-A., LONGUET, M., 1978: Characterization of proteins from nucleolar preribosomes of mouse leukaemia cells by two-dimensional polyacrylamide gel electrophoresis. Eur. J. Biochem. **85**, 105—114.

— TAVITIAN, A., 1979: Ribosomal proteins synthesis and exchange in the absence of 28 S and 18 S ribosomal RNA synthesis in L 5178 cells. Biochim. Biophys. Acta **563**, 129—142.

— HAMELIN, R., TAVITIAN, A., 1978: Influence of toyocamycin on the assembly and processing of preribosomal ribonucleoproteins in the nucleolus of mammalian cells. Biochim. Biophys. Acta **521**, 241—250.

— LONGUET, M., TAVITIAN, A., 1979: Kinetic studies on ribosomal proteins assembly in preribosomal particles and ribosomal subunits of mammalian cells. Biochim. Biophys. Acta **563**, 113—128.

BACH, R., ALLET, B., CRIPPA, M., 1981 a: Sequence organization of the spacer in the ribosomal genes of *Xenopus clivii* and *Xenopus borealis*. Nucl. Acids Res. **9**, 5311—5330.

BACH, R., GRUMMT, I., Allet, B., 1981 b: The nucleotide sequence of the initiation region of the ribosomal transcription unit from mouse. Nucl. Acids Res. **9**, 1559—1569.

BACHELLERIE, J. P., MARTIN-PREVEL, C., ZALTA, J., 1971: Cinétique de l'incorporation d'uridine [^{3}H] dans les fractions subnucléolaires de cellules d'hépatome ascitique du rat. Biochim. **53**, 383—389.

— NICOLOSO, M., ZALTA, J. P., 1975: Early nucleolar preribosomal RNA-Protein in mammalian cells. Eur. J. Biochem. **55**, 119—129.

— — — 1977 a: Nucleolar chromatin in CHO cells. Topographical distribution of rDNA sequences and isolation of ribosomal transcription complexes. Eur. J. Biochem. **79**, 23—32.

— AMALRIC, F., NICOLOSO, M., ZALTA, J. P., SIMARD, R., 1977 b: Nucleolar chromatin in mammalian cells. I. An *in vitro* dissociation system of isolated nucleoli: biochemical and ultrastructural study. Biol. Cell. **28**, 1—8.

BAER, B. W., KORNBERG, R. D., 1979: Random location of nucleosomes on gene for 5 SrRNA. J. Biol. Chem. **254**, 9678—9681.

BAILEY, G. S., NIXON, J. E., HENDRICKS, J. D., SINNHUBER, R. O., VANHOLDE, K. E., 1980: Carcinogen aflatoxin B_1 is located preferentially in internucleosomal DNA following exposure *in vivo* in rainbow trout. Biochemistry **19**, 5836—5842.

BAKKEN, A., MORGAN, G., SOLLNER-WEBB, B., ROAN, J., BUSBY, S., REEDER, R. H., 1982: Mapping of transcription initiation and termination signals on *Xenopus laevis* ribosomal DNA. Proc. Nat. Acad. Sci. U.S.A. **79**, 56—60.

BALLAL, N. R., CHOI, Y. C., MOUCHE, R., BUSCH, H., 1977: Fidelity of synthesis of preribosomal RNA in isolated nucleoli and nucleolar chromatin. Proc. Nat. Acad. Sci. U.S.A. **74**, 2446—2450.

— SAMAL, B., CHOI, Y. C., BUSCH, H., 1979: Studies on the specificity of preribosomal RNA transcription in nucleoli after selective deproteinization. Nucl. Acids Res. **7**, 919—934.

— DASKAL, Y., SAMAL, B., BUSCH, H., 1980: Characterization of nucleolar residue fraction that specifically transcribes pre-rRNA. Cell Biol. Int. Rep. **4**, 997—1008.

BALTIMORE, D., 1981: Gene conversion: some implications for immunoglobulin genes. Cell **24**, 592—594.

BANDMAN, E., GURNEY, T., 1975: Differences in the cytoplasmic distribution of newly synthesized poly(A) in serum-stimulated and resting cultures of Balb/c 3T3 cells. Exp. Cell Res. **90**, 150—168.

BARNETT, T., RAE, P. M. M., 1979: A 9.6 kb intervening sequence in *D. virilis* rDNA and sequence homology in rDNA interruptions of diverse species of *Drosophila* and other *Diptera*. Cell **16**, 763—775.

BARSOTTI, P., DERENZINI, M., NOVELLO, F., PESSION-BRIZZI, A., MARINOZZI, V., 1980: Pretreatment of rats with cycloheximide prevents hepatocyte nucleolar fragmentation induced by α-amanitin. Biol. Cell **39**, 159—162.

BASERGA, R., IDE, T., WHELLY, S., 1978: Stimulation of rRNA synthesis in isolated nuclei and nucleoli by SV 40 T-antigen. Cold Spring Harbor Symp. Quant. Biol. **42**, 685—691.

BATISTONI, R., ANDRONICO, F., NARDI, I., BARSACCHI-PILONE, G., 1978: Chromosome location of the ribosomal genes in *Triturus vulgaris meridionalis* (*Amphibia Urodela*) III. Inheritance of the chromosomal sites for 18 S + 28 S ribosomal RNA. Chromosoma **65**, 231—240.

BÄUMLEIN, H., WOBUS, U., 1976: Chromosomal localization of ribosomal 5 S rRNA genes in *Chironomus thummi* by *in situ* hybridization of iodinated 5 S RNA. Chromosoma **57**, 199—204.

BAYEV, A. A., GEORGIEV, O. I., HADJIOLOV, A. A., KERMEKCHIEV, M. B., NIKOLAEV, N., SKRYABIN, K. G., ZAKHARYEV, V. M., 1980: The structure of the yeast ribosomal RNA genes. 2. The nucleotide sequence of the initiation site for ribosomal RNA transcription. Nucl. Acids Res. **8**, 4919—4925.

— — — NIKOLAEV, N., SKRYABIN, K. G., ZAKHARYEV, V. M., 1981: The structure of the yeast ribosomal RNA genes. 3. Precise mapping of the 18 S and 25 S rRNA genes and structure of the adjacent regions. Nucl. Acids Res. **9**, 789—799.

BECKINGHAM, K., 1981: The ribosomal DNA of *Calliphora erythrocephala*. The cistron classes of total genomic DNA. J. Mol. Biol. **149**, 141—169.

— WHITE, R., 1980: The ribosomal DNA of *Calliphora erythrocephala*; an analysis of hybrid plasmids containing rDNA. J. Mol. Biol. **137**, 349—373.

BEEBEE, T. Y. C., BUTTERWORTH, P. H. W., 1980: Eukaryotic DNA dependent RNA polymerases: an evaluation of their role in the regulation of gene expression. In: Eukaryotic Gene Regulation (KOLODNY, G., ed.), pp. 1—55. Boca Raton, Fla.: CRC Press.

— KORNER, A., BOND, R. P. M., 1972: Differential inhibition of mammalian RNA polymerases by an exotoxin from *Bacillus thuringiensis*. Biochem. J. **127**, 619—624.

BELL, G. I., VALENZUELA, P., RUTTER, W. J., 1976: Phosphorylation of yeast RNA polymerases. Nature **261**, 429—431.

— DEGENNARO, L. J., GELFAND, D. H., BISHOP, R. J., VALENZUELA, P., RUTTER, W. J., 1977: Ribosomal RNA genes of *Saccharomyces cerevisiae*. I. Physical map of the repeating unit and location of the regions coding for 5 S, 5.8 S, 18 S, and 25 S rRNA. J. Biol. Chem. **252**, 8118—8125.

BENCZE, J. L., BRASCH, K., WHITE, B. N., 1979: The location of 5 S rRNA genes in lampbrush polytene chromosomes from *Drosophila*. Exp. Cell Res. **120**, 365—372.

BENECKE, B. J., FERENCZ, A., SEIFART, K. H., 1973: Resistance of hepatic RNA polymerases to compounds affecting RNA and protein synthesis *in vivo*. FEBS Lett. **31**, 53—58.

BEREZNEY, R., 1979: Dynamic properties of the nuclear matrix. In: The Cell Nucleus, Vol. VII (BUSCH, H., ed.), pp. 413—456. New York: Academic Press.

— COFFEY, D. S., 1977: Nuclear matrix. Isolation and characterization of a framework structure from rat liver nuclei. J. Cell Biol. **73**, 616—637.

BERGER, E., 1977: The ribosomes of *Drosophila*. V. Normal and defective ribosome biosynthesis in *Drosophila* cell cultures. Mol. Gen. Genet. **155**, 35—40.

— SCHWEIGER, H. G., 1975 a: An apparent lack of nontranscribed spacers in rDNA of a green alga. Mol. Gen. Genet. **139**, 269—275.

— — 1975 b: Ribosomal DNA in different members of a family of green algae: an electron microscopical study. Planta **127**, 49—62.

BERGER, E., ZELLMER, D. M., KLOPPSTECH, K., RICHTER, G., DILLARD, W. L., SCHWEIGER, H. G., 1978: Alternating polarity in rRNA genes. Cell Biol. Int. Rep. **2**, 41—50.

BERNHARD, W., 1966: Ultrastructural aspects of the normal and pathological nucleolus in mammalian cells. Nat. Cancer Inst. Monogr. **23**, 13—38.

— GRANBOULAN, N., 1968: Electron microscopy of the nucleolus in vertebrate cells. In: Ultrastructure in Biological Systems (DALTON, A. J., HAGUENAU, F., eds.), pp. 81—149. New York: Academic Press.

— FRAYSSINET, C., LAFARGE, C., LEBRETON, E., 1965: Lésions nucléolaires précoces provoquées par l'aflatoxine dans les cellules hépatiques du rat. Compt. Rend. Acad. Sci. **261**, 1785—1789.

BERNIER, R., IGLESIAS, R., SIMARD, R., 1972: Detection of DNA by tritiated actinomycin D on ultrathin frozen sections. J. Cell Biol. **53**, 798—808.

BHULLAR, B. S., HEWITT, J., CANDIDO, E. P. M., 1981: The large high mobility group proteins of rainbow trout are localized predominantly in the nucleus and nucleoli of a cultured trout cell line. J. Biol. Chem. **256**, 8801—8806.

BIRD, A. P., 1980: Gene reiteration and gene amplification. In: Cell Biology, Vol. 3 (GOLDSTEIN, L., PRESCOTT, D. M., eds.), pp. 62—113. New York: Academic Press.

— SOUTHERN, E. M., 1978: Use of restriction enzymes to study eucaryotic DNA methylation. J. Mol. Biol. **118**, 27—47.

— TAGGART, M. H., GEHRING, CH. A., 1981 a: Methylated and unmethylated ribosomal rRNA genes in the mouse. J. Mol. Biol. **152**, 1—17.

— — MACLEOD, D., 1981 b: Loss of rDNA methylation accompanies the onset of ribosomal gene activity in early development of *X. laevis*. Cell **26**, 381—390.

BIRKENHEIMER, E. H., BROWN, D. D., JORDAN, E., 1978: A nuclear extract of *Xenopus* oocytes that accurately transcribes 5 S DNA. Cell **15**, 1077—1086.

BIRNSTIEL, M. L., WALLACE, H., SIRLIN, J. L., FISCHBERG, M., 1966: Localization of the ribosomal DNA complements in the Nucleolar Organizer Region of *Xenopus laevis*. Nat. Cancer Inst. Monogr. **23**, 431—448.

— CHIPCHASE, M., SPEIRS, J., 1971: The ribosomal RNA cistrons. Progr. Nucl. Acid Res. Mol. Biol. **11**, 351—389.

— SELLS, B. H., PURDOM, I. F., 1972: Kinetic complexity of RNA molecules. J. Mol. Biol. **63**, 21—39.

BISWAS, B. B., GANGULY, A., DAS, A., 1975: Eukaryotic RNA polymerases and the factors that control them. Progr. Nucl. Acid Res. Mol. Biol. **15**, 145—184.

BLACKBURN, E. H., GALL, J. G., 1978: A tandemly repeated sequence at the termini of the extrachromosomal ribosomal rRNA genes in *Tetrahymena*. J. Mol. Biol. **120**, 33—53.

BLATTI, S. P., INGLES, C. J., LINDELL, T. J., MORRIS, P. W., WEAVER, R. F., WEINBERG, F., RUTTER, W. J., 1970: Structure and regulatory properties of eucaryotic RNA polymerase. Cold Spring Harbor Symp. Quant. Biol. **35**, 649—658.

BOCTOR, A., GRASSMAN, A., 1973: Differential sensitivity of rat liver and rat hepatoma cells to α-amanitin. Biochem. Pharmacol. **22**, 17—28.

BOGENHAGEN, D. F., BROWN, D. D., 1981: Nucleotide sequences in *Xenopus* 5 S DNA required for transcription termination. Cell **24**, 261—270.

BOGENHAGEN, D. F., SAKONJU, S., BROWN, D. D., 1980: A control region in the center of the 5 S rRNA gene directs specific initiation of transcription. II. The 3′ border of the region. Cell **19**, 27—35.

— WORMINGTON, W. M., BROWN, D. D., 1982: Stable transcription complexes of *Xenopus* 5 S rRNA genes: A means to maintain the differentiated state. Cell **28**, 413—421.

BOLLA, R., ROTH, H. E., WEISSBACH, H., BROT, N., 1977: Effect of ribosomal proteins on synthesis and assembly of preribosomal particles in isolated rat liver nuclei. J. Biol. Chem. **252**, 721—725.

BOLLEN, G. H. P. M., MAGER, W. H., JENNESKENS, L. W., PLANTA, R. J., 1980: Small-size mRNAs code for ribosomal proteins in yeast. Eur. J. Biochem. **105**, 75—80.

— COHEN, L. H., MAGER, W. H., KLAASEN, A. W., PLANTA, R. J., 1981: Isolation of cloned ribosomal protein genes from the yeast *Saccharomyces carlsbergensis*. Gene **14**, 279—287.

— MOLENAAR, C. M. T., COHEN, L. H., RAAMSDONK-DUIN, M. M. C., MAGER, W. H., PLANTA, R. J., 1982: Ribosomal protein genes of yeast contain intervening sequences. Gene **18**, 29—37.

BOLOGNARI, A., CONTINI, A., 1979: The role of the nucleolus in carcinogenesis. Rivista di biol. norm. e patol. **7**, fasc. II, 43—68.

BOMBIK, B. M., HUANG, C. H., BASERGA, R., 1977: Isolation of transcriptionally active chromatin from mammalian nucleoli. Proc. Nat. Acad. Sci. U.S.A. **74**, 69—73.

BONCINELLI, E., FURIA, M., 1979: Patterns of the reversion to *bb* conditions of magnified *bb* loci in *D. melanogaster*. Mol. Gen. Genet. **176**, 81—85.

BONNER, W. M., 1978: Protein migration and accumulation in nuclei. In: The Cell Nucleus, Vol. VI (BUSCH, H., ed.), pp. 97—148. New York: Academic Press.

BORCHSENIUS, S., BONVEN, B., LEER, J. C., WESTERGAARD, O., 1981: Nuclease-sensitive regions on the extrachromosomal r-chromatin from *Tetrahymena pyriformis*. Eur. J. Biochem. **117**, 245—250.

BOSELEY, P. G., TUYNS, A., BIRNSTIEL, M. L., 1978: Mapping of the *Xenopus laevis* 5.8 S rDNA by restriction and DNA sequencing. Nucl. Acids Res. **5**, 1121—1137.

— MOSS, T., MACHLER, M., PORTMANN, R., BIRNSTIEL, M. L., 1979: Sequence organization of the spacer DNA in a ribosomal gene unit of *Xenopus laevis*. Cell **17**, 19—31.

BOTCHAN, P., REEDER, R. H., DAWID, I. B., 1977: Restriction analysis of the nontranscribed spacers of *Xenopus laevis* ribosomal DNA. Cell **11**, 599—607.

BOURGEOIS, C. A., HERNANDEZ-VERDUN, D.; HUBERT, J., BOUTEILLE, M., 1979: Silver staining of NORs in electron microscopy. Exp. Cell Res. **123**, 449—452.

BOUTEILLE, M., HERNANDEZ-VERDUN, D., 1979: Localization of a gene: the nucleolar organizer. Biomedicine **30**, 282—287.

— LAVAL, M., DUPUY-COIN, A. M., 1974: Localization of nuclear functions as revealed by ultrastructural autoradiography and cytochemistry. In: The Cell Nucleus, Vol. 1 (BUSCH, H., ed.), pp. 5—64. New York: Academic Press.

— HERNANDEZ-VERDUN, D., DUPUY-COIN, A. M., BOURGEOIS, C. A., 1982: Nucleoli and nucleolar-related structures in normal, infected and drug-treated cells. In: The Nucleolus (JORDAN, E. G., CULLIS, C. A., eds.), pp. 179—211. Cambridge: Cambridge Univ. Press.

BOUVIER, D., HUBERT, J., SEVE, A. P., BOUTEILLE, M., 1982: RNA is responsible for the three-dimensional organization of nuclear matrix proteins in HeLa cells. Biol. Cell **43**, 143—146.

BOWMAN, L. H., EMERSON, C. P., JR., 1977: Post-transcriptional regulation of ribosome accumulation during myoblast differentiation. Cell **10**, 587—596.

— GOLDMAN, W. E., GOLDBERG, G. I., HEBERT, M. B., SCHLESSINGER, D., 1983: Location of the initial cleavage sites in mouse pre-rRNA. Mol. Cell Biol. **3**, 1501—1510.

— RABIN, B., SCHLESSINGER, D., 1981: Multiple ribosomal RNA cleavage pathways in mammalian cells. Nucl. Acids Res. **9**, 4951—4966.

BOWMAN, P. D., MEEK, R. L., DANIEL, C. W., 1976: Decreased synthesis of nucleolar RNA in aging human cells *in vitro*. Exp. Cell Res. **101**, 434—437.

BOZZONI, I., BALDARI, C. T., AMALDI, F., BOUNGIORNO-NARDELLI, M., 1981 a: Replication of ribosomal DNA in *Xenopus laevis*. Eur. J. Biochem. **118**, 585—590.

— BECCARI, E., LUO, Z. X., AMALDI, F., 1981 b: *Xenopus laevis* ribosomal protein genes: isolation of recombinant cDNA clones and study of the genomic organization. Nucl. Acids Res. **9**, 1069—1086.

— TOGNONI, A., PIERANDREI-AMALDI, P., BECCARI, E., BUONGIORNO-NARDELLI, M., AMALDI, F., 1982: Isolation and structural analysis of ribosomal protein genes in *Xenopus laevis*. J. Mol. Biol. **161**, 353—371.

BRACHET, J., 1947: Embryologie chimique, pp. 522. Paris: Masson.

— 1957: Biochemical Cytology, p. 535. New York: Academic Press.

BRAGA, E. A., YUSSIFOV, T. N., NOSIKOV, V. V., 1982: Structural organization of rat ribosomal genes. Restriction endonuclease analysis of genomic and cloned ribosomal DNA. Gene **20**, 145—156.

BRANDHORST, B. P., MCCONKEY, E. H., 1974: Stability of RNA in mammalian cells. J. Mol. Biol. **85**, 451—463.

BRANLANT, C., KROL, A., MACHATT, M. A., POUYET, J., EBEL, J. P., EDWARDS, K., KÖSSEL, H., 1981: Primary and secondary structure of *Escherichia coli* MRE 600 23 S ribosomal RNA. Comparison with models of secondary structure for maize chloroplast 23 S rRNA and for large portions of mouse and human 16 S mitochondrial rRNAs. Nucl. Acids Res. **9**, 4303—4324.

BRAM, R. J. R., YOUNG, R. A., STEITZ, J. A., 1980: The ribonuclease III site flanking 23 S sequences in the 30 S ribosomal precursor RNA of *E. coli*. Cell **19**, 393—401.

BRAND, C. R., KLOOTWIJK, J., SIBOM, C. P., PLANTA, R. J., 1979: Pseudouridylation of yeast ribosomal precursor RNA. Nucl. Acids Res. **7**, 121—134.

BRAUN, R., EVANS, T. E., 1969: Replication of nuclear satellite and mitochondrial DNA of *Physarum*. Biochim. Biophys. Acta **182**, 511—522.

BREANT, B., BUHLER, J. M., SENTENAC, A., FROMAGEOT, P., 1983: On the phosphorylation of yeast RNA polymerases A and B. Eur. J. Biochem. **130**, 247—251.

BREATHNACH, R., CHAMBON, P., 1981: Organization and expression of eucaryotic split genes coding for proteins. Ann. Rev. Biochem. **50**, 349—383.

BRIMACOMBE, R., STOFFLER, G., WITTMANN, H. G., 1978: Ribosome structure. Ann. Rev. Biochem. **47**, 217—249.

BROWN, A., PAN, C.-J., MARZLUFF, W. F., 1982: Methylation of RNA in a cell-free system from mouse myeloma cells. Biochemistry **21**, 4303—4310.

BROWN, D. D., 1982: How a simple animal gene works. The Harvey Lectures, Ser. **76**, 27—45.

— DAWID, I. B., 1968: Specific gene amplification in oocytes. Oocyte nuclei contain extrachromosomal replicas of the genes for rRNA. Science **160**, 272—274.

— GURDON, J. B., 1964: Absence of ribosomal RNA synthesis in the anucleolate mutant of *Xenopus laevis*. Proc. Nat. Acad. Sci. U.S.A. **51**, 139—146.

— — 1977: High-fidelity transcription of 5 S DNA injected into *Xenopus* oocytes. Proc. Nat. Acad. Sci. U.S.A. **74**, 2064—2068.

— — 1978: Cloned single repeating units of 5 S DNA direct accurate transcription of 5 S RNA when injected into *Xenopus* oocytes. Proc. Nat. Acad. Sci. U.S.A. **75**, 2849—2853.

— WENSINK, P. C., JORDAN, E., 1971: Purification and some characteristics of 5 S DNA from *Xenopus laevis*. Proc. Nat. Acad. Sci. U.S.A. **68**, 3175—3179.

— — — 1972: A comparison of the ribosomal DNAs of *Xenopus laevis* and *Xenopus mulleri*: The evolution of tandem genes. J. Mol. Biol. **63**, 57—73.

— CARROLL, D., BROWN, R. D., 1977: The isolation and characterization of a second oocyte 5 S DNA from *Xenopus laevis*. Cell **12**, 1045—1056.

BUCCI, S., NARDI, I., MANCINO, G., FIUME, L., 1971: Incorporation of tritiated uridine in nuclei of *Triturus* oocytes treated with α-amanitin. Exp. Cell Res. **69**, 462—465.

BUCHER, N. L. R., MALT, R., 1971: Regeneration of Liver and Kidney. Boston: Little, Brown & Co.

BUHLER, J.-M., IBORRA, F., SENTENAC, A., FROMAGEOT, P., 1976: The presence of phosphorylated subunits in yeast RNA polymerases A and B. FEBS Lett. **71**, 37—41.

— HUET, J., DAVIES, K. E., SENTENAC, A., FROMAGEOT, P., 1980: Immunological studies of yeast nuclear RNA polymerases at the subunit level. J. Biol. Chem. **255**, 9949—9954.

BUONGIORNO-NARDELLI, M., AMALDI, F., LAVA-SANCHEZ, P. A., 1976: Electron microscope analysis of amplifying rRNA from *Xenopus laevis*. Exp. Cell Res. **98**, 95—103.

BURNS, M. W., OHNUKI, Y., ROUNDS, D. E., OLSON, R. S., 1970: Modification of nucleolar expression following laser microirradiation of chromosomes. Exp. Cell Res. **60**, 133—142.

BUSBY, S. J., REEDER, R. H., 1982: Fate of amplified nucleoli in *Xenopus laevis* embryos. Dev. Biol. **91**, 458—467.

— — 1983: Spacer sequences regulate transcription of ribosomal gene plasmids injected into *Xenopus* embryos. Cell **34**, 989—996.

BUSBY, W. F., WOGAN, G. N., 1979: Food-borne mycotoxins and alimentary mycotoxicoses. In: Food-borne Infections and Intoxications, 2nd ed. (RIEMANN, H. P., BRYAN, F. L., eds.), pp. 519—610. New York: Academic Press.

BUSCH, H. (ed.), 1974: The Molecular Biology of Cancer, p. 637. New York: Academic Press.

Busch, H. (ed.), 1979: The molecular biology of cancer: the cancer cell and its functions. In: Effects of Drugs on the Cell Nucleus (Busch, H., Yeoman, G., Daskal, Y., eds.), pp. 35—37. New York: Academic Press.
— Smetana, K., 1970: The Nucleolus, p. 626. New York: Academic Press.
— — 1974: The nucleus of the cancer cell. In: The Molecular Biology of Cancer (Busch, H., ed.), pp. 41—80. New York: Academic Press.
— Byvoet, P., Smetana, K., 1963: The nucleolus of the cancer cell. Cancer Res. **23**, 313—339.
— Ballal, N. R., Rao, M. R. S., Choi, Y. C., Rothblum, L. I., 1978: Factors affecting nucleolar rDNA readouts. In: The Cell Nucleus, Vol. V, B (Busch, H., ed.), pp. 416—468. New York: Academic Press.
— Daskal, Y., Györkey, F., Smetana, K., 1979: Silver staining of nucleolar granules in tumor cells. Cancer Res. **39**, 857—863.
— Busch, R. K., Chan, P.-K., Kelsey, D., Takahashi, K., 1982: Nucleolar antigens of human tumors. Methods Cancer Res. **19**, 109—178.

Busiello, E., DiGirolamo, M., 1975: RNA metabolism in nuclei isolated from HeLa cells. Eur. J. Biochem. **55**, 61—70.

Butler, M. J., Davies, K. E., Walker, I. O., 1978: The structure of nucleolar chromatin in *Physarum polycephalum*. Nucl. Acids Res. **5**, 667—678.

Caboche, M., Bachellerie, J.-P., 1977: RNA methylation and control of eucaryotic RNA biosynthesis. Eur. J. Biochem. **74**, 19—29.

Campbell, G. R., Littau, V. C., Melera, P. W., Allfrey, V. G., Johnson, E. M., 1979: Unique sequence arrangement of ribosomal genes in the palindromic rDNA molecule of *Physarum polycephalum*. Nucl. Acids Res. **6**, 1433—1447.

Capco, D. G., Wan, K. M., Penman, S., 1982: The nuclear matrix: Three-dimensional architecture and protein composition. Cell **29**, 847—858.

Carin, M., Jensen, B. F., Jentsch, K. D., Leer, J. C., Nielsen, O. F., Westergaard, O., 1980: In vitro splicing of the ribosomal RNA precursor in isolated nucleoli from *Tetrahymena*. Nucl. Acids Res. **8**, 5551—5566.

Carter, C. J., Cannon, M., 1980: Maturation of ribosomal precursor RNA in *Saccharomyces cerevisiae*. J. Mol. Biol. **153**, 179—199.

Caspersson, T. O., 1950: Cell Growth and Cell Function, p. 185. New York: W. W. Norton & Co.

Cassidy, B. G., Subrahmanyam, C. S., Rothblum, L. I., 1982: The nucleotide sequence of the 5′-region of rat rDNA and adjoining spacer. Biochem. Biophys. Res. Commun. **107**, 1571—1576.

Cassidy, D. M., Blackler, A. W., 1974: Repression of nucleolar organizer activity in an interspecific hybrid of the genus *Xenopus*. Devel. Biol. **41**, 84—96.

Cave, M. D., 1976: Absence of rDNA amplification in the uninucleolate oocyte of the cockroach *Blattella germanica*. J. Cell. Biol. **71**, 49—58.

Cazillis, M., Houssais, J.-F., 1981: The ribosomal proteins of L-cells. Eur. J. Biochem. **114**, 355—363.

Cech, Th. R., Brehm, S. L., 1981: Replication of the extrachromosomal ribosomal RNA genes of *Tetrahymena thermophilia*. Nucl. Acids Res. **9**, 3531—3543.
— Rio, D. C., 1979: Localization of transcribed regions on extrachromosomal

rRNA genes of *Tetrahymena thermophila* by R-loop mapping. Proc. Nat. Acad. Sci. U.S.A. **76**, 5051—5055.

Cech, Th. R., Tanner, N. K., Tinoko, I., Weir, B. R., Zuker, M., Perlman, P. S., 1983: Secondary structure of *Tetrahymena* ribosomal RNA intervening sequence: structural homology with fungal mitochondrial intervening sequences. Proc. Nat. Acad. Sci. U.S.A. **80**, 3903—3907.

Chambers, J. Ch., Watanable, Sh. Taylor, J. H., 1982: Dissection of a replication origin of *Xenopus* DNA. Proc. Nat. Acad. Sci. U.S.A. **79**, 5572—5576.

Chambon, P., 1975: Animal RNA polymerases. Ann. Rev. Biochem. **44**, 613—638.

Chesterton, C. J., Coupar, B. E. H., Butterworth, P. H. W., Buss, J., Green, M. H., 1975: Studies on the control of rRNA synthesis of HeLa cells. Eur. J. Biochem. **57**, 79—83.

Childs, G., Maxson, R., Cohn, R. H., Kedes, L., 1981: Orphons: Dispersed genetic elements derived from tandem repetitive genes of eucaryotes. Cell **23**, 651—663.

Chiu, S.-M., Oleinick, N. L., Friedman, L. R., Stambrook, P. J., 1982: Hypersensitivity of DNA in transcriptionally active chromatin to ionizing radiation. Biochim. Biophys. Acta **699**, 15—21.

Chooi, W. Y., 1976: RNA transcription and ribosomal protein assembly in *Drosophila melanogaster*. In: Handbook of Genetics (King, R. C., ed.), Vol. 5, pp. 219—265. New York: Plenum Press.

— Leiby, K. R., 1981: An electron microscopic method for localization of ribosomal proteins during transcription of ribosomal DNA: A method for studying protein assembly. Proc. Nat. Acad. Sci. U.S.A. **78**, 4823—4827.

Chouinard, L. A., 1971: A light- and electron-microscope study of the nucleolus during growth of the oocyte in the prepubertal mouse. J. Cell Sci. **9**, 637—663.

Cihak, A., Pitot, H. C., 1970: Incorporation of label from 5-fluoroorotate into non-ribosomal cytoplasmic RNA in rat liver. FEBS Lett. **6**, 206—212.

Cihlar, R. L., Sypherd, P. S., 1980: The organization of the ribosomal RNA genes in the fungus *Mucor racemosus*. Nucl. Acids Res. **8**, 793—804.

Clark, C. G., Gerbi, S. A., 1982: Ribosomal RNA evolution by fragmentation of the 23 S progenitor: maturation pathway parallels evolutionary emergence. J. Mol. Evol. **18**, 329—336.

Clark-Walker, G. D., Azad, A. A., 1980: Hybridizable sequences between cytoplasmic ribosomal RNAs and 3 micron circular DNAs of *Saccharomyces cerevisiae* and *Torulopsis glabrata*. Nucl. Acids Res. **8**, 1009—1022.

Clissold, P., Cole, R. J., 1973: Regulation of rRNA synthesis during mammalian myogenesis in culture. Exp. Cell Res. **80**, 159—169.

Cochet-Meilhac, M., Nuret, P., Courvalin, J. C., Chambon, P., 1974: Animal DNA-dependent RNA polymerases. Determination of the cellular number of RNA polymerase B molecules. Biochim. Biophys. Acta **353**, 185—192.

Cockburn, A. F., Taylor, W. C., Firtel, R. A., 1978: *Dictyostelium* rDNA consists of non-chromosomal palindromic dimers containing 5 S and 36 S coding regions. Chromosoma **70**, 19—29.

Cohen, M., jr, 1976: Evolution of 5 S rRNA genes in the chromosomes of the *Virilis* group of *Drosophila*. Chromosoma **55**, 359—371.

Comings, D. E., Peters, K. E., 1981: Two-dimensional gel electrophoresis of nuclear particles. In: The Cell Nucleus, Vol. 9 (Busch, H., ed.), pp. 89—118. New York: Academic Press.

Cooke, A., Kay, J. E., 1973: Effect of phytohaemagglutinin on the nuclear RNA polymerase activity of human lymphocytes. Exp. Cell Res. **79**, 179—185.

Cooper, C. S., Quincey, R. V., 1979: The role of subunits in yeast DNA-dependent RNA polymerase A. Biochem. J. **181**, 301—308.

Cooper, H. L., 1969: Ribosomal RNA wastage in resting and growing lymphocytes. J. Biol. Chem. **244**, 5590—5596.

— 1970: Control of synthesis and wastage of ribosomal RNA in lymphocytes. Nature (London) **227**, 1105—1107.

— 1973: Degradation of 28 S RNA late in rRNA maturation in non-growing lymphocytes and its reversal after growth stimulation. J. Cell Biol. **59**, 250—254.

— Gibson, E. M., 1971: Control of synthesis and wastage of rRNA in lymphocytes. J. Biol. Chem. **246**, 5059—5066.

Cooper, K. W., 1959: Cytogenetic analysis of major heterochromatic elements (especially Xh and Y) in *Drosophila melanogaster* and the theory of "heterochromatin". Chromosoma **10**, 535—588.

Cory, S., Adams, J. M., 1977: A very large repeating unit of mouse DNA containing the 18 S, 28 S, and 5.8 S rRNA genes. Cell **11**, 795—805.

Coupar, B. E. H., Chesterton, C. J., 1975: Purification of forms A_I and A_{II} DNA-dependent RNA polymerases from rat liver nucleoli using low-ionic-strength extraction conditions. Eur. J. Biochem. **59**, 25—30.

— — 1977: The mechanism by which heparin stimulates transcription by isolated rat liver nuclei: polyribonucleotide elongation rates and the number of transcribing RNA polymerase molecules. Eur. J. Biochem. **79**, 525—533.

— Davies, J. A., Chesterton, C. J., 1978: Quantification of hepatic transcribing RNA polymerase molecules, polyribonucleotide elongation rates and messenger RNA complexity. Eur. J. Biochem. **84**, 611—623.

Cox, B. J., Turnock, C., 1973: Synthesis and processing of ribosomal RNA in cultured plant cells. Eur. J. Biochem. **37**, 367—376.

Cox, R. A., 1977: Structure and function of procaryotic and eucaryotic ribosomes. Progr. Biophys. Mol. Biol. **32**, 193—231.

— Kelly, J. M., 1981: Mature 23 S rRNA of procaryotes appears homologous with the precursor of 25–28 S rRNA of eucaryotes. Comments on the evolution of 23–28 S rRNA. FEBS Lett. **130**, 1—10.

— Peden, K., 1979: A study of the organization of the ribosomal RNA gene cluster of *Neurospora crassa* by means of restriction endonuclease analysis and cloning bacteriophage lambda. Molec. Gen. Genetics **174**, 17—24.

Cox, R. F., 1976: Quantitation of elongating form A and B RNA polymerases in chick oviduct nuclei and effects of estradiol. Cell **7**, 455—465.

Craig, N. C., 1974: Ribosomal RNA synthesis in eucaryotes and its regulation. In: MTP International Review of Science, Series 1, Biochemistry, 6: Biochemistry of Nucleic Acids (Burton, K., ed.), pp. 255—288. London: Butterworths.

— Perry, R. P., 1971: Persistent cytoplasmic synthesis of ribosomal proteins during the selective inhibition of rRNA synthesis. Nature (London), New Biol. **229**, 75—80.

CRAIG, N. C., KELLY, D. E., PERRY, R. P., 1971: Lifetime of messenger RNA which codes for ribosomal protein in L-cells. Biochim. Biophys. Acta **246**, 493—498.

CRAMER, J. H., FARRELLY, F. W., ROWND, R. H., 1976: Restriction endonuclease analysis of rDNA from *Saccharomyces cerevisiae*. Mol. Gen. Genet. **148**, 233—241.

— — BARNITZ, J. T., ROWND, R. H., 1977: Construction and restriction endonuclease mapping of hybrid plasmids containing *Saccharomyces cerevisiae* rDNA. Mol. Gen. Genet. **151**, 229—244.

CRISTOFALO, V. J., 1970: Metabolic aspects of aging in diploid human cells. In: Aging in Cell and Tissue Culture (HOLECKOVA, M., CRISTOFALO, V. J., eds.), pp. 83—119. New York: Plenum Press.

CROUSE, H. V., GERBI, S. A., LIANG, C. M., MAGNUS, L., MERSER, I. M., 1977: Localization of ribosomal DNA within the proximal X heterochromatin of *Sciara coprophila* (*Diptera, Sciaridae*). Chromosoma (Berl.) **64**, 305—318.

CROY, R. G., WOGAN, G. N. 1981: Temporal patterns of covalent DNA adducts in rat liver after single and multiple doses of aflatoxin B_1. Cancer Res. **41**, 197—203.

CULLIS, C. A., 1979: Quantitative variation of ribosomal RNA genes in flax genotrophs. Heredity **42**, 237—246.

— CHARLTON, L., 1981: The induction of ribosomal DNA changes in flax. Plant Sci. Lett. **20**, 213—217.

DABAUVALLE, M.-C., FRANKE, W. W., 1982: Karyophilic proteins: polypeptides synthesized *in vitro* accumulate in the nucleus on microinjection into the cytoplasm of amphibian oocytes. Proc. Nat. Acad. Sci. U.S.A. **79**, 5302—5306.

DABEVA, M. D., DUDOV, K. P., 1982 a: Quantitative alterations in the nucleolar and nucleoplasmic rRNA in regenerating rat liver. Biochem. J. **204**, 179—183.

— — 1982 b: Transcriptional control of ribosome production in regenerating rat liver. Biochem. J. **208**, 101—108.

— — HADJIOLOV, A. A., EMANUILOV, I. A., TODOROV, B. N., 1976: Intranuclear maturation pathways of rat liver rRNA. Biochem. J. **160**, 495—503.

— — — STOYKOVA, A. S., 1978: Ouantitative analysis of rat liver nucleolar and nucleoplasmic rRNA. Biochem. J. **171**, 367—374.

D'ALESSIO, J. M., HARRIS, G. H., PERNA, P. J., PAULE, M. R., 1981: Ribosomal RNA repeat unit of *Acanthamoeba castellanii*: cloning and restriction endonuclease map. Biochemistry **20**, 3822—3827.

DARLIX, J.-L., ROCHAIX, J.-D., 1981: Nucleotide sequence and structure of cytoplasmic 5 S RNA and 5.8 S RNA of *Chlamydomonas reinhardii*. Nucl. Acids Res. **9**, 1291—1299.

DARNELL, J. E., 1982: Variety in the level of gene control in eucaryotic cells. Nature **297**, 365—371.

DASKAL, Y., 1979: Drug effects on nucleolar and extranucleolar chromatin. In: Effects of Drugs on the Cell Nucleus (BUSCH, H. *et al.* eds.), pp. 107—125. New York: Academic Press.

— PRESTAYKO, A. W., BUSCH, H., 1974: Ultrastructural and biochemical studies of the isolated fibrillar component of nucleoli from Novikoff hepatoma ascites cells. Exp. Cell Res. **88**, 1—14.

DASKAL, Y., MERSKI, J. A., HUGHES, J. B., BUSCH, H., 1975: The effect of cycloheximide on the ultrastructure of ratt liver cells. Exp. Cell Res. **93**, 395—401.

— SMETANA, K., BUSCH, H., 1980: Evidence from studies on segregated nucleoli that nucleolar silver staining proteins C 23 and B 23 are in the fibrillar component. Exp. Cell Res. **127**, 285—291.

DAUPHINAIS, C., 1981: The control of ribosomal RNA transcription in lymphocytes. Eur. J. Biochem. **114**, 487—492.

DAWID, I. B., BOTCHAN, P., 1977: Sequences homologous to ribosomal insertions occur in the *Drosophila* genome outside the nucleolus organizer. Proc. Nat. Acad. Sci. U.S.A. **74**, 4233—4237.

— WELLAUER, P. K., 1976: A reinvestigation of 5′-3′ polarity in 40 S precursor to rRNA in *Xenopus laevis*. Cell **8**, 443—448.

— — LONG, E. O., 1978: Ribosomal DNA in *Drosophila melanogaster*. I. Isolation and characterization of cloned fragments. J. Mol. Biol. **126**, 749—768.

— LONG, E. O., DINOCERA, P. P., PARDUE, M. L., 1981: Ribosomal insertion-like elements in *Drosophila melanogaster* are interspersed with mobile sequences. Cell **25**, 399—408.

DE KLOET, S. R., 1966: Ribonucleic acid synthesis in yeast. Biochem. J. **99**, 566—581.

— 1968: Effects of 5-fluorouracil and 6-azauracil on the synthesis of RNA and protein in *Saccharomyces carlsbergensis*. Biochem. J. **106**, 167—178.

DE LA TORRE, C., GIMENEZ-MARTIN, G., 1982: The nucleolar cycle. In: The Nucleolus (JORDAN, E. G., CULLIS, C. A., eds.), pp. 153—178. Cambridge: Cambridge Univ. Press.

DELIHAS, N., ANDERSEN, J., 1982: Generalized structures of the 5 S ribosomal RNAs. Nucl. Acids Res. **10**, 7323—7344.

DENG, J. S., TAKASAKI, Y., TAN, E. M., 1981: Nonhistone nuclear antigens reactive with autoantibodies. Immunofluorescent studies of distribution in synchronized cells. J. Cell Biol. **91**, 654—660.

DENIS, H., 1977: Accumulation du RNA dans les oocytes des vertèbres infèrieurs. Biol. Cell **28**, 87—92.

DENOYA, C., COSTA GIOMI, P., SCODELLER, E. A., VASQUEZ, C., LA TORRE, J. L., 1981: Processing of naked 45 S rRNA precursor *in vitro* by an RNA-associated endoribonuclease. Eur. J. Biochem. **115**, 375—383.

DENTON, T. E., LIEM, S. L., CHENG, K. M., BARRETT, J. V., 1981: The relationship between aging and ribosomal gene activity in humans as evidenced by silver staining. Mech. Ageing Develop. **15**, 1—7.

DERENZINI, M., BONETTI, E., 1975: Cycloheximide-induced ultrastructural changes in hepatocyte nuclei in partially hepatectomized rats. Virchow's Arch. Cell Path. **19**, 115—125.

— PESSION, A., BETTS-EUSEBI, C. M., NOVELLO, F., 1983: Relationship between the extended, non-nucleosomal intranucleolar chromatin *in situ* and rRNA synthesis. Exp. Cell Res. **145**, 127—143.

DETKE, S., KELLER, J. M., 1982: Comparison of the proteins present in HeLa interphase nucleoskeletons and metaphase chromosome scaffolds. J. Biol. Chem. **257**, 3905—3911.

D'Eustachio, P., Meyuhas, O., Ruddle, F., Perry, R. P., 1981: Chromosomal distribution of ribosomal protein genes in the mouse. Cell **24**, 307—312.

Dev, V. G., Miller, D. A., Miller, O. J., Rechsteiner, M., 1979: Time of suppression of human rRNA genes in mouse-human hybrid cells. Exp. Cell Res. **123**, 47—54.

Dice, J. F., 1982: Altered degradation of proteins microinjected into senescent human fibroblasts. J. Biol. Chem. **257**, 14624—14627.

Dimova, R. N., Gajdardjieva, K. G., Dabeva, M. D., Hadjiolov, A. A., 1979: Early effects of D-galactosamine on rat liver nucleolar structures. Biol. Cellulaire **35**, 1—10.

— Markov, D. V., Gajdardjieva, K. C., Dabeva, M. D., Hadjiolov, A. A., 1982: Electron microscopic localization of silver staining NOR-proteins in rat liver nucleoli upon D-galactosamine block of transcription. Eur. J. Cell Biol. **28**, 272—277.

Din, N., Engberg, J., 1979: Extrachromosomal rRNA genes in *Tetrahymena*: structure and evolution. J. Mol. Biol. **134**, 555—574.

— — Kaffenberger, W., Eckert, W., 1979: The intervening sequence in the 26 S rRNA coding region of *T. thermophila* is transcribed within the largest stable precursor for rRNA. Cell **18**, 525—532.

— — Gall, J. G., 1982: The nucleotide sequence at the transcription termination site of the ribosomal RNA gene in *Tetrahymena thermophila*. Nucl. Acids Res. **10**, 1503—1515.

Dingwall, C., Sharnick, S. V., Laskey, R. A., 1982: A polypeptide domain that specifies migration of nucleoplasmin into the nucleus. Cell **30**, 449—458.

Dix, D., Cohen, P., 1980: On the role of aging in cancer incidence. J. Theor. Biol. **83**, 163—173.

Doenecke, D., McCarthy, B. J., 1975: The nature of protein association with chromatin. Biochemistry **14**, 1373—1378.

Doerschung, E. B., Miksche, J. P., Stern, H., 1976: DNA variation and rDNA constancy in two *Crepis* species and the interspecific hybrid exhibiting nucleolar-organizer suppression. Heredity **37**, 441—450.

Duceman, B. W., Jacob, S. T., 1980: Transcriptionally active RNA polymerases from Morris hepatomas and rat liver. Biochem. J. **190**, 781—789.

Dudov, K. P., Dabeva, M. D., 1983: Post-transcriptional regulation of ribosome formation in the nucleus of regenerating rat liver. Biochem. J. **210**, 83—92.

— — Hadjiolov, A. A., Todorov, B. N., 1978: Processing and migration of ribosomal RNA in the nucleolus and nucleoplasm of rat liver nuclei. Biochem. J. **171**, 375—383.

— Hadjiolova, K. V., Kermekchiev, M. B., Stanchev, B. S., Hadjiolov, A. A., 1983: A 12 S precursor to 5.8 S rRNA associated with rat liver nucleolar 28 S rRNA. Biochim. Biophys. Acta **739**, 79—84.

Duhamel-Maestracci, N., Simard, R., 1980: Nucleolus organizers regions and ribosomal genes in Chinese hamster genome. Biol. Cell **39**, 1—6.

Dulbecco, R., 1973: Genes in cancer. Folia biologica (Praha) **29**, 9—17.

Dupuy-Coin, A. M., Ege, T., Bouteille, M., Ringertz, N. R., 1976: Ultrastructure of chick erythrocyte nuclei undergoing reactivation in heterokaryons and enucleated cells. Exp. Cell Res. **101**, 355—369.

ECKERT, W. A., FRANKE, W. W., SCHEER, U., 1975: Nucleocytoplasmic translocation of RNA in *Tetrahymena pyriformis* and its inhibition by actinomycin D and cycloheximide. Exp. Cell Res. **94**, 31—46.

EDSTRÖM, J.-E., LÖNN, U., 1976: Cytoplasmic zone analysis. RNA flow studies by micromanipulation. J. Cell Biol. **56**, 562—572.

— GRAMPP, W., SCHOR, N., 1961: The intracellular distribution and heterogeneity of ribonucleic acid in starfish oocytes. J. Biophys. Biochem. Cytol. **11**, 549—557.

EGAWA, K., CHOI, Y. C., BUSCH, H., 1971: Studies on the role of 23 S nucleolar RNA as an intermediate in the synthesis of 18 S ribosomal RNA. J. Mol. Biol. **56**, 565—577.

EICHLER, D. C., EALES, S. J., 1982: Isolation and characterization of a single-stranded specific endoribonuclease from Ehrlich cell nucleoli. J. Biol. Chem. **257**, 14384—14389.

— TATAR, T. F., 1980: Properties of a purified nucleolar ribonuclease from Ehrlich ascites carcinoma cells. Biochemistry **19**, 3016—3022.

ELGIN, S. C. R., 1981: DNase I-hypersensitive sites of chromatin. Cell **27**, 413—415.

ELICEIRI, G. L., 1976: Turnover of ribosomal RNA in liver. Biochim. Biophys. Acta **447**, 391—394.

— GREEN, H., 1969: Ribosomal RNA synthesis in mouse-human hybrid cells. J. Mol. Biol. **41**, 253—260.

ELSEVIER, S. M., RUDDLE, F. H., 1975: Location of genes coding for 18 S and 28 S ribosomal RNA within the genome of *Mus musculus*. Chromosoma **52**, 219—228.

EMANUILOV, I., NICOLOVA, R. C., DABEVA, M. D., HADJIOLOV, A. A., 1974: Quantitative ultrastructural study of the action of α-amanitin on mouse liver cell nucleoli. Exp. Cell Res. **86**, 401—403.

EMERSON, C. P., 1971: Regulation of the synthesis and stability of ribosomal RNA during contact inhibition of growth. Nature (London), New Biol. **232**, 101—106.

ENGBERG, J., 1983: Strong sequence conservation of a 38 bp region near the center of the extrachromosomal rDNA palindrome in different *Tetrahymena* species. Nucl. Acids Res. **11**, 4939—4946.

— PEARLMAN, R., 1972: The amount of rRNA genes in *Tetrahymena pyriformis* in different physiological states. Eur. J. Biochem. **26**, 393—400.

— CHRISTIANSEN, G., LEICK, V., 1974: Autonomous rDNA molecules containing single copies of rRNA genes in the macronucleus of *Tetrahymena pyriformis*. Biochem. Biophys. Res. Commun. **59**, 1356—1365.

— ANDERSSON, P., LEICK, V., COLLINS, J., 1976: rDNA molecules from *Tetrahymena pyriformis* GL are giant palindromes. J. Mol. Biol. **104**, 455—470.

— DIN, N., ECKERT, W. A., KAFFENBERGER, W., PEARLMAN, R. E., 1980: Detailed transcription map of the extrachromosomal ribosomal RNA genes in *Tetrahymena thermophila*. J. Mol. Biol. **142**, 289—313.

ENGELKE, D. R., NG, S.-Y., SHASTRY, B. S., ROEDER, R. G., 1980: Specific interaction of a purified transcription factor with an internal control region of 5 S rRNA genes. Cell **19**, 717—728.

ENNIS, H. L., 1966: Synthesis of ribonucleic acid in L cells during inhibition of protein synthesis by cycloheximide. Mol. Pharmacol. **2**, 543—557.

ERDMANN, V. A., 1982: Collection of published 5 S and 5.8 S RNA sequences and their precursors. Nucl. Acids Res. **10**, r93—r115.

— HUYSMANS, E., VANDENBERGHE, A., DE WACHTER, R., 1983: Collection of published 5 S and 5.8 S ribosomal RNA sequences. Nucl. Acids Res. **11**, r105—r133.

ESPONDA, P., GIMENEZ-MARTIN, G., 1972: Ultrastructural morphology of the nucleolar organizing region. J. Ultrastr. Res. **39**, 509—519.

ESTABLE, C., 1966: Morphology structure and dynamics of the nucleolonema. Nat. Cancer Inst. Monogr. **23**, 91—106.

— SOTELO, J. R., 1951: Una nueva estructura celular: el nucleolonema. Publ. Inst. Inv. Cien. Biol. **1**, 105—126.

EVANS, H. J., BUCKLAND, R. A., PARDUE, M. L., 1974: Location of the genes coding for 18 S and 28 S ribosomal DNA in the human genome. Chromosoma **48**, 405—426.

FABER, A. J., COOK, A., HANCOCK, R., 1981: Characterization of a chromatin fraction bearing pulse-labeled RNA. Eur. J. Biochem. **120**, 357—361.

FABIJANSKI, S., PELLEGRINI, M., 1982a: Isolation of a cloned DNA segment containing a ribosomal protein gene of *Drosophila melanogaster*. Gene **18**, 267—276.

— — 1982b: A *Drosophila* ribosomal protein gene is located near repeated sequences including rDNA sequences. Nucl. Acids Res. **10**, 5979—5991.

FAKAN, S., 1978: High resolution autoradiography studies on chromatin functions. In: The Cell Nucleus (BUSCH, H., ed.), Vol. V, pp. 3—53. New York: Academic Press.

— 1980: Ultrastructural visualization of transcription at the cellular and molecular level. Biol. Cell **39**, 113—116.

— NOBIS, P., 1978: Ultrastructural localization of transcription sites and of RNA distribution during the cell cycle of synchronized CHO cells. Exp. Cell Res. **113**, 327—337.

— PUVION, E., 1980: The ultrastructural visualization of nucleolar and extranucleolar RNA synthesis and distribution. Intern. Rev. Cytol. **65**, 255—299.

FALIKS, D., MEYUHAS, O., 1982: Coordinate regulation of ribosomal protein mRNA level in regenerating rat liver. Study with the corresponding mouse cloned cDNAs. Nucl. Acids Res. **10**, 789—801.

FAN, H., PENMAN, S., 1971: Regulation of synthesis and processing of nucleolar components in metaphase-arrested cells. J. Mol. Biol. **59**, 27—42.

FODOR, I., BERIDZE, T., 1980: Structural organization of plant ribosomal DNA. Biochem. Internat. **1**, 493—501.

FOE, V. E., 1978: Modulation of ribosomal RNA synthesis in *Oncopeltus fasciatus*. An electron microscopic study of the relationship between changes in chromatin structure and transcriptional activity. Cold Spring Harbor Symp., Quant. Biol. **42**, 732—739.

— WILKINSON, L. E., LAIRD, C. D., 1976: Comparative organization of active transcription units in *Oncopeltus fasciatus*. Cell **9**, 131—146.

FONG, W.-F., FUCHS, M. S., 1976: The long term effect of α-amanitin of RNA synthesis in adult female *Aedes aegypti*. Insect Biochem. **6**, 123—130.

FORD, P. J., SOUTHERN, E. M., 1973: Different sequences for 5 S RNA in kidney cells and ovaries of *Xenopus laevis*. Nature (London), New Biol. **241**, 7—12.

FRANKE, W. W., SCHEER, U., SPRING, H., TRENDELENBURG, M. F., KROHNE, G., 1976: Morphology of transcriptional units of rDNA. Exp. Cell Res. **100**, 233—244.

— — TRENDELENBURG, M., ZENTGRAF, H., SPRING, H., 1978: Morphology of transcriptionally active chromatin. Cold Spring Harbor Symp. Quant. Biol. **42**, 755—772.

— — SPRING, H., TRENDELENBURG, M. F., ZENTGRAF, H., 1979: Organization of nucleolar chromatin. In: The Cell Nucleus (BUSCH, H., ed.), Vol. VII, pp. 49—95. New York: Academic Press.

— KLEINSCHMIDT, J. A., SPRING, H., KROHNE, G., GRUND, CH., TRENDELENBURG, M. F., STOEHR, M., SCHEER, U., 1981: A nucleolar skeleton of protein filaments demonstrated in amplified nucleoli of *Xenopus*. J. Cell Biol. **90**, 289—299.

FRANZ, G., KUNZ, W., 1981: Intervening sequences in ribosomal RNA genes and *bobbed* phenotype in *Drosophila hydei*. Nature **292**, 638—640.

FREE, S. J., RICE, P. W., METZENBERG, R. L., 1979: Arrangement of the genes coding for ribosomal ribonucleic acids in *Neurospora crassa*. J. Bacteriol. **137**, 121—1226.

FRIED, H. M., WARNER, J. R., 1982: Molecular cloning and analysis of yeast gene for cycloheximide resistance and ribosomal protein L 29. Nucl. Acids Res. **10**, 3133—3148.

— PEARSON, N. J., KIM, CH. H., WARNER, J. R., 1981: The genes for fifteen ribosomal proteins of *Saccharomyces cerevisiae*. J. Biol. Chem. **256**, 10176—10183.

FARBER, J. L., FARMAR, R., 1973: Differential effects of cycloheximide on protein and RNA synthesis as a function of dose. Biochem. Biophys. Res. Commun. **51**, 626—631.

FEDOROFF, N. V., 1979: On spacers. Cell **16**, 697—710.

— BROWN, D. D., 1978: The nucleotide sequence of oocyte 5 S DNA in *Xenopus laevis*: 1. The AT-rich spacer. Cell **13**, 701—716.

FERENCZ, A., SIEFART, K. H., 1975: Comparative effect of heparin on RNA synthesis of isolated rat liver nucleoli and purified RNA polymerase A. Eur. J. Biochem. **53**, 605—613.

FILES, J. G., HIRSH, D., 1981: Ribosomal DNA of *Caenorhabditis elegans*. J. Mol. Biol. **149**, 223—240.

FINANCSEK, I., MIZUMOTO, K., MURAMATSU, M., 1982 a: Nucleotide sequence of the transcription initiation region of a rat ribosomal RNA gene. Gene **18**, 115—122.

— — MISHIMA, Y., MURAMATSU, M., 1982 b: Human ribosomal RNA gene: Nucleotide sequence of the transcription initiation region and comparison of three mammalian genes. Proc. Nat. Acad. Sci. U.S.A. **79**, 3092—3096.

FINDLY, R. C., GALL, J. G., 1978: Free ribosomal RNA genes in *Paramecium* are tandemly repeated. Proc. Nat. Acad. Sci. U.S.A. **75**, 3312—3316.

Findly, R. C., Gall, J. G., 1980: Organization of ribosomal genes in *Paramecium tetraurelia*. J. Cell. Biol. **84**, 547—559.

Finkelstein, D. B., Blamire, J., Marmur, J., 1972: Location of rRNA cistrons in yeast. Nature (London), New Biol. **240**, 279—281.

Fiume, L., Laschi, L., 1965: Lesioni ultrastrutturali prodotte nelle cellule parenchimali epatiche dalla falloidina e dalla α-amanitina. Sper. Arch. Biol. Norm. Patol. **115**, 288—297.

— Wieland, T., 1970: Amanitins. Chemistry and action. FEBS Lett. **8**, 1—5.

Flavell, R. B., Martini, G., 1982: The genetic control of nucleolus formation with special reference to common breadwheat. In: The Nucleolus (Jordan, E. G., Cullis, C. A., eds.), pp. 113—128. Cambridge: Cambridge Univ. Press.

French, C. K., Fouts, D. L., Manning, J. E., 1979: Sequence arrangement of the rDNA of *Sarcophaga Bullata*. J. Cell Biol. **83**, 193 a.

Friedrich, H., Hemleben, V., Meagher, R., Key, J., 1979: Purification and restriction endonuclease mapping of soybean 18 S and 25 S ribosomal RNA genes. Planta **146**, 467—473.

Fujisawa, T., Abe, S., Kawaga, T., Satake, M., Ogata, K., 1973: Studies on the processing of 45 S RNA in rat liver nucleolus with specific reference to 29.5 S RNA. Biochim. Biophys. Acta **324**, 226—240.

— Imai, K., Tanaka, V., Ogata, K., 1979: Studies on the protein components of 110 S and total ribonucleoprotein particles of rat liver. J. Biochem. (Tokyo) **85**, 277—286.

Gabrielsen, O. S., Øyen, T. B., 1982: Yeast RNA polymerase I binds preferentially to A + T rich linkers in rDNA. Nucl. Acids Res. **10**, 5893—5904.

Gaetani, S., Mengheri, E., Scapin, S., Spadoni, M., 1977: Long-term protein deficiency and rat liver ribosome cycle. J. Nutr. **107**, 1035—1043.

Gajdardjieva, K. C., Dabeva, M. D., Chelibonova-Lorer, H., Hadjiolov, A. A., 1977: The use of D-galactosamine for a pulse-chase study of rRNA maturation in rat liver. FEBS Lett. **84**, 48—52.

— — Hadjiolov, A. A., 1980: Maturation and nucleo-cytoplasmic transport of rat liver rRNA upon D-galactosamine inhibition of transcription. Eur. J. Biochem. **104**, 451—458.

— — Markov, D. V., Hadjiolov, A. A., 1980: Isolation and characterization of D-galactosamine-induced nucleolar fibrillar remnants from rat liver. Compt. Rend. Acad. Sci. Bulgarie **33**, 977—980.

— Markov, D. V., Dimova, R. N., Kermekchiev, M. B., Todorov, I. T., Dabeva, M. D., Hadjiolov, A. A., 1982: Isolation and initial characterization of nucleolar fibrillar remnants from the liver of rats treated with D-galactosamine. Exp. Cell Res. **140**, 95—104.

Gajkowska, B., Puvion, E., Bernhard, W., 1977: Unusual perinucleolar accumulation of ribonucleoprotein granules induced by camptothecin in isolated liver cells. J. Ultrastr. Res. **60**, 335—347.

Galcheva-Gargova, Z., Petrov, P., Dessev, G. N., 1982: Effect of chromatin decondensation on the intranuclear matrix. Eur. J. Cell Biol. **28**, 155—159.

Galibert, F., Tiollais, P., Eladari, M. E., 1975: Fingerprint studies of the rRNA in mammalian cells. Eur. J. Biochem. **55**, 239—245.

GALL, J. G., 1968: Differential synthesis of the genes for rRNA during amphibian oogenesis. Proc. Nat. Acad. Sci. U.S.A. **60**, 553—559.

— 1969: The genes for rRNA during oogenesis. Genetics **61** (Suppl.) 121—132.

— 1974: Free ribosomal RNA genes in the macronucleus of *Tetrahymena*. Proc. Nat. Acad. Sci. U.S.A. **71**, 3078—3081.

— PARDUE, M. L., 1969: Formation and detection of RNA-DNA hybrid molecules in cytological preparations. Proc. Nat. Acad. Sci. U.S.A. **63**, 378—383.

— ROCHAIX, J. D., 1974: The amplified rDNA of *Dytiscid* beetles. Proc. Nat. Acad. Sci. U.S.A. **71**, 1819—1823.

GARIGLIO, P., BUSS, J., GREEN, M. H., 1974: Sarkosyl activation of RNA polymerase activity in mitotic mouse cells. FEBS Lett. **44**, 330—333.

GAZARYAN, K. G., ANANYANZ, T. G., FEDINA, A. B., ANDREEVA, N. B., 1973: Studies on the mechanism of genome inactivation in avian erythropoiesis. II. RNA polymerase and template activity of nuclei and chromatin (russ.). Mol. Biologiya **7**, 73—83.

GELFANT, S., 1981: Cycling-noncycling cell transitions in tissue aging, immunological surveillance, transformation and tumor growth. Int. Rev. Cytol. **70**, 1—25.

GENCHEV, D. D., KERMEKCHIEV, M. B., HADJIOLOV, A. A., 1980: Free pyrimidine nucleotide pool of Ehrlich ascites tumour cells. Biochem. J. **188**, 85—90.

GEORGIEV, O. I., NIKOLAEV, N., HADJIOLOV, A. A., SKRYABIN, K. G., ZAKHARYEV, V. M., BAYEV, A. A., 1981: The structure of the yeast rRNA genes. 4. Complete sequence of the 25 S rRNA gene from *Saccharomyces cerevisiae*. Nucl. Acids Res. **9**, 6853—6858.

— DUDOV, K. P., HADJIOLOV, A. A., SKRYABIN, K. G., 1983: Evidence for interaction of 5.8 S rRNA with the 5′ and 3′ terminal segments of *Saccharomyces cerevisiae* 25 S rRNA. Folia biol. (Praha) **29**, 510—522.

GERLACH, W. L., BEDBROOK, J. R., 1979: Cloning and characterization of rRNA genes from wheat and barley. Nucl. Acids Res. **7**, 1869—1879.

— DYER, T. A., 1980: Sequence organization of the repeating units in the nucleus of wheat which contain 5 S rRNA genes. Nucl. Acids Res. **8**, 4851—4865.

GEUSKENS, M., BERNHARD, W., 1966: Cytochimie ultrastructurale du nucléole. III. Action de l'actinomycine D sur le métabolisme du RNA nucléolaire. Exp. Cell Res. **44**, 579—598.

GEYER, P. K., MEYUHAS, O., PERRY, R., JOHNSON, L. F., 1982: Regulation of ribosomal protein mRNA content and translation in growth-stimulated mouse fibroblasts. Mol. Cell. Biol. **2**, 685—693.

GIMENEZ-MARTIN, G., RISUENO, M. C., FERNANDEZ-GOMEZ, M. E., AHMADIAN, P., 1973: Effect of cordycepin on the fine structure of interphase nucleoli in plant cells. Cytobiologie **7**, 181—192.

— TORRE, C. DE LA, LOPEZ-SAEZ, J. F., ESPONDA, P., 1977: Plant nucleolus: Structure and physiology. Cytobiologie **14**, 421—462.

GIRI, CH. P., GOROVSKY, M. A., 1980: DNase I sensitivity of ribosomal genes in isolated nucleosome core particles. Nucl. Acids Res. **8**, 197—214.

GIVENS, J. F., PHILLIPS, R. L., 1976: The nucleolus organizer region of maize (*Zea mays* L.). Chromosoma **57**, 103—117.

GLÄTZER, K. H., 1975: Visualization of gene transcription in spermatocytes of *Drosophila hydei*. Chromosoma **53**, 371—379.

— 1980: Regular substructures within homologous transcripts of spread *Drosophila* germ cells. Exp. Cell Res. **125**, 519—523.

GLAZER, R. L., 1977: The action of N-hydroxy-2-acetylaminofluorene on the synthesis of ribosomal and poly(A)-RNA in normal and regenerating liver. Biochim. Biophys. Acta **475**, 492—500.

GLOTZ, C., ZWIEB, C., BRIMACOMBE, R., EDWARDS, K., KÖSSEL, H., 1981: Secondary structure of the large subunit rRNA from *E. coli, Zea Mays* chloroplast and human and mouse mitochondrial ribosomes. Nucl. Acids Res. **9**, 3287—3306.

GLOVER, D. M., 1977: Cloned segment of *Drosophila melanogaster* containing new types of sequence insertion. Proc. Nat. Acad. Sci. U.S.A. **74**, 4932—4936.

— HOGNESS, D. S., 1977: A novel arrangements of the 18 S and 28 S sequences in a repeating unit of *Drosophila melanogaster*. Cell **10**, 167—176.

— WHITE, R. L., FINNEGAN, D. J., HOGNESS, D. C., 1975: Characterization of six cloned DNAs from *Drosophila melanogaster* including one that contains the genes for rRNA. Cell **5**, 149—157.

GOESSENS, G., 1976: High resolution autoradiographic studies of Ehrlich tumor cell nucleoli. Exp. Cell Res. **100**, 88—94.

— LEPOINT, A., 1974: The fine structure of the nucleolus during interphase and mitosis in Ehrlich ascites tumor cells cultivated *in vitro*. Exp. Cell Res. **87**, 63—72.

— — 1979: The nucleolus organizing regions (NOR's): Recent data and hypothesis. Biol. Cell **35**, 211—220.

— — 1982: Localization of AgNOR proteins in Ehrlich tumor cell nucleoli. Biol. Cell **43**, 139—142.

GOLDBERG, I. H., FRIEDMAN, P. A., 1971: Antibiotics and nucleic acids. Ann. Rev. Biochem. **40**, 775—805.

GOLDBLATT, P. J., ARCHER, J., EASTWOOD, C., 1975: The effect of high and low doses of cycloheximide on nucleolar RNA synthesis. Lab. Invest. **33**, 117—123.

GOLDMAN, W. E., GOLDBERG, G., BOWMAN, L. H., STEINMETZ, D., SCHLESSINGER, D., 1983: Mouse rDNA: Sequences and evolutionary analysis of spacer and mature RNA regions. Mol. Cell. Biol. **3**, 1488—1500.

GOLDSBROUGH, P. B., CULLIS, C. A., 1981: Characterization of the genes for ribosomal RNA in flax. Nucl. Acids Res. **9**, 1301—1309.

— ELLIS, T. H. N., CULLIS, C. A., 1981: Organization of 5 S rRNA genes in flax. Nucl. Acids Res. **9**, 5895—5904.

— — LOMONOSSOFF, G. P., 1982: Sequence variation and methylation of the flax 5 S rRNA genes. Nucl. Acids Res. **10**, 4501—4514.

GOLDSTEIN, L., KO, C., 1981: Distribution of proteins between nucleus and cytoplasm of *Amoeba proteus*. J. Cell Biol. **88**, 516—525.

GOODPASTURE, C., BLOOM, S. E., 1975: Visualization of nuclear organizer region in mammalian chromosomes using silver staining. Chromosoma **53**, 37—50.

GORCHAKOVA, G. A., SIDORENKO, A. D., 1976: Nuclear ribonucleases and post-transcriptional processing of RNA (russ.). Biokhimiya **41**, 630—638.

GORENSTEIN, C. G., WARNER, J. R., 1976: Coordinate regulation of the synthesis of eucaryotic ribosomal proteins. Proc. Nat. Acad. Sci. U.S.A. **73**, 1547—1550.

GORENSTEIN, C. G., WARNER, J. R., 1977: Synthesis and turnover of ribosomal proteins in the absence of 60 S subunit assembly in *Saccharomyces cerevisiae*. Mol. Gen. Genet. **157**, 327—333.

— — 1979: The monocistronic nature of ribosomal protein genes in yeast. Current Genet. **1**, 9—15.

GOTOH, S., NIKOLAEV, N., BATTANER, E., BIRGE, C. H., SCHLESSINGER, D., 1974: *E. coli* RNAse III cleaves HeLa cell nuclear RNA. Biochem. Biophys. Res. Commun. **59**, 972—978.

GOTTESFELD, J., BLOOMER, L. S., 1982: Assembly of transcriptionally active 5 S rRNA gene chromatin *in vitro*. Cell **28**, 781—791.

GOURSE, R. L., GERBI, S. A., 1980 a: Fine structure of rRNA. III. Location of evolutionarily conserved regions within rDNA. J. Mol. Biol. **140**, 321—339.

— — 1980 b: Fine structure of ribosomal RNA. IV. Extraordinary evolutionary conservation in sequences that flank introns in rDNA. Nucl. Acids Res. **8**, 3623—3637.

GRAHAME-SMITH, D. G., ISAAC, P., HEAL, D. J., BOND, R. P. M., 1975: Inhibition of adenyl cyclase by an exotoxin of *Bacillus thuringiensis*. Nature (London) **253**, 58—60.

GRAINGER, R. M., MAIZELS, N., 1980: *Dictyostelium* ribosomal RNA is processed during transcription. Cell **20**, 619—623.

— OGLE, R. C., 1978: Chromatin structure of rRNA genes in *Physarum polycephalum*. Chromosoma **65**, 115—126.

GRANBOULAN, N., GRANBOULAN, P., 1965: Cytochimie ultrastructurale du nucléole. II. Etude des sites de synthèse du RNA dans le nucléole et le noyau. Exp. Cell Res. **38**, 604—619.

GRAZIANI, F., CAIZZI, R., GARGANO, S., 1977: Circular ribosomal DNA during ribosomal magnification in *Drosophila melanogaster*. J. Mol. Biol. **112**, 49—63.

GREEN, CH. J., KAMMEN, H. O., PENHOET, E. E., 1982: Purification and properties of a mammalian tRNA pseudouridine synthase. J. Biol. Chem. **257**, 3045—3052.

GREEN, M. H., BUSS, J., GARIGLIO, P., 1975: Activation of nuclear RNA polymerase by Sarkosyl. Eur. J. Biochem. **53**, 217—225.

GROOPMAN, J. D., BUSBY, W. F., JR., WOGAN, G. N., 1980: Nuclear distribution of aflatoxin B_1 and its interaction with histones in rat liver *in vivo*. Cancer Res. **40**, 4343—4351.

GROSS, R. H., RINGLER, J., 1979: Ribonucleic acid synthesis in isolated *Drosophila* nuclei. Biochemistry **18**, 4923—4927.

GRUCA, S., KRZYZOWSKA-GRUCA, S., VORBRODT, A., KRAWCZYK, Z., 1978: Intranucleolar localization of the RNA polymerase A activity in isolated nuclei of regenerating rat liver. Exp. Cell Res. **114**, 462—467.

GRUISSEM, W., SEIFART, K. H., 1982: Transcription of 5 S RNA genes *in vitro* is feedback inhibited by HeLa 5 S RNA. J. Biol. Chem. **257**, 1468—1472.

— KOTZERKE, M., SEIFART, K. H., 1981: Transcription of the cloned genes for ribosomal 5 S RNA in a system reconstituted *in vitro* from HeLa cells. Eur. J. Biochem. **117**, 407—415.

GRUMMT, F., GRUMMT, I., MAYER, E., 1979: Ribosome biosynthesis is not necessary for initiation of DNA replication. Eur. J. Biochem. **97**, 37—42.

GRUMMT, I., 1977: The effects of histidine starvation on the methylation of ribosomal RNA. Eur. J. Biochem. **79**, 133—141.
— 1978: *In vitro* synthesis of pre-rRNA in isolated nucleoli. In: The Cell Nucleus (BUSCH, H., ed.), Vol. V, pp. 373—414. New York: Academic Press.
— 1981 a: Mapping of a mouse ribosomal DNA promoter by *in vitro* transcription. Nucl. Acids Res. **9**, 6093—6102.
— 1981 b: Specific transcription of mouse ribosomal DNA in a cell-free system that mimics control *in vivo*. Proc. Nat. Acad. Sci. U.S.A. **78**, 727—731.
— 1982: Nucleotide sequence requirements for specific initiation of transcription by RNA polymerase I. Proc. Nat. Acad. Sci. U.S.A. **79**, 6908—6911.
— GROSS, H. J., 1980: Structural organization of mouse rDNA: Comparison of transcribed and non-transcribed regions. Molec. Gen. Genet. **177**, 223—229.
— SMITH, V. A., GRUMMT, F., 1976: Amino acid starvation affects the initiation frequency of nucleolar RNA polymerase. Cell **7**, 439—445.
— SOELLNER, C., SCHOLZ, I., 1979 a: Characterization of a cloned ribosomal fragment from mouse which contains the 18 S coding region and adjacent spacer sequences. Nucl. Acids Res. **6**, 1351—1369.
— HALL, S. H., CROUCH, R. J., 1979 b: Localization of an endonuclease specific for double-stranded RNA within the nucleolus and its implication in processing ribosomal transcripts. Eur. J. Biochem. **94**, 437—443.
— ROTH, E., PAULE, M., 1982: Ribosomal RNA transcription *in vitro* is species specific. Nature **296**, 173—174.
GRUNDMANN, E., FECHLER, W., 1965: RNS-Gehalt und Volumen der Nucleolen in der Rattenleber während experimenteller Carcinogenesis durch Diäthylnitrosamin. Z. Krebsforschg. **67**, 80—92.

HACKETT, P. B., EGBERTS, E., TRAUB, P., 1978: Characterization of Ehrlich ascites tumor cell messenger RNA specifying ribosomal proteins by translation *in vitro*. J. Mol. Biol. **119**, 253—267.
HADJIOLOV, A. A., 1966: Turnover and messenger activity of rat liver ribonucleic acids. Biochim. Biophys. Acta **119**, 547—556.
— 1977: Patterns of ribosome biogenesis in eukaryotes. Trends Biochem. Sci. **2**, 84—86.
— 1980: Biogenesis of ribosomes in eukaryotes. In: Subcellular Biochemistry (ROODYN, D. B., ed.), Vol. 7, pp. 1—80. New York: Plenum Press.
— COX, R. A., 1973: A spectrophotometric study of secondary structure of pre-rRNA from ascites tumor cells. Biochem. J. **135**, 349—351.
— HADJIOLOVA, K. V., 1979: The effect of 5-fluoropyrimidines on the processing of ribonucleic acids in liver. In: Antimetabolites in Biochemistry, Biology, and Medicine (SKODA, J., LANGEN, P., eds.), pp. 77—85. New York: Pergamon Press.
— MILCHEV, G. I., 1974: Synthesis and maturation of rRNA in isolated HeLa cell nuclei. Biochem. J. **142**, 263—272.
— NIKOLAEV, N., 1976: Maturation of ribosomal ribonucleic acids and the biogenesis of ribosomes. Progr. Biophys. Mol. Biol. **31**, 95—144.
— TENCHEVA, Z. S., 1958: Über die Veränderungen der Nukleinsäuren in der Leber

von Ratten bei der Cancerogenese durch 4-Dimethylaminoazobenzol. Z. Krebsforschg. **62**, 361—369.

Hadjiolov, A. A., Venkov, P. V., Tsanev, R. G., 1966: Ribonucleic acids fractionation by density-gradient centrifugation and by agar gel electrophoresis: A comparison. Anal. Biochem. **17**, 263—267.

— Dabeva, M. D., Mackedonski, V. V., 1974a: The action of α-amanitin *in vivo* on the synthesis and maturation of mouse liver ribonucleic acids. Biochem. J. **138**, 321—334.

— Hadjiolova, K. V., Nikolova, R., Emanuilov, I., 1974b: Evidence that the synthesis and nucleo-cytoplasmic transfer of liver messenger-like RNA is independent of rRNA maturation. Int. J. Biochem. **5**, 353—358.

— Dabeva, M. D., Dudov, K. P., Gajdardjieva, K. C., Georgiev, O. I., Nikolaev, N., Stoyanova, B. B., 1978: Control of ribosomal RNA processing in eukaryotes. FEBS Symposia **51**, 319—328.

— Georgiev, O. I., Nosikov, V. V., Yavachev, L. P., 1984: Primary and secondary structure of rat 28 S ribosomal RNA. Nucl. Acids Res. **12**, 3677—3693.

Hadjiolova, K. V., Golovinsky, E. V., Hadjiolov, A. A., 1973: The site of action of 5-fluoroorotic acid on the maturation of mouse liver ribonucleic acids. Biochim. Biophys. Acta **319**, 373—382.

— Naydenova, Z. G., Hadjiolov, A. A., 1981: Inhibition of rRNA maturation in Friend erythroleukemia cells by 5-fluorouridine and toyocamycin. Biochem. Pharmacol. **30**, 1861—1863.

— Georgiev, O. I., Hadjiolov, A. A., 1984a: Excess 5′-terminal sequences in the rat nucleolar 28 S ribosomal RNA. Exp. Cell Res. **153**, 266—270.

— — Nosikov, V. V., Hadjiolov, A. A., 1984b: Localization and structure of endonuclease cleavage sites involved in the processing of the rat 32 S precursor to rRNA. Biochem. J. **220**, 105—116.

Haim, L., Iapalucci-Espinoza, S., Conde, R., Franze-Fernandez, M. T., 1983: Control of activation of liver RNA polymerase I occurring after re-feeding of protein-depleted mice. Biochem. J. **210**, 837—844.

Hall, L., Braun, R., 1977: The organization of genes for tRNA and rRNA in amoebae and plasmodia of *Physarum polycephalum*. Eur. J. Biochem. **76**, 165—174.

Hall, L. M. C., Maden, B. E. H., 1980: Nucleotide sequence through the 18 S–28 S intergene region of a vertebrate ribosomal transcription unit. Nucl. Acids Res. **8**, 5993—6005.

Hall, T. J., Cummings, M. R., 1975: *In vitro* synthesis and processing of rRNA in the housefly ovary. Dev. biol. **46**, 233—242.

Hallberg, R. L., Brown, D. D., 1969: Coordinated synthesis of some ribosomal proteins and rRNA in embryos of *Xenopus laevis*. J. Mol. Biol. **46**, 393—411.

Hancock, R., 1982: Topological organization of interphase DNA: the nuclear matrix and other skeletal structures. Biol. Cell **46**, 105—122.

— Boulikas, T., 1982: Functional organization of the nucleus. Int. Rev. Cytol. **79**, 165—214.

Harford, A. G., Zuchowski, C. I., 1977: The effect of X chromosome heterozygosity on the structure of ribosomal genes in *Drosophila melanogaster*. Cell **11**, 389—394.

Harrington, C. A., Chikaraishi, D. M., 1983: Identification and sequence of the initiation site for rat 45 S ribosomal RNA synthesis. Nucl. Acids Res. **11**, 3317—3332.

Hastie, N. D., Mahy, B. W. J., 1973: Effects of α-amanitin *in vivo* on RNA polymerase activity of cultured chick fibroblast cell nuclei: resistance of ribosomal RNA synthesis to the drug. FEBS Lett. **32**, 95—99.

Hatlen, L., Attardi, G., 1971: Proportion of the HeLa genome complementary to tRNA and 5 S RNA. J. Mol. Biol. **56**, 535—553.

Hay, E. D., Gurdon, J. B., 1967: Fine structure of the nucleolus in normal and mutant *Xenopus* embryos. J. Cell Sci. **2**, 151—162.

Hayashi, Y., Kominami, R., Muramatsu, M., 1977: Effect of cycloheximide on the synthesis and processing of 5 S rRNA in HeLa. J. Biochem. (Tokyo) **81**, 451—459.

Hayflick, L., 1965: The limited *in vitro* lifetime of human diploid cell strains. Exp. Cell Res. **37**, 614—636.

— Moorhead, P. S., 1961: The serial cultivation of human diploid cell strains. Exp. Cell Res. **25**, 585—621.

Heady, J. E., McConkey, E. H., 1971: Completion of nascent HeLa ribosomal proteins in a cell free system. Biochem. Biophys. Res. Commun. **40**, 30—36.

Heitz, E., 1931: Nukleolen und Chromosomen in der Gattung *Vicia*. Planta **15**, 495—505.

— 1933: Über totale und partielle somatische Heteropyknose, sowie strukturelle Geschlechtschromosomen bei *Drosophila funebris*. Z. Zellforschg. Mikr. Anat. **19**, 720—742.

Henderson, A. S., Warburton, D., Atwood, K. C., 1972: Location of rDNA in the human chromosome complement. Proc. Nat. Acad. Sci. U.S.A. **69**, 3394—3398.

— Echer, E. M., Vu, M. T., Atwood, K. C., 1974: The chromosomal location of ribosomal DNA in the mouse. Chromosoma **49**, 155—160.

— Atwood, K. C., Warburton, D., 1976 a: Chromosomal distribution of rDNA in *Pan paniscus, Gorilla gorilla beringei*, and *Symphalangus syndactylus*: comparison to related primates. Chromosoma **59**, 147—155.

— — Yu, M. T., Warburton, D., 1976 b: The site of 5 S RNA genes in primates. Chromosoma **56**, 29—32.

Henshaw, E. C., Guiney, D. C., Hirsch, C. A., 1973: The ribosome cycle in mammalian protein synthesis. I. The place of monomeric ribosomes and ribosomal subunits in the cycle. J. Biol. Chem. **248**, 4367—4376.

Hentschel, C. C., Tata, J. R., 1977: Differential activation of free and template-engaged RNA polymerase I and II during the resumption of development of dormant *Artemia* gastrulae. Dev. Biol. **57**, 293—304.

Herlan, G., Eckert, W. A., Kaffenberger, W., Wunderlich, F., 1979: Isolation and characterization of an RNA-containing nuclear matrix from *Tetrahymena* macronuclei. Biochemistry **18**, 1782—1788.

Hernandez-Verdun, D., 1983: The nucleolus organizer regions. Biol. Cell **49**, 191—202.

HERNANDEZ-VERDUN, D., BOUTEILLE, M., 1979: Nucleologenesis in chick erythrocyte nuclei reactivated by cell fusion. J. Ultrastr. Res. **69**, 164—179.

— HUBERT, J., BOURGEOIS, C., BOUTEILLE, M., 1978: Identification ultrastructurale de l'organisateur nucléolaire par la technique à l'argent. C. R. Acad. Sci. (Paris) **287**, 1421—1423.

— BOUTEILLE, M., EGE, T., RINGERTZ, N. R., 1979: Fine structure of nucleoli in micronucleated cells. Exp. Cell Res. **124**, 223—235.

— HUBERT, J., BOURGEOIS, C. A., BOUTEILLE, M., 1980: Ultrastructural localization of Ag-NOR stained proteins in the nucleolus during the cell cycle and in other nucleolar structures. Chromosoma (Berl.) **79**, 349—362.

— DERENZINI, M., BOUTEILLE, M., 1982: The morphological relationship in electron microscopy between NOR-silver proteins. Chromosoma (Berl.) **85**, 461—473.

HERSHEY, N. D., CONRAD, S. E., SODJA, A., YEN, P. H., COHEN, M., DAVIDSON, N., 1977: The sequence arrangement of *Drosophila melanogaster* 5 S DNA cloned in recombinant plasmids. Cell **11**, 585—598.

HERZOG, J., FARBER, J. L., 1975: Fibrillar nucleolar remnants do not contain macromolecular precursors of ribosomal RNA. Demonstration by the effects of D-galactosamine. Exp. Cell Res. **93**, 502—505.

HIGASHI, K., HANASAKI, N., NAKANISHI, A., SHIMOMOURA, E., HIRANO, H., GOTOH, S., SAKAMOTO, Y., 1978: Difference in susceptibility to sonication of chromatins containing transcriptionally active and inactive ribosomal genes. Biochim. Biophys. Acta **520**, 612—622.

HIGASHINAKAGAWA, T., WAHN, H., REEDER, R. H., 1977: Isolation of ribosomal gene chromatin. Dev. Biol. **55**, 375—386.

— SAIGA, H., SHINTANI, N., NARUSHIMA-IIO, M., MITA, T., 1981: Localization of putative transcription initiation site on the cloned rDNA fragment of *Tetrahymena pyriformis*. Nucl. Acids Res. **9**, 5905—5916.

HILDEBRANDT, A., SAUER, H. W., 1976: Levels of RNA polymerases during the mitotic cycle of *Physarum polycephalum*. Biochim. Biophys. Acta **425**, 316—321.

HIPSKIND, R. A., REEDER, R. H., 1980: Initiation of ribosomal RNA chains in homogenates of oocyte nuclei. J. Biol. Chem. **255**, 7896—7906.

HIRSCH, C. A., HIATT, H. H., 1966: Turnover of liver ribosomes in fed and in fasted rats. J. Biol. Chem. **241**, 5936—5948.

HIRSCH, J., MARTELO, O. J., 1976: Phosphorylation of rat liver RNA polymerase I by nuclear protein kinases. J. Biol. Chem. **251**, 5408—5413.

HOLLINGER, T. G., SMITH, L. D., 1976: Conservation of RNA polymerase during maturation of the *Rana pipiens* oocyte. Dev. Biol. **51**, 86—97.

HONJO, T., REEDER, R. H., 1973: Preferential transcription of *Xenopus laevis* rRNA in interspecies hybrids between *X. laevis* and *X. mulleri*. J. Mol. Biol. **80**, 217—225.

HORI, H., OSAWA, S., 1979: Evolutionary change in 5 S RNA secondary structure and a phylogenic tree of 54 5 S RNA species. Proc. Nat. Acad. Sci. U.S.A. **76**, 381—385.

HOROWITZ, B., GOLDFINGER, B. A., MARMUR, J., 1976: Effect of cordycepin triphosphate on the nuclear DNA-dependent RNA polymerases and poly (A) polymerase from the yeast *Saccharomyces cerevisiae*. Arch. Biochem. Biophys. **172**, 143—148.

Hoshikawa, Y., Iida, Y., Iwabuchi, M., 1983: Nucleotide sequence of the transcriptional initiation region of *Dictyostelium discoideum* rRNA gene and comparison of the initiation regions of three lower eucaryotes' genes. Nucl. Acids Res. **11**, 1725—1733.

Hourcade, D., Dressler, D., Wolfson, J., 1973: The amplification of rRNA genes involves a rolling circle intermediate. Proc. Nat. Acad. Sci. U.S.A. **70**, 2926—2930.

Howell, X. M., 1977: Visualization of ribosomal gene activity: silver stains proteins associated with rRNA transcribed from oocyte chromosomes. Chromosoma **62**, 361—367.

— Denton, T. E., Diamond, J. R., 1975: Differential staining of the satellite regions of human acrocentric chromosomes. Experientia **31**, 260—262.

Hsu, T. C., Arrighi, F. E., Klevecz, R. R., Brinkley, B. R., 1965: The nucleoli in mitotic division of mammalian cells *in vitro*. J. Cell Biol. **26**, 539—553.

— Brinkley, B. R., Arrighi, F. E., 1967: The structure and behaviour of the nucleolus organizers in mammalian cells. Chromosoma **23**, 137—153.

— Spirito, S. E., Pardue, M. L., 1975: Distribution of 18 + 28 S ribosomal genes in mammalian genomes. Chromosoma **53**, 25—36.

Hubbell, H. R., Rothblum, L. I., Hsu, T. C., 1979: Identification of a silver binding protein associated with the cytological silver staining of actively transcribing nucleolar regions. Cell Biol. Int. Rep. **3**, 615—622.

Huet, J., Phalente, L., Buttin, G., Sentenac, A., Fromageot, P., 1982 a: Probing yeast RNA polymerase A subunits with monospecific antibodies. EMBO J. **1**, 1193—1198.

— Sentenac, A., Fromageot, P., 1982 b: Spot-immunodetection of conserved determinants in eucaryotic RNA polymerases. J. Biol. Chem. **257**, 2613—2618.

Hughes, D. G., Maden, B. E. H., 1978: The pseudouridine contents of the rRNA of three vertebrate species. Biochem. J. **171**, 781—786.

Hunt, J. A. 1976: Ribonucleic acid synthesis in rabbit erythroid cells. Analysis of rates of synthesis of nuclear and cytoplasmic RNA. Biochem. J. **160**, 727—744.

Hwang, K. M., Yang, L. C., Carrico, C. K., Schulz, R. A., Schenkman, J. B., Sartorelli, A. C., 1974: Production of membrane whorls in rat liver by some inhibitors of protein synthesis. J. Cell Biol. **62**, 20—31.

Hyman, L. E., McCusker, J., Hereford, L., Haber, J. E., Rosbash, M., 1980: Mapping of ribosomal protein genes in the yeast *Saccharomyces cerevisiae*. Eur. J. Cell Biol. **22**, 11—18.

Ide, T., Whelly, S., Baserga, R., 1977: Stimulation of RNA synthesis in isolated nuclei by partially purified preparations of SV 40 T-antigen. Proc. Nat. Acad. Sci. U.S.A. **74**, 3189—3192.

Ignotz, G. G., Hokari, S., DePhilip, R. M., Tsukada, K., Lieberman, I., 1981: Lodish model and regulation of ribosomal protein synthesis by insulin-deficient chick embryo fibroblasts. Biochemistry **20**, 2550—2558.

INDIK, Z. K., TARTOF, K. D., 1982: Glutamate tRNA genes are adjacent to 5 S RNA genes in *Drosophila* and reveal a conserved upstream sequence (the ACT-TA box). Nucl. Acids Res. **10**, 4159—4172.

ISHIKAWA, H., 1975: Polynucleotide fragments from the 28 S ribosomal RNA of insects. Nucl. Acids Res. **2**, 87—100.

— 1979: Re-joining of the 18 S fragments dissociated from the 28 S ribosomal RNA of insect: A structural role of 5.8 S RNA. Biochem. Biophys. Res. Commun. **90**, 417—424.

ISRAELEWSKI, N., SCHMIDT, E. R., 1982: Spacer size heterogeneity in rDNA of *Chironomus thummi* is due to a 120 bp repeat homologous to a predominantly centromeric repeated sequence. Nucl. Acids Res. **10**, 7689—7700.

JACOB, S. T., 1973: Mammalian RNA polymerases. Progr. Nucl. Acid Res. Mol. Biol. **13**, 93—126.

— ROSE, K. M., 1980: Basic enzymology of transcription in procaryotes and eucaryotes. In: Cell Biology (GOLDSTEIN, L., PRESCOTT, D. M., eds.), Vol. 3, pp. 113—152. New York: Academic Press.

— SAJDEL, E. M., MUECKE, W., MUNRO, H. N., 1970: Soluble RNA polymerases of rat liver nuclei: properties, template specificity and amanitin response *in vitro* and *in vivo*. Cold Spring Harbor Symp. Quant. Biol. **35**, 681—691.

JACQ, B., 1981: Sequence homologies between eucaryotic 5.8 S rRNA and the 5'-end of procaryotic 23 S rRNA: evidence for a common evolutionary origin. Nucl. Acids Res. **9**, 2913—2920.

— JOURDAN, R., JORDAN, B. R., 1977: Structure and processing of precursor 5 S RNA in *Drosophila melanogaster*. J. Mol. Biol. **117**, 785—795.

— MILLER, J. R., BROWNLEE, G. G., 1977: A pseudogene structure in 5 S DNA of *Xenopus laevis*. Cell **12**, 109—120.

JAEHNING, J. A., STEWART, C. C., ROEDER, R. G., 1975: DNA-dependent RNA polymerase levels during the response of human peripheral lymphocytes to phytohaemagglutinin. Cell **4**, 51—57.

JAYE, M., WU, F.-S., LUCAS-LENARD, J., 1980: Inhibition of synthesis of ribosomal proteins and of ribosome assembly after infection of L cells with vesicular stomatitis virus. Biochim. Biophys. Acta **606**, 1—12.

JELINEK, W., GOLDSTEIN, L., 1973: Isolation and characterization of some of the proteins that shuttle between cytoplasm and nucleus in *Amoeba proteus*. J. Cell Physiol. **81**, 181—197.

JENDRISAK, J., 1980: The use of α-amanitin to inhibit *in vivo* RNA synthesis and germination in wheat embryos. J. Biol. Chem. **255**, 8529—8533.

JOHN, H. A., BIRNSTIEL, M. L., JONES, K. W., 1969: RNA-DNA hybrids at the cytological level. Nature **223**, 582—587.

JOHNSON, E. M., 1980: A family of inverted repeat sequences and specific single-strand gaps at the termini of the *Physarum* rDNA palindrome. Cell **22**, 875—886.

— ALLFREY, V. G., BRADBURY, E. M., MATTEWS, H. R., 1978 a: Altered nucleosome structure containing DNA sequences complementary to 19 S and 26 S rRNA in *Physarum polycephalum*. Proc. Nat. Acad. Sci. U.S.A. **75**, 1116—1120.

JOHNSON, E. M., MATTEWS, H. R., LITTAU, V. C., LOTHSTEIN, L., BRADBURY, E. M., ALLFREY, V. G., 1978 b: The structure of chromatin containing DNA complementary to 19 S and 26 S rRNA in active and inactive stages of *Physarum polycephalum*. Arch. Biochem. Biophys. **191**, 537—550.

— CAMPBELL, G. R., ALLFREY, V. G., 1979: Different nucleosome structures on transcribing and nontranscribing ribosomal gene sequences. Science **206**, 1192—1194.

JOHNSON, L. D., HENDERSON, A. S., ATWOOD, K. C., 1974: Location of the genes for 5 S RNA in the human chromosome complement. Cytogenet. Cell Genet. **13**, 103—105.

JOHNSON, L. F., WILLIAMS, J. G., ABELSON, H. T., GREEN, H., PENMAN, S., 1975: Changes in RNA in relation to growth of the fibroblast. III. Posttranscriptional regulation of mRNA formation in resting and growing cells. Cell **4**, 69—75.

JOHNSON, T. C., HOLLAND, J. J., 1965: Ribonucleic acid and protein synthesis in mitotic HeLa cells. J. Cell Biol. **27**, 565—574.

JOHNSON, T. R., KUMAR, A., 1977: Ribosome processing in HeLa cells. Studies on structural aspects of precursor and mature ribosomes. J. Cell Biol. **73**, 419—427.

JOLLY, D. J., THOMAS, C. A., JR., 1980: Nuclear RNA transcripts from *Drosophila melanogaster* rRNA genes containing introns. Nucl. Acids Res. **8**, 67—84.

JONES, K. W., 1965: The role of the nucleolus in the formation of ribosomes. J. Ultrastr. Res. **13**, 257—265.

JONES, R. W., 1978 a: Preparation of chromatin containing rDNA from the macronucleus of *Tetrahymena pyriformis*. Biochem. J. **173**, 145—154.

— 1978 b: Histone composition of a chromatin fraction containing rDNA isolated from the macronucleus of *Tetrahymena pyriformis*. Biochem. J. **173**, 155—164.

JORDAN, E. G., MCGOVERN, J., 1981: The quantitative relationship of the fibrillar centres and other nucleolar components to changes in growth conditions, serum deprivation and low doses of actinomycin D in cultured diploid human fibroblasts (strain MRC-5). J. Cell Sci. **52**, 373—389.

JORDAN, B. R., LATIL-DAMOTTE, M., JOURDAN, R., 1980 a: Sequence of the 3′-terminal portion of *Drosophila melanogaster* 18 S rRNA and of the adjoining spacer. FEBS Lett. **117**, 227—231.

— — — 1980 b: Coding and spacer sequences in the 5.8 S–2 S region of *Sciara coprophila* rDNA. Nucl. Acids Res. **8**, 3565—3573.

KABACK, D. B., DAVIDSON, N., 1980: Organization of the ribosomal RNA gene cluster in the yeast *Saccharomyces cerevisiae*. J. Mol. Biol. **138**, 745—754.

— BHARGAVA, M. M., HALVORSON, H. O., 1973: Location and arrangement of genes coding for rRNA in *Saccharomyces cerevisiae*. J. Mol. Biol. **79**, 735—739.

KAFTORY, A., HERSHKO, A., FRY, M., 1978: Protein turnover in senescent cultured chick embryo fibroblasts. J. Cell Physiol. **94**, 147—160.

KAN, N. C., GALL, J. G., 1981: Sequence homology near the center of the extrachromosomal rDNA palindrome in *Tetrahymena*. J. Mol. Biol. **153**, 1151—1155.

KARAGYOZOV, L. K., HADJIOLOV, A. A., 1981: Isolation of active transcription complexes from animal cell nuclei by nitrocellulose binding. J. Biochem. Biophys. Methods **5**, 329—339.

KARAGYOZOV, L. K., VALKANOV, M., HADJIOLOV, A. A., 1978: Transcription of DNA-histone complexes by yeast RNA polymerase B. Nucl. Acid Res. **5**, 1907—1917.

— STOYANOVA, B. B., HADJIOLOV, A. A., 1980: Effect of cycloheximide on the *in vivo* and *in vitro* synthesis of ribosomal RNA in rat liver. Biochim. Biophys. Acta **607**, 295—303.

KARASAKI, S., 1965: Electron microscopic examination of the sites of nuclear RNA synthesis during amphibian embryogenesis. J. Cell Biol. **26**, 937—958.

— 1969: The fine structure of proliferating cells in preneoplastic rat livers during azo-dye carcinogenesis. J. Cell Biol. **40**, 322—335.

KARRER, K., GALL, J., 1976: The macronuclear rDNA of *Tetrahymena pyriformis* is a palindrome. J. Mol. Biol. **104**, 421—453.

KAUFMANN, S. H., COFFEY, D. S., SHAPER, J. H., 1981: Considerations in the isolation of rat liver nuclear matrix, nuclear envelope and pore complex lamina. Exp. Cell Res. **132**, 105—123.

KAY, J. E., 1968: Early effects of phytohaemagglutinin in lymphocyte RNA synthesis. Eur. J. Biochem. **4**, 225—232.

KEDINGER, C., SIMARD, R., 1974: The action of α-amanitin on RNA synthesis in Chinese hamster ovary cells. J. Cell Biol. **63**, 831—842.

— GNIAZDOWSKI, M., MANDEL, J. L., GISSINGER, F., CHAMBON, P., 1970: α-Amanitin: a specific inhibitor of one of two DNA dependent RNA polymerase activities from calf thymus. Biochem. Biophys. Res. Commun. **38**, 165—171.

KELLY, J. M., COX, R. A., 1981: The nucleotide sequence at the 3′-end of *Neurospora crassa* 25 S rRNA and the location of a 5.8 S rRNA binding site. Nucl. Acids Res. **9**, 1111—1118.

KEPPLER, D., RUDIGER, J., BISCHOFF, E., DECKER, K., 1970: The trapping of uridine phosphates by D-galactosamine, D-glucosamine and 2-deoxy-D-galactose. Eur. J. Biochem. **17**, 246—253.

— PAUSCH, J., DECKER, K., 1974: Selective uridine triphosphate deficiency induced by D-galactosamine in liver and reversed by pyrimidine nucleotide precursors. J. Biol. Chem. **249**, 211—216.

KHAN, M. S. N., MADEN, B. E. H., 1976: Nucleotide sequences within the ribosomal ribonucleic acid of HeLa cells, *Xenopus laevis*, and chick embryo fibroblasts. J. Mol. Biol. **101**, 235—254.

— SALIM, M., MADEN, B. E. H., 1978: Extensive homologies between the methylated nucleotide sequences in several vertebrate rRNA. Biochem. J. **169**, 531—542.

KHANN, M. M., LINDELL, T. J., 1980: Actinomycin D binds with highest affinity to nonribosomal DNA. J. Biol. Chem. **255**, 3581—3584.

KIDD, S. J., GLOVER, D. M., 1980: A DNA segment from *D. melanogaster* which contains five tandemly repeating units homologous to the major rDNA insertion. Cell **19**, 103—110.

— — 1981: *Drosophila melanogaster* ribosomal DNA containing type II insertions is variably transcribed in different strains and tissues. J. Mol. Biol. **151**, 645—662.

KIEF, D. R., WARNER, J. R., 1981: Coordinate control of synthesis of rRNA and ribosomal proteins during nutritional shift-up in *Saccharomyces cerevisiae*. Mol. Cell. Biol. **1**, 1007—1015.

KIERSZENBAUM, A. L., TRES, L. L., 1975: Structural and transcriptional features of the mouse spermatid genome. J. Cell Biol. **65**, 258—270.

KIMMEL, A. R., GOROVSKY, M. A., 1978: Organization of the 5 S RNA genes in macro and micronuclei of *Tetrahymena pyriformis*. Chromosoma **67**, 1—20.

KIMURA, M., OHTA, T., 1973: Eukaryotes-prokaryotes divergence estimated by 5 S ribosomal RNA sequences. Nature. New Biol. **243**, 199—200.

KIRSCHNER, R. H., RUSLI, M., MARTIN, T. E., 1977: Characterization of the nuclear envelope, pore complexes and dense lamina of mouse liver nuclei by high resolution scanning electron microscopy. J. Cell Biol. **72**, 118—132.

KISS, G. B., PEARLMAN, E., 1981: Extrachromosomal rDNA of *Tetrahymena thermophila* is not a Perfect Palindrome. Gene **13**, 281—287.

KISTER, H. P., MULLER, B., ECKERT, W. A., 1983: Complex endonucleolytic cleavage pattern during early events in the processing of pre-rRNA in the lower eucaryote *Tetrahymena thermophila*. Nucl. Acids Res. **11**, 3487—3502.

KLEIN, H. L., PETES, T. D., 1981: Intrachromosomal gene conversion in yeast. Nature **289**, 144—148.

KLEMENZ, R., GEIDUSCHEK, E. P., 1980: The 5′-terminus of the precursor ribosomal RNA of *Saccharomyces cerevisiae*. Nucl. Acids Res. **8**, 2679—2689.

KLOETZEL, P.-M., WHITFIELD, W., SOMMERVILLE, J., 1981: Analysis and reconstruction of an RNP particle which stores 5 S RNA and tRNA in amphibian oocytes. Nucl. Acids Res. **9**, 605—621.

KLOOTWIJK, J., PLANTA, R. J., 1973 a: Analysis of the methylation sites in yeast ribosomal RNA. Eur. J. Biochem. **39**, 325—330.

— — 1973 b: Modified sequences in yeast ribosomal RNA. Mol. Biol. Rep. **1**, 187—191.

— DEJONGE, P., PLANTA, R. J., 1979: The primary transcript of the ribosomal repeating unit in yeast. Nucl. Acid Res. **6**, 27—33.

KNIBIEHLER, B., NAVARRO, A., MIRRE, C., STAHL, A., 1977: Localization of ribosomal cistrons in the quail oocyte during meiotic prophase I. Exp. Cell Res. **110**, 153—157.

KNIGHT, E., DARNELL, J. E., 1967: Distribution of 5 S RNA in HeLa cells. J. Mol. Biol. **28**, 491—500.

KOCH, J., CRUCEANU, A., 1971: Hormone-induced gene amplification in somatic cells. Z. Physiol. Chem. **352**, 137—142.

KÖCHEL, H. G., KÜNTZEL, H., 1981: Nucleotide sequence of the *Aspergillus nidulans* mitochondrial gene coding for the small ribosomal subunit RNA: homology to *E. coli* 16 S rRNA. Nucl. Acids Res. **9**, 5689—5996.

KOMINAMI, R., MURAMATSU, M., 1977: Heterogeneity of 5′-termini of nucleolar 45 S, 32 S, and 28 S RNAs in mouse hepatoma. Nucl. Acids Res. **4**, 229—240.

— HAMADA, H., FUJII-KURIYAMA, Y., MURAMATSU, M., 1978: 5′-terminal processing of ribosomal 28 S RNA. Biochemistry **17**, 3965—3970.

— URANO, Y., MISHIMA, Y., MURAMATSU, M., 1981: Organization of ribosomal RNA gene repeats of the mouse. Nucl. Acids Res. **9**, 3219—3233.

— MISHIMA, Y., URANO, Y., SAKAI, M., MURAMATSU, M., 1982: Cloning and determination of the transcription termination site of ribosomal RNA gene of the mouse. Nucl. Acids Res. **10**, 1963—1979.

KOPAC, M. J., MATEJKO, G. M., 1958: Malignant nucleoli: cytological studies and perspectives. Ann. N.Y. Acad. Sci. **73**, 237—282.

KORN, L. J., BROWN, D. D., 1978: Nucleotide sequence of *Xenopus borealis* oocyte 5 S DNA: comparison of sequences that flank several related eucaryotic genes. Cell **15**, 1145—1156.

— GURDON, J. B., 1981: The reactivation of developmentally inert 5 S RNA genes in somatic nuclei injected into *Xenopus* oocytes. Nature **289**, 461—465.

— BIRKENHEIMER, E. H., BROWN, D. D., 1979: Transcription initiation of *Xenopus* 5 S ribosomal RNA genes *in vitro*. Nucl. Acids Res. **7**, 947—958.

KOSHIBA, K., THIRUMALACHARY, C., DASKAL, Y., BUSCH, H., 1971: Ultrastructural and biochemical studies on ribonucleoprotein particles from isolated nucleoli of thioacetamide-treated rat liver. Exp. Cell Res. **68**, 235—246.

KRAMER, R. A., PHILIPPSEN, P., DAVIS, R. W., 1978: Divergent transcription in the yeast rRNA coding region as shown by hybridization to separated strands and sequence analysis of cloned DNA. J. Mol. Biol. **123**, 405—416.

KRAUSE, B., SEIFART, K. H., 1981: Transcription of ribosomal 5 S RNA from HeLa chromatin by homologous and heterologous eucaryotic RNA polymerases. Biochem. Internat. **2**, 201—210.

KRAUTER, K. S., SOEIRO, R., NADAL-GINARD, B., 1979: Transcriptional regulation of ribosomal RNA accumulation during L_6E_9 myoblast differentiation. J. Mol. Biol. **134**, 727—741.

KRAWCZYK, Z., CHORÁZY, M., 1978: Changes in cellular concentration of DNA-dependent RNA polymerases A and B during regeneration of rat liver. Acta Biochimica Polonica **25**, 257—271.

KRUGER, K., GRABOWSKI, P. J., ZAUG, A. J., SANDS, J., GOTTSCHLING, D. E., CECH, T. R., 1982: Self-splicing RNA: autoexcision and autocyclization of the rRNA intervening sequence of *Tetrahymena*. Cell **31**, 147—157.

KRUISWIJK, T., PLANTA, R. J., KROP, J. M., 1978: The course of the assembly of ribosomal subunits in yeast. Biochim. Biophys. Acta **517**, 378—389.

KRYSTAL, M., ARNHEIM, N., 1978: Length heterogeneity in a region of the human ribosomal gene spacer is not accompanied by extensive population polymorphism. J. Mol. Biol. **126**, 91—104.

— D'EUSTACHIO, P., RUDDLE, F. H., ARNHEIM, N., 1981: Human nucleolus organizers on nonhomologous chromosomes can share the same ribosomal gene variants. Proc. Nat. Acad. Sci. U.S.A. **78**, 5744—5748.

KULKER, H. C., POGO, A. O., 1980: The stringent and relaxed phenomena in *Saccharomyces cerevisiae*. J. Biol. Chem. **255**, 1526—1535.

KUMAR, A., SUBRAHMANIAN, A. R., 1975: Ribosome assembly in HeLa cells: Labeling pattern of ribosomal proteins by two-dimensional resolution. J. Mol. Biol. **94**, 409—423.

— WARNER, J. R., 1972: Characterization of ribosomal precursor particles from HeLa cell nucleoli. J. Mol. Biol. **63**, 233—246.

— WU, R. S., 1973: Role of ribosomal RNA transcription in ribosome processing in HeLa cells. J. Mol. Biol. **80**, 265—276.

KÜNTZEL, H., HEIDRICH, M., PIECHULLA, B., 1981: Phylogenetic tree derived from bacterial, cytosomal and organelle 5 S rRNA sequences. Nucl. Acids Res. **9**, 1451—1461.

KUNZ, W., GLÄTZER, K. H., 1979: Similarities and differences in the gene structures of the three nucleoli of *Drosophila hydei*. Z. Physiol. Chem. **360**, 313—320.

KUPRIJANOVA, N., POPENKO, N., EISNER, G., VENGEROV, Y., TIMOFEEVA, M., TIKHONENKO, A., SKRYABIN, K. G., BAYEV, A. A., 1982: Organization of loach ribosomal genes (*Misgurnus fossilis* L.). Mol. Biol. Repts. **8**, 143—148.

KURATA, S., MISUMI, Y., SAKAGUCHI, B., SHIOKAWA, K., YAMANA, K., 1978: Does the rate of ribosomal RNA synthesis vary depending on the number of nucleoli in a nucleus? Exp. Cell Res. **115**, 415—419.

KUTER, D. J., RODGERS, A., 1976: The protein composition of HeLa ribosomal subunits and nucleolar precursor particles. Exp. Cell Res. **102**, 205—212.

KUWANO, M., IKEHARA, Y., 1973: Inhibition by α-amanitin of messenger RNA formation in cultured fibroblasts: potentiation by amphotericin B. Exp. Cell Res. **82**, 454—457.

KWAN, C. N., GOTOH, S., SCHLESSINGER, D., 1974: Nucleases in HeLa cells nucleoplasm and nucleoli. Biochim. Biophys. Acta **349**, 428—441.

LABHART, P., KOLLER, TH., 1982: Structure of the active nucleolar chromatin of *Xenopus laevis* oocytes. Cell **28**, 279—292.

— BANZ, E., NESS, P. J., PARISH, R. W., KOLLER, TH., 1984: A structural concept for nucleoli of *Dictyostelium discoideum* deduced from dissociation studies. Chromosoma (Berl.) **89**, 111—120.

LACROUTE, F., 1973: RNA and protein elongation rates in *Saccharomyces cerevisiae*. Mol. Gen. Genet. **125**, 319—327.

LAFARGE, C., FRAYSSINET, C., 1970: The reversibility of inhibition of RNA and DNA synthesis induced by aflatoxin in rat liver. A tentative explanation for carcinogenic mechanism. Int. J. Cancer **6**, 74—83.

LAIRD, C. D., CHOOI, W. Y., 1976: Morphology of transcription units in *Drosophila melanogaster*. Chromosoma **58**, 193—218.

— WILKINSON, L. E., FOE, V. E., CHOOI, W. Y., 1976: Analysis of chromatin-associated fiber arrays. Chromosoma **58**, 169—192.

LAMPERT, A., FEIGELSON, P., 1974: A short lived polypeptide component of one of two discrete functional pools of hepatic nuclear α-amanitin resistant RNA polymerases. Biochem. Biophys. Res. Commun. **58**, 1030—1038.

LANGEN, P., 1975: Antimetabolites of Nucleic Acid Metabolism, p. 273. New York: Gordon and Breach.

LARIONOV, V. L., SHOUBOCHKINA, E. A., 1982: Construction of hybrid plasmids containing yeast replication origin (russ.). Mol. Biologiya **16**, 948—955.

— GRISHIN, A. V., SMIRNOV, M. N., 1980: 3 μm DNA- and extrachromosomal ribosomal DNA in the yeast *Saccharomyces cerevisiae*. Gene **12**, 41—49.

LARKIN, J. C., WOOLFORD, J. L., 1983: Molecular cloning and analysis of the CRY 1 gene: a yeast ribosomal protein gene. Nucl. Acids Res. **11**, 403—419.

LASTICK, S. M., 1980: The assembly of ribosomes in HeLa cell nucleoli. Eur. J. Biochem. **113**, 175—182.

— MCCONKEY, E. H., 1976: Exchange and stability of HeLa ribosomal proteins *in vivo*. J. Biol. Chem. **251**, 2867—2875.

LAVAL, M., BOUTEILLE, M., MOULÉ, Y., 1976: Effect of the ionic environment on the transcriptional activity of rat liver nucleoli. Exp. Cell Res. **102**, 365—375.

LAVAL, M., HERNANDEZ-VERDUN, D., BOUTEILLE, M., 1981: Remnant nucleolar structures and residual RNA synthesis in chick erythrocytes. Exp. Cell Res. **132**, 157—167.

LEBOWITZ, P., WEISSMAN, S. M., 1979: Organization and transcription of the Simian virus 40 genome. Curr. Topics Microbiol. Immunol. **87**, 44—151.

LEER, J. C., NIELSEN, O. F., PIPER, P. W., WESTERGAARD, O., 1976: Isolation of the ribosomal RNA gene from *Tetrahymena* in the state of transcriptionally active chromatin. Biochem. Biophys. Res. Commun. **72**, 720—731.

— TIRYAKI, D., WESTERGAARD, O., 1979: Termination of transcription in nucleoli isolated from *Tetrahymena*. Proc. Nat. Acad. Sci. U.S.A. **76**, 5563—5566.

LEER, R. J., RAAMSDONK-DUIN, M. C., MOLENAAR, C. M., COHEN, L. H., MAGER, W. H., PLANTA, R., 1982: The structure of the gene coding for the phosphorylated ribosomal protein S 10 in yeast. Nucl. Acids Res. **10**, 5869—5878.

LEICK, W., 1969: Formation of subribosomal particles in the macronuclei of *Tetrahymena pyriformis*. Eur. J. Biochem. **8**, 221—228.

LEON, P. E., 1976: Molecular hybridization of iodinated 4 S, 5 S, and 18 S + 28 S RNA to salamander chromosomes. J. Cell Biol. **69**, 287—300.

LEONARD, TH. B., JACOB, S. T., 1977: Alterations in DNA-dependent RNA polymerases I and II from rat liver by thioacetamide: preferential increase in the level of chromatin-associated nucleolar RNA polymerase I B. Biochemistry **16**, 4538—4544.

LEPOINT, A., GOESSENS, G., 1978: Nucleologenesis in Ehrlich tumor cells. Exp. Cell Res. **117**, 89—94.

— — 1982: Quantitative analysis of Ehrlich tumor cell nucleoli during interphase. Exp. Cell Res. **137**, 456—459.

LEVIS, R., 1978: Kinetic analysis of 5 S RNA formation in *Drosophila* cells. J. Mol. Biol. **122**, 279—283.

— PENMAN, S., 1978: Processing steps and methylation in the formation of the rRNA in cultured *Drosophila* cells. J. Mol. Biol. **121**, 219—238.

LIAU, M. C., HURLBERT, R. B., 1975: Interrelationship between synthesis and methylation of rRNA in isolated Novikoff tumor nucleoli. Biochemistry **14**, 127—134.

— PERRY, R. P., 1969: Ribosome precursor particles in nucleoli. J. Cell Biol. **42**, 272—283.

— HUNT, M. E., HURLBERT, R. B., 1976: Role of rRNA methylases in the regulation of ribosome production in mammalian cells. Biochemistry **15**, 3158—3164.

LIEBHABER, S. A., WOLF, S., SCHLESSINGER, D., 1978: Differences in rRNA metabolism of primary and SV-40 transformed human fibroblasts. Cell **13**, 121—127.

LIKOVSKY, Z., SMETANA, K., 1978: On the presence of micronucleoli in mature avian erythrocytes. Folia biologica (Praha) **24**, 304—308.

LIMA-DE-FARIA, A., 1973: Equations defining the position of ribosomal cistrons in the eucaryotic chromosome. Nature (New Biol.) **241**, 136—139.

— 1976: The chromosome field. I. Prediction of the location of ribosomal cistrons. Hereditas **83**, 1—22.

LIMA-DE-FARIA, A., 1980: Classification of genes, rearrangements and chromosomes according to the chromosome field. Hereditas **93**, 1—46.

LINDELL, T. J., 1976: Evidence for an extranucleolar mechanism of actinomycin D action. Nature **263**, 347—350.

— O'MALLEY, A. F., PUGLISHI, B., 1978: Inhibition of nucleoplasmic transcription of rapidly labeled nuclear proteins by low concentration of actinomycin D *in vivo*. Proposed role of messenger RNA in ribosomal RNA transcription. Biochemistry **17**, 1154—1160.

LIPPS, H. J., STEINBRÜCK, G., 1978: Free genes for rRNAs in the macronuclear genome of the ciliate *Stylonychia mytilus*. Chromosoma **69**, 21—26.

LISCHWE, M. A., SMETANA, K., OLSON, M. O. J., BUSCH, H., 1979: Protein C 23 and B 23 are the major nucleolar silver staining proteins. Life Sci. **25**, 701—708.

— RICHARDS, R. L., BUSCH, R. K., BUSCH, H., 1981: Localization of phosphoprotein C 23 to nucleolar structures and to the nucleolus organizer regions. Exp. Cell Res. **136**, 101—109.

LLOYD, D., POOLE, R. K., EDWARDS, S. W., 1982: The Cell Division Cycle. New York: Academic Press.

LOEB, J. N., HOWELL, R. R., TOMKINS, G. M., 1965: Turnover of ribosomal RNA in rat liver. Science **149**, 1093—1095.

LOENING, U. E., 1967: The fractionation of high molecular weight RNA by polyacrylamide gel electrophoresis. Biochem. J. **102**, 251—260.

— 1975: The mechanism of synthesis of ribosomal RNA. FEBS Symposia **33**, 151—157.

— BAKER, A. M., 1976: Isolation of nucleoli. In: Subnuclear Components (BIRNIE, G. D., ed.), pp. 107—128. London: Butterworths.

— JONES, K., BIRNSTIEL, M. L., 1969: Properties of the rRNA precursor in *Xenopus laevis*: Comparison to the precursor in mammals and in plants. J. Mol. Biol. **45**, 353—366.

LOHR, D., 1983: Chromatin structure differs between coding and upstream flanking sequences of the yeast 35 S ribosomal genes. Biochemistry **22**, 927—934.

LONG, E. O., DAWID, I. B., 1979 a: Restriction analysis of spacers in ribosomal DNA of *Drosophila melanogaster*. Nucl. Acids Res. **7**, 205—215.

— — 1979 b: Expression of ribosomal DNA insertions in *Drosophila melanogaster*. Cell **18**, 1185—1196.

— — 1980: Repeated genes in eucaryotes. Ann. Rev. Biochem. **49**, 727—764.

— REBBERT, M. L., DAWID, I. B., 1981 a: Nucleotide sequence of the initiation site for ribosomal RNA transcription in *Drosophila melanogaster*: Comparison of genes with and without insertions. Proc. Nat. Acad. Sci. U.S.A. **78**, 1513—1517.

— COLLINS, M., KIEFER, B. I., DAWID, I. B., 1981 b: Expression of the ribosomal DNA insertions in bobbed mutants of *Drosophila melanogaster*. Mol. Gen. Genet. **182**, 377—384.

LÖNN, U., EDSTRÖM, J.-E., 1976: Mobility restriction *in vivo* of the heavy ribosomal subunit in a secretory cell. J. Cell Biol. **70**, 573—580.

— — 1977 a: Movements and associations of ribosomal subunits in a secretory cell during growth inhibition by starvation. J. Cell Biol. **73**, 696—704.

Lönn, U., Edström, J.-E., 1977b: Protein synthesis inhibitors and export of ribosomal subunits. Biochim. Biophys. Acta **475**, 677—679.

Lord, A., Nicole, L., Lafontaine, J. B., 1977: Ultrastructural and radioautographic investigation of the nucleolar cycle in *Physarum polycephalum*. J. Cell Sci. **23**, 25—42.

Love, R., Soriano, R. Z., 1971: Correlation of nucleolini with fine structural nucleolar constituents of cultured normal and neoplastic cells. Cancer Res. **31**, 1030—1037.

Lusby, E. W., McLaughlin, C. S., 1980: The effect of amino acid starvation on a major acid soluble compound in *Saccharomyces cerevisiae*. Mol. Gen. Genet. **179**, 699—701.

MacCarty, W. C., 1936: The value of the macronucleolus in the cancer problem. Amer. J. Cancer **26**, 529—532.

Macgregor, H. C., 1972: The nucleolus and its genes in amphibian oogenesis. Biol. Rev. **47**, 177—210.

— 1982: Ways of amplifying ribosomal genes. In: The Nucleolus (Jordan, E. G., Cullins, C. A., eds.), pp. 129—153. Cambridge: Cambridge Univ. Press.

Macieira-Coelho, A., Loria, E., 1976: Changes in RNA synthesis during the life span of human fibroblasts *in vitro*. Gerontology **22**, 79—88.

Mackedonski, V. V., Hadjiolov, A. A., 1970: Selective inhibition of ribosomal RNA synthesis in Ehrlich ascites tumor cells by a non-specific inhibitor, 2,4-dinitrophenol. Biochim. Biophys. Acta **204**, 462—469.

— Nikolaev, N., Sebesta, K., Hadjiolov, A. A., 1972: Inhibition of ribonucleic acid biosynthesis in mice liver by exotoxin of *Bacillus thuringiensis*. Biochim. Biophys. Acta **272**, 56—66.

Macleod, D., Bird, A., 1982: DNAse I sensitivity and methylation of active versus inactive rRNA genes in *Xenopus* species hybrids. Cell **29**, 211—218.

Maden, B. E. H., 1971: The structure and formation of ribosomes in animal cells. Progr. Biophys. Mol. Biol. **22**, 127—177.

— 1980: Methylation map of *Xenopus laevis* ribosomal RNA. Nature **288**, 293—296.

— 1982: 18 S ribosomal DNA and 18 S ribosomal RNA in *Xenopus laevis* and other eucaryotes. In: The Cell Nucleus (Busch, H., ed.), Vol. X, p. 319—350. New York: Academic Press.

— Robertson, J. S., 1974: Demonstration of the "5.8 S" ribosomal sequence in HeLa cell ribosomal precursor RNA. J. Mol. Biol. **87**, 227—235.

— Salim, M., 1974: The methylated nucleotide sequences in HeLa cell ribosomal RNA and its precursors. J. Mol. Biol. **88**, 133—164.

— Vaughan, M. H., Warner, J. R., Darnell, J. E., 1969: Effects of valine deprivation on ribosome formation in HeLa cells. J. Mol. Biol. **45**, 265—275.

— Salim, M., Summers, D. F., 1972: Maturation pathway for rRNA in the HeLa cell nucleolus. Nature (London), New Biol. **237**, 5—9.

— — Robertson, J. S., 1974: Progress in the structural analysis of mammalian 45 S and ribosomal RNA. In: Ribosomes (Nomura, M., Tissieres, A., Lengyel, P., eds.), pp. 829—839. Cold Spring Harbor, N.Y.: Cold Spring Harbor Laboratory.

MADEN, B. E. H., KHAN, M. S. N., HUGHES, D. G., GODDARD, J. P., 1977: Inside 45 S ribonucleic acid. Biochem. Soc. Symp. **42**, 165—179.

— HALL, L. M. C., SALIM, M., 1982a: Ribosome formation in the eucaryotic nucleolus: recent advances from sequence analysis. In: The Nucleolus (JORDAN, E. G., CULLIS, C. A., eds.), pp. 87—101. Cambridge: Cambridge Univ. Press.

— MOSS, M., SALIM, M., 1982b: Nucleotide sequence of an external transcribed spacer in *Xenopus laevis* rDNA: sequences flanking the 5′ and 3′ ends of 18 S rRNA are noncomplementary. Nucl. Acids Res. **10**, 2387—2398.

MAGER, W. H., PLANTA, R. J., 1976: Yeast ribosomal proteins are synthesized on small polysomes. Eur. J. Biochem. **62**, 193—197.

MAISEL, J. C., MCCONKEY, E. H., 1971: Nucleolar protein metabolism in actinomycin D treated HeLa cells. J. Mol. Biol. **61**, 251.

MAIZELS, N., 1976: *Dictyostelium* 17 S, 25 S, and 5 S rDNAs lie within a 38,000 base pair repeated unit. Cell **9**, 431—438.

MAMRACK, M. D., OLSON, M. O. J., BUSCH, H., 1979: Amino acid sequence and sites of phosphorylation in a highly acidic region of nucleolar nonhistone protein C23. Biochemistry **18**, 3381—3386.

MANDAL, R. K., 1969: RNA synthesis in ascites tumor cells during inhibition of protein synthesis. Biochim. Biophys. Acta **182**, 375—381.

MANDAL, K., DAWID, I. B., 1981: The nucleotide sequence at the transcription termination site of ribosomal RNA in *Drosophila melanogaster*. Nucl. Acids Res. **9**, 1801—1811.

MANNING, R. F., SAMOLS, D. R., GAGE, L. P., 1978: The genes for 18 S, 5.8 S, and 28 S ribosomal RNA of *Bombix mori* are organized into tandem repeats of uniform length. Gene **4**, 153—166.

MAO, J., APPEL, B., SCHAAK, J., SHARP, S., YAMADA, H., SÖLL, D., 1982: The 5 S RNA genes in *Schizosaccharomyces pombe*. Nucl. Acids Res. **10**, 487—500.

MARINOZZI, V., FIUME, L., 1971: Effects of α-amanitin on mouse and rat liver cell nuclei. Exp. Cell Res. **67**, 311—322.

MARKOV, G. G., DESSEV, G. N., RUSSEV, G. C., TSANEV, R. G., 1975: Effects of γ-irradiation on biosynthesis of different types of ribonucleic acids in normal and regenerating rat liver. Biochem. J. **146**, 41—51.

MARTIN, T. E., OKAMURA, C. S., 1981: Immunochemistry of nuclear hnRNP complexes. In: The Cell Nucleus (BUSCH, H., ed.), Vol. IX, pp. 119—144. New York: Academic Press.

MARUSHIGE, K., BONNER, J., 1971: Fractionation of liver chromatin. Proc. Nat. Acad. Sci. U.S.A. **68**, 2941—2944.

MARZLUFF, W. F., MURPHY, E. C., HUANG, R. C. C., 1973: Transcription of Ribonucleic Acid in Isolated Mouse Myeloma Nuclei. Biochemistry **12**, 3440—3446.

— — — 1974: Transcription of the genes for 5 S RNA and transfer RNA in isolated mouse myeloma cell nuclei. Biochemistry **13**, 3689—3696.

— WHITE, E., BENJAMIN, R., HUANG, R. C. C., 1975: Low molecular weight RNA species from chromatin. Biochemistry **14**, 3715—3724.

MATHIS, D. J., GOROVSKY, M. L., 1976: Subunit structure of rDNA containing chromatin. Biochemistry **15**, 750—755.

— — 1978: Structure of rDNA-containing chromatin of *Tetrahymena pyriformis* analyzed by nuclease digestion. Cold Spring Harbor Symp. Quant. Biol. **42**, 773—778.

Mathis, D. J., Oudet, P., Chambon, P., 1980: Structure of transcribing chromatin. Progr. Nucleic Acid Res. Mol. Biol. **24**, 1—54.

Matsui, T., Onishi, T., Muramatsu, M., 1976: Nucleolar DNA-dependent RNA polymerase from rat liver. 2. Two forms and their physiological significance. Eur. J. Biochem. **71**, 361—368.

— Fuke, M., Busch, H., 1977: Fidelity of rRNA synthesis by nucleoli and nucleolar chromatin. Biochemistry **16**, 39—45.

— Weinfeld, H., Sandberg, A., 1979: Quantitative conservation of chromatin-bound RNA polymerase I and II in mitosis. J. Cell Biol. **80**, 451—464.

Matsuura, S., Morimoto, T., Tashiro, Y., Higashinakagawa, T., Muramatsu, M., 1974: Ultrastructural and biochemical studies on the precursor ribosomal particles isolated from rat liver nuclei. J. Cell Biol. **63**, 629—640.

Mattaj, I. W., Mackenzie, A., Jost, J.-P., 1982: Cloning in a cosmid vector of complete 37 kb and 25 kb rDNA repeat units from the chicken. Biochim. Biophys. Acta **698**, 204—210.

Mauck, J. C., Green, H., 1973: Regulation of RNA synthesis in fibroblasts during transition from resting to growing state. Proc. Nat. Acad. Sci. U.S.A. **70**, 2819—2823.

Maul, G. G., Hamilton, T. H., 1967: The intranuclear localization of two DNA-dependent RNA polymerase activities. Proc. Nat. Acad. Sci. U.S.A. **57**, 1371—1378.

Maxam, A. M., Tizard, R., Skryabin, K. G., Gilbert, W., 1977: Promotor region for yeast 5 S rRNA. Nature (London) **267**, 643—645.

Mayo, V. S., Andrean, B. A. G., DeKloet, S. R., 1968: Effects of cycloheximide and 5-fluorouracil on the synthesis of ribonucleic acid in yeast. Biochim. Biophys. Acta **169**, 297—305.

McClintock, B., 1934: The relation of a particular chromosomal element to the development of the nucleoli in *Zea mays*. Z. Zellforsch. Mikr. Anat. **21**, 294—328.

McConkey, E. H., Bielka, H., Gordon, J., Lastick, S. M., Lin, A., Ogata, K., Reboud, J. P., Traut, R. R., Warner, J. R., Welfle, H., Wool, I., 1979: Proposed uniform nomenclature for mammalian ribosomal proteins. Mol. Gen. Genet. **169**, 1—6.

McKnight, S. L., Miller, O. L., jr., 1976: Ultrastructural patterns of RNA synthesis during early embryogenesis of *Drosophila melanogaster*. Cell **8**, 305—319.

— Bustin, M., Miller, O. L., jr., 1978: Electron microscopic analysis of chromosome metabolism in the *Drosophila melanogaster* embryo. Cold Spring Harbor Symp. Quant. Biol. **42**, 741—754.

— Hipskind, R. A., Reeder, R., 1980: Ultrastructural analysis of ribosomal gene transcription *in vitro*. J. Biol. Chem. **255**, 7907—7911.

Meedel, T. H., Levine, E. M., 1978: Regulation of protein synthesis in human diploid fibroblasts: reduced initiation efficiency in resting cultures. J. Cell Physiol. **94**, 229—242.

Mercer, W. E., Avignolo, C., Galanti, N., Rose, K. M., Hyland, J. K., Jacob, S.

T., Baserga, R., 1984: Cellular DNA replication is independent of the synthesis and accumulation of ribosomal RNA. Exp. Cell Res. **150**, 118—130.

Meyer, G. F., Hennig, W., 1974: The nucleolus in primary spermatocytes of *Drosophila hydei*. Chromosoma **46**, 121—144.

Meyerink, J. H., Retel, J., 1977: Topographical analysis of yeast rDNA by cleavage with restriction endonucleases. Nucl. Acids Res. **3**, 2697—2707.

— — Raue, H., Planta, R., Ende, A. van der, van Bruggen, E., 1978: Genetic organization of the ribosomal transcription units of the yeast *Saccharomyces carlsbergensis*. Nucl. Acids Res. **5**, 2801—2808.

— Klootwijk, J., Planta, R. J., van Ende, A., van Bruggen, E. F., 1979: Extrachromosomal circular ribosomal DNA in the yeast *Saccharomyces carlsbergensis*. Nucl. Acids Res. **7**, 69—76.

Meyuhas, O., Perry, R. P., 1980: Construction and identification of cDNA clones for mouse ribosomal proteins: application for the study of r-protein gene expression. Gene **10**, 113—129.

Miassod, R., Cecchini, J.-P., Lares, L., Richard, J., 1973: Maturation of ribosomal ribonucleic acids in suspensions of higher plant cells. A DNA-RNA hybridization study. FEBS Lett. **35**, 71—75.

Michot, B., Bachellerie, J.-P., Raynal, F., 1982a: Sequence and secondary structure of mouse 28 S rRNA 5′terminal domain. Organization of the 5.8 S–28 S rRNA complex. Nucl. Acids Res. **10**, 5273.

— — — Renalier, M.-H., 1982b: Homology of the 5′-terminal sequence of 28 S rRNA of mouse with yeast and *Xenopus*. FEBS Lett. **140**, 193—197.

— — — — 1982c: Sequence of the 3′-terminal domain of mouse 18 S rRNA. FEBS Lett. **142**, 260—266.

— — — 1983: Structure of mouse rRNA precursors. Complete sequence and potential folding of the spacer regions between 18 S and 28 S rRNA. Nucl. Acids Res. **11**, 3375—3391.

Miesfeld, R., Arnheim, N., 1982: Identification of the *in vivo* and *in vitro* origin of transcription in human rDNA. Nucl. Acids Res. **10**, 3933—3949.

Miller, D. A., Dev, V., Tantravahi, R., Miller, O. J., 1976: Suppression of human nucleolus organizer activity in mouse-human somatic hybrid cells. Exp. Cell Res. **101**, 235—243.

Miller, J. R., Brownlee, G. G., 1978: Is there a correction mechanism in the 5 S multigene system? Nature (London) **275**, 556—558.

— Melton, D. A., 1981: A transcriptionally active pseudogene in *Xenopus laevis* oocyte 5 S DNA. Cell **24**, 829—835.

— Cartwright, E. M., Brownlee, G., Fedoroff, N., Brown, D., 1978: The nucleotide sequence of oocyte 5 S DNA in *Xenopus laevis*. II. GC-rich region. Cell **13**, 717—725.

Miller, K. G., Sollner-Webb, B., 1981: Transcription of mouse rRNA genes by RNA polymerase I: *in vitro* and *in vivo* initiation and processing sites. Cell **27**, 165—174.

Miller, L., 1973: Control of 5 S RNA synthesis during early development of anucleolate and partial nucleolate mutants of *Xenopus laevis*. J. Cell Biol. **59**, 624—632.

Miller, O. J., Tantravahi, R., Miller, D. A., Yu, L. C., Szabo, P., Prensky, W.,

1979: Marked increase in ribosomal RNA gene multiplicity in a rat hepatoma cell line. Chromosoma **71**, 183—195.

MILLER, O. L., JR., 1981: The nucleolus, chromosomes and visualization of genetic activity. J. Cell Biol. **91**, 15 s—27 s.

— BAKKEN, A. H., 1972: Morphological studies of transcription. Acta Endocrinol. **168** (Suppl.), 155—177.

— BEATTY, B. R., 1969 a: Visualization of nucleolar genes. Science **164**, 955—957.

— — 1969 b: Extrachromosomal nucleolar genes in amphibian oocytes. Genetics **61** (Suppl.) 134—143.

— HAMKALO, B. A., 1972 a: Electron microscopy of active genes. FEBS Symposia **23**, 367—378.

— — 1972 b: Visualization of RNA synthesis on chromosomes. Int. Rev. Cytol. **33**, 1—25.

MIRAULT, M.-E., SCHERRER, K., 1971: Isolation of preribosomes from HeLa cells and their characterization by electrophoresis on uniform and exponential-gradient-polyacrylamide gels. Eur. J. Biochem. **23**, 372—386.

— — 1972: *In vitro* processing of HeLa cell preribosomes by a nucleolar endoribonuclease. FEBS Lett. **20**, 233—239.

MIRRE, C., KNIBIEHLER, B., 1981: Ultrastructural autoradiographic localization of the rRNA transcription sites in the quail nucleolar components using two RNA antimetabolites. Biol. Cell **42**, 73—78.

— — 1982: A re-evaluation of the relationships between the fibrillar centers and the nucleolus-organizing regions in reticulated nucleoli. J. Cell Sci. **55**, 247—259.

— STAHL, A., 1978 a: Peripheral RNA synthesis of fibrillar center in nucleoli of Japanese quail oocytes and somatic cells. J. Ultrastr. Res. **64**, 377—387.

— — 1978 b: Ultrastructure and activity of the nucleolar organizer in the mouse oocyte during meiotic prophase. J. Cell Sci. **31**, 79—100.

— — 1981: Ultrastructural organization, sites of transcription and distribution of fibrillar centers in the nucleolus of the mouse oocyte. J. Cell Sci. **48**, 105—126.

MISHIMA, Y., KOMINAMI, R., HONJO, T., MURAMATSU, M., 1980: Cloning and determination of a putative promoter region of a mouse ribosomal deoxyribonucleic acid fragment. Biochemistry **19**, 3780—3786.

— YAMAMOTO, O., KOMINAMI, R., MURAMATSU, M., 1981: *In vitro* transcription of a cloned mouse ribosomal RNA gene. Nucl. Acids Res. **9**, 6773—6785.

MOHAN, J., RITOSSA, F. M., 1970: Regulation of rRNA synthesis and its bearing on the bobbed phenotype of *Drosophila melanogaster*. Dev. Biol. **22**, 495—512.

MOLENAAR, I., SILLEVIS SMITH, W. W., ROZIJN, T. H., TONINO, G. M., 1970: Biochemical and electron microscopic study of isolated yeast nuclei. Exp. Cell Res. **60**, 148—156.

MOLGAARD, H. V., MATTHEWS, H. R., BRADBURY, E. M., 1976: Organization of genes for rRNA in *Physarum polycephalum*. Eur. J. Biochem. **68**, 541—549.

MONK, R. J., MEYUHAS, O., PERRY, R. P., 1981: Mammals have multiple genes for individual ribosomal proteins. Cell **24**, 301—306.

MONNERON, A., BURGLEN, J., BERNHARD, W., 1970: Action of toyocamicin on nucleolar fine structure and function. J. Ultrastr. Res. **32**, 370—389.

MOORE, G. P. M., RINGERTZ, N. R., 1973: Localization of DNA-dependent RNA polymerase activities in fixed human fibroblasts by autoradiography. Exp. Cell Res. **76**, 223—228.

MORA, M., DARZYNKIEWICZ, Z., BASERGA, R., 1980: DNA synthesis and cell division in a mammalian cell mutant temperature sensitive for the processing of ribosomal RNA. Exp. Cell Res. **125**, 241—249.

MOROI, Y., HARTMANN, A. L., NAKANE, P. K., TAN, E. M., 1981: Distribution of kinetochore (centromere) antigen in mammalian cell nuclei. J. Cell Biol. **90**, 254—259.

MOSS, T., 1982: Transcription of cloned *Xenopus laevis* ribosomal DNA microinjected into *Xenopus* oocytes, and the identification of an RNA polymerase I promoter. Cell **30**, 835—842.

— BIRNSTIEL, M. L., 1979: The putative promoter of *Xenopus laevis* ribosomal gene is reduplicated. Nucl. Acids Res. **6**, 3733—3743.

— — 1982: The structure and function of the ribosomal gene spacer. In: The Nucleolus (JORDAN, E. G., CULLIS, C. A., eds.), pp. 73—86. Cambridge: Cambridge Univ. Press.

— BOSELEY, P. G., BIRNSTIEL, M., 1980: More ribosomal spacer sequences from *Xenopus laevis*. Nucl. Acids Res. **8**, 467—485.

MOYNE, G., NASH, R. E., PUVION, E., 1977: Perichromatin granules in isolated rat hepatocytes treated with cortisol and cycloheximide. Biol. Cell **30**, 5—16.

MOYZIS, R. K., GRADY, D. L., LI, D. W., MIRVIS, S. E., TS'O, P. O., 1980: Extensive homology of nuclear RNA and polysomal poly (A) mRNA between normal and neoplastically transformed cells. Biochemistry **19**, 821—832.

MURAMATSU, M., SHIMADA, N., HIGASHINAKAGAWA, T., 1970: Effect of cycloheximide on the nucleolar RNA synthesis in rat liver. J. Mol. Biol. **53**, 91—106.

— MATSUI, T., ONISHI, T., MISHIMA, Y., 1979: Nucleolar RNA polymerase and transcription of nucleolar chromatin. In: The Cell Nucleus (BUSCH, H., ed.), Vol. VII, pp. 123—161. New York: Academic Press.

NABESHIMA, Y., OGATA, K., 1980: Stimulation of the synthesis of ribosomal proteins in regenerating rat liver with special reference to the increase in the amounts of effective mRNAs for ribosomal proteins. Eur. J. Biochem. **107**, 323—329.

— TSURUGI, K., OGATA, K., 1975: Preferential biosynthesis of ribosomal structural proteins by free and loosely bound polysomes from regenerating rat liver. Biochim. Biophys. Acta **414**, 30—43.

— IMAI, K., OGATA, K., 1979: Biosynthesis of ribosomal proteins by poly(A)-containing mRNAs from rat liver in a wheat germ cell-free system and sizes of mRNAs coding ribosomal proteins. Biochim. Biophys. Acta **564**, 105—121.

NARAYAN, K., BIRNSTIEL, M., 1969: Biochemical and ultrastructural characteristics of ribonucleoprotein particles isolated from rat liver cell nucleoli. Biochim. Biophys. Acta **190**, 470—485.

NATH, K., BOLLON, A. P., 1977: Organization of the yeast rRNA gene cluster via cloning and restriction analysis. J. Biol. Chem. **252**, 6562—6571.

— — 1978: Restriction analysis of tandemly repeated yeast rRNA genes. Mol. Gen. Genet. **160**, 235—245.

NAVASHIN, M., 1934: Chromosome alterations caused by hybridization and their bearing upon certain general genetic problems. Cytologia **5**, 169—203.

NAZAR, R. N., 1977: Studies on the 5′-termini of Novikoff ascites hepatoma ribosomal precursor RNA. Biochemistry **16**, 3215—3219.

— 1980: A 5.8 S rRNA-like sequence in prokaryotic 23 S rRNA. FEBS Lett. **119**, 212—214.

— SITZ, TH., 1980: Role of the 5′-terminal sequence in the RNA binding site of yeast 5.8 SrRNA. FEBS Lett. **115**, 71—76.

— — BUSCH, H., 1975: Structural analysis of mammalian ribosomal RNA and its precursors. Nucleotide sequence of 5.8 S rRNA. J. Biol. Chem. **250**, 8591—8597.

NEAL, G. E., 1972: The effect of aflatoxin B_1 on normal and cortisol-stimulated rat liver RNA synthesis. Biochem. J. **130**, 619—629.

NG, S. Y., PARKER, C. S., ROEDER, R. G., 1979: Transcription of cloned *Xenopus* 5 S RNA genes by *X. laevis* RNA polymerase III in reconstituted systems. Proc. Nat. Acad. Sci. U.S.A. **76**, 136—140.

NICOLOFF, H., ANASTASSOVA-KRISTEVA, M., RIEGER, R., KÜNZEL, G., 1979: Nucleolar dominance as observed in barley translocation lines with specifically reconstructed SAT chromosomes. Theor. Appl. Genet. **55**, 247—251.

NIESSING, J., SCHNIEDERS, B., KUNZ, W., SEIFART, K. H., SEKERIS, C. E., 1970: Inhibition of RNA synthesis by α-amanitin *in vivo.* Z. Naturforschg. **25b**, 1119—1125.

NIKOLAEV, N., GEORGIEV, O. I., VENKOV, P. V., HADJIOLOV, A. A., 1979: The 37 S precursor to rRNA is the primary transcript of rRNA genes in *Saccharomyces cerevisiae*. J. Mol. Biol. **127**, 297—307.

— BIRGE, C. H., GOTOH, S., GLAZIER, K., SCHLESINGER, D., 1975: Primary processing of high molecular weight preribosomal RNA in *Escherichia coli* and HeLa cells. Brookhaven Symp. Biol. **26**, 175—193.

NIKOLOV, E. N., DABEVA, M. D., NIKOLOV, T. K., 1983: Turnover of ribosomes in regenerating rat liver. Int. J. Biochem. 1255—1260.

NILES, E. G., 1978: Isolation of a high specific activity 35 S rRNA precursor from *Tetrahymena pyriformis* and identification of its 5′-terminus pppAp. Biochemistry **16**, 3215—3219.

— SUTIPHONG, J., HAQUE, S., 1981 a: Structure of the *Tetrahymena pyriformis* rRNA gene. Nucleotide sequence of the transcription initiation region. J. Biol. Chem. **256**, 12849—12856.

— CUNNINGHAM, K., JAIN, R., 1981 b: Structure of the *Tetrahymena pyriformis* rRNA gene. Nucleotide sequence of the transcription termination region. J. Biol. Chem. **256**, 12857—12860.

NISSEN-MEYER, J., EIKHOM, T. S., 1976 a: Effect of the growth conditions on the ratio between native 40 S and 60 S ribosomal subunits in various cell types. Exp. Cell Res. **98**, 41—46.

— — 1976 b: An excess of small ribosomal subunits and a higher rate of turnover of the 60 S than of the 40 S ribosomal subunits in L cells grown in suspension culture. J. Mol. Biol. **101**, 211—221.

NOEL, J. S., DEWEY, W. C., ABEL, J. H., THOMPSON, R. P., 1971: Ultrastructure of the nucleolus during the chinese hamster cell cycle. J. Cell Biol. **49**, 830—847.

NOLLER, H. F., KOP, J., WHEATON, V., BROSIUS, J., GUTELL, R., KOPYLOV, A., DOHME, F., HERR, W., STAHL, D. A., GUPTA, P., WOESE, C. R., 1981: Secondary structure model for 23 S ribosomal RNA. Nucl. Acids Res. **9**, 6167—6189.

NOMIYAMA, H., KUHARA, S., KUKITA, T., OTSUKA, T., SAKAKI, Y., 1981: Nucleotide sequence of the ribosomal RNA gene of *Physarum polycephalum*: intron 2 and its flanking regions of the 26 S rRNA gene. Nucl. Acids Res. **9**, 5507—5520.

NOMURA, M., MORGAN, E. A., JASKUNAS, S. R., 1977: Genetics of bacterial ribosomes. Annu. Rev. Genet. **11**, 297—347.

— YATES, J. L., DEAN, D., POST, L. E., 1980: Feedback regulation of ribosomal protein gene expression in *E. coli*: structural homology of rRNA and ribosomal protein mRNA. Proc. Nat. Acad. Sci. U.S.A. **77**, 7084—7088.

— DEAN, D., YATES, J. L., 1982: Feedback regulation of ribosomal protein synthesis in *Escherichia coli*. Trends Biochem. Sci. **7**, 92—95.

NORDGREN, H., STENRAM, U., 1972: Decreased half-life of the RNA of free and membrane-bound ribosomes in the liver of protein-deprived rats. Z. Physiol. Chem. **353**, 1832—1836.

NOSE, K., OKAMOTO, H., 1980: Transcriptional activity of nuclei from WI 38 cells at various passages. J. Cell Physiol. **102**, 51—54.

NUSSINOV, R., 1980: Some rules in the ordering of nucleotides in the DNA. Nucl. Acids Res. **8**, 4545—4562.

OBORA, M., HIRANO, H., HIGASHI, K., 1982: Purification and characterization of nucleolar RNA methylase from Ehrlich ascites tumor cells of mice. Biochemistry **21**, 1374—1380.

OHTA, N., NEWTON, A., 1981: Isolation and mapping of ribosomal RNA genes of *Caulobacter crescentus*. J. Mol. Biol. **153**, 291—303.

OLERT, J., SAWATSKI, G., KLING, H., GEBAUER, J., 1979: Cytological and histochemical studies on the mechanism of the selective silver staining of nucleolus organizer regions. (NORs). Histochem. **60**, 91—99.

OLSEN, G. J., SOGIN, M. L., 1982: Nucleotide sequence of *Dictyostelium discoideum* 5.8 S rRNA: Evolutionary and secondary structural implications. Biochemistry **21**, 2335—2340.

OLSEN, M. O., GUETZOW, K., BUSCH, H., 1981: Localization of phosphoprotein C 23 in nucleoli by immunological methods. Exp. Cell Res. **135**, 259—265.

ONISHI, T., MURAMATSU, M., 1978: Techniques of *in vitro* RNA synthesis in isolated nucleoli. In: Methods of Cell Biology (PRESCOTT, D., ed.), Vol. 19, pp. 301—315. New York: Academic Press.

— MATSUI, T., MURAMATSU, M., 1977: Effect of cycloheximide on the nucleolar RNA synthesis in rat liver. Changes in RNA polymerase I and nucleolar template activity. J. Biochem. (Tokyo) **82**, 1109—1119.

ORGANTINI, J. E., JOSEPH, C., FARBER, J., 1975: Increases in the activity of the solubilized rat liver nuclear RNA polymerases following partial hepatectomy. Arch. Biochem. Biophys. **170**, 485—491.

ØYEN, T. B., 1973: Chromosome I as a possible site for some rRNA cistrons in *Saccharomyces cerevisiae*. FEBS Lett. **30**, 53—56.

PACE, N., WALKER, TH., SCHROEDER, E., 1977: Structure of the 5.8 S RNA component of the 5.8 S–28 S rRNA junction complex. Biochemistry **16**, 5321—5327.

PALADE, G. E., 1958: Microsomes and ribonucleoprotein particles. In: Microsomal Particles and Protein Synthesis (ROBERTS, R. B., ed.), pp. 36—61. London: Pergamon Press.

PAN, W.-CH., BLACKBURN, E. H., 1981: Single extrachromosomal ribosomal RNA gene copies are synthesized during amplification of the rDNA in *Tetrahymena*. Cell **23**, 459—466.

— ORIAS, E., FLACKS, M., BLACKBURN, E. H., 1982: Allele-specific, selective amplification of a rRNA gene in *Tetrahymena thermophila*. Cell **28**, 595—604.

PARDOLL, D. M., VOGELSTEIN, B., 1980: Sequence analysis of nuclear matrix associated DNA from rat liver. Exp. Cell Res. **128**, 466—470.

PARDUE, M. L., GALL, J. G., 1969: Molecular hybridization of radioactive RNA to the DNA of cytological preparations. Proc. Nat. Acad. Sci. U.S.A. **64**, 600—608.

— BROWN, D. D., BIRNSTIEL, M. L., 1973: Localization of genes for 5 S rRNA in *Xenopus laevis*. Chromosoma **42**, 191—203.

— HSU, T. C., 1975: Locations of 18 S and 28 S ribosomal genes on the chromosomes of the Indian muntjac. J. Cell Biol. **64**, 251—254.

— GERBI, S. A., ECKHARDT, R. A., GALL, J. G., 1978: Cytological localization of DNA complementary to rRNA in polytene chromosomes of *Diptera*. Chromosoma **29**, 269—290.

PARKER, C. S., JAEHNING, J. A., ROEDER, R. G., 1977: Faithful gene transcription by eucaryotic RNA polymerases in reconstructed systems. Cold Spring Harbor Symp. Quant. Biol. **42**, 577—587.

PAVLAKIS, G. N., JORDAN, B. R., WURST, R. M., VOURNAKIS, J. N., 1979: Sequence and secondary structure of *D. melanogaster* 5.8 S and 2 S rRNAs and of the processing site between them. Nucl. Acids Res. **7**, 2213—2238.

PEACOCK, W. J., APPELS, R., ENDOW, S., GLOVER, D., 1981: Chromosomal distribution of the major insert in *Drosophila melanogaster* 28 S rRNA genes. Genet. Res. (Cambr.) **37**, 209—214.

PEARSON, M. L., 1980: Muscle differentiation in cell culture. In: The Molecular Genetics of Development (LEIGHTON, T., LOOMIS, W. F., eds.), pp. 361—418. New York: Academic Press.

PEARSON, N. J., FRIED, H. M., WARNER, J. R., 1982: Yeast use transcriptional control to compensate for extra copies of a ribosomal protein gene. Cell **29**, 347—355.

PEBUSQUE, M.-J., SEITE, R., 1981: Electron microscopic studies of silver-stained proteins in nucleolar organizer regions: location in nucleoli of rat sympathetic neurons during light and dark periods. J. Cell Sci. **51**, 85—94.

— VIO, M., SEITE, R., 1981: Ultrastructural location of Ag-NOR stained proteins in nucleoli of rat sympathetic neurons during the dark period. Biol. Cell **40**, 151—154.

PELHAM, H. R. B., BROWN, D. D., 1980: A specific transcription factor that can bind either the 5 S gene or 5 S RNA. Proc. Nat. Acad. Sci. U.S.A. **77**, 4170—4174.

Pellegrini, M., Manning, J., Davidson, N., 1977: Sequence arrangement of the rDNA of *Drosophila melanogaster*. Cell **10**, 213—224.

Perry, R. P., 1962: The cellular sites of synthesis of ribosomal and 4 S RNA. Proc. Nat. Acad. Sci. U.S.A. **48**, 2179—2186.

— 1963: Selective effects of actinomycin D on the intracellular distribution of RNA synthesis in tissue culture cells. Exp. Cell Res. **29**, 400—406.

— 1967: The nucleolus and the synthesis of ribosomes. Progr. Nucl. Acids Res. Mol. Biol. **6**, 219—257.

— 1973: The regulation of ribosome content in eucaryotes. Biochem. Soc. Symp. **37**, 114—135.

— 1976: Processing of RNA. Annu. Rev. Biochem. **45**, 605—629.

— 1981: RNA processing comes of age. J. Cell Biol. **91**, 28 s—38 s.

— Kelley, D. E., 1970: Inhibition of RNA synthesis by actinomycin D: Characteristic dose-response of different RNA species. J. Cell Physiol. **76**, 127—139.

— Hell, A., Errera, M., 1961: The role of the nucleolus in ribonucleic acid and protein synthesis. I. Incorporation of cytidine into normal and nucleolar inactivated HeLa cells. Biochim. Biophys. Acta **49**, 47—57.

Peters, K. E., Comings, D. E., 1980: Two-dimensional gel electrophoresis of rat liver nuclear washes, nuclear matrix and hnRNA proteins. J. Cell Biol. **86**, 135—155.

Peters, M. A., Walker, Th., Pace, N., 1982: Independent binding sites in mouse 5.8 S rRNA for 28 S rRNA. Biochemistry **21**, 2329—2333.

Peterson, J. L., McConkey, E. H., 1976: Non-histone chromosomal proteins from HeLa cells. A survey by high resolution, two-dimensional electrophoresis. J. Biol. Chem. **251**, 548—554.

Peterson, R. C., Doening, J. L., Brown, D., 1980: Characterization of two *Xenopus* somatic 5 S DNAs and one minor oocyte-specific 5 S DNA. Cell **20**, 131—141.

Petes, Th. D., 1979 a: Yeast ribosomal RNA genes are located on chromosome XII. Proc. Nat. Acad. Sci. U.S.A. **76**, 410—414.

— 1979 b: Meiotic mapping of yeast ribosomal deoxyribonucleic acid on chromosome XII. J. Bacteriol. **138**, 185—192.

— 1980: Unequal meiotic recombination within tandem arrays of yeast ribosomal RNA genes. Cell **19**, 765—774.

— Botstein, D., 1977: Simple Mendelian inheritance of the reiterated ribosomal DNA of yeast. Proc. Nat. Acad. Sci. U.S.A. **74**, 5091—5095.

— Hereford, L. M., Skryabin, K. G., 1978: Characterization of two types of yeast ribosomal RNA genes. J. Bacteriol. **134**, 295—305.

Petrov, P. T., Sekeris, C., 1971: Early action of α-amanitin on extranucleolar ribonucleoproteins, as revealed by electron microscopic observation. Exp. Cell Res. **69**, 393—401.

Philippsen, P., Thomas, M., Kramer, R., Davis, R., 1978: Unique arrangement of coding sequences for 5 S, 5.8 S, 18 S, and 28 S rRNA in *Saccharomyces cerevisiae* as determined by R-loop and hybridization analysis. J. Mol. Biol. **123**, 387—404.

PHILIPS, D. M., PHILIPS, S. G., 1973: Repopulation of post mitotic nucleoli by preformed RNA. II. Ultrastructure. J. Cell Biol. **58**, 54—60.

PHILIPS, R. L., 1978: Molecular cytogenetics of the nucleolus organizer region. In: Genetics and Breeding of Maize (WALDEN, D. B., ed.), pp. 711—741. New York: J. Wiley.

PHILIPS, S. G., PHILIPS, D. M., 1969: Sites of nucleolus production in cultured hamster cells. J. Cell Biol. **40**, 248—268.

— — 1979: Nucleolus-like bodies in micronuclei of cultured *Xenopus* cells. Exp. Cell Res. **120**, 295—306.

PHILLIPS, W. F., MCCONKEY, E. H., 1976: Relative stoichiometry of ribosomal proteins in HeLa cell nucleoli. J. Biol. Chem. **251**, 2876—2881.

PICARD, B., WEGNEZ, M., 1979: Isolation of a 7 S particle from *Xenopus laevis* oocytes: A 5 S RNA-protein complex. Proc. Nat. Acad. Sci. U.S.A. **76**, 241—245.

PICCOLETTI, R., ALETTI, M. G., CAIRO, G., BERNELLI-ZAZZERA, A., 1982: Number of transcribing RNA polymerase molecules and polyribonucleotide elongation rates in regenerating rat liver. Effect of cycloheximide treatment. Cell Biol. Intern. Rep. **6**, 669—674.

PIERANDREI-AMALDI, P., BECCARI, E., 1980: Messenger RNA for ribosomal proteins in *Xenopus laevis* oocytes. Eur. J. Biochem. **106**, 603—611.

— CAMPIONI, N., BECCARI, E., BOZZONI, I., AMALDI, F., 1982: Expression of ribosomal protein genes in *Xenopus laevis* development. Cell **30**, 163—171.

PILONE, G. B., NARDI, I., BATISTONI, R., ANDRONICO, F., BACCARI, E., 1974: Chromosome location of the genes for 28 S, 18 S, and 5 S rRNA in *Triturus marmoratus*. Chromosoma **49**, 135—153.

PLANTA, R. J., MEYERINK, J. H., KLOOTWIJK, J., 1978: Organization and transcription of genes coding for ribosomal components in eucaryotes. FEBS Symposia **51**, 401—411.

PLOTON, D., BENDAYAN, M., ADNET, J. J., 1983: Ultrastructural localization of Ag-NOR proteins and nucleic acids in reticulated nucleoli. Biol. Cell **49**, 29—34.

PLUTA, A. F., SPEAR, B. B., 1981: Localization of genes for ribosomal RNA in the nuclei of *Oxytricha fallax*. Exp. Cell Res. **135**, 387—392.

POGO, A. O., LITTAU, V. C., ALLFREY, V. G., MIRSKY, A. E., 1967: Modification of ribonucleic acid synthesis in nuclei isolated from normal and regenerating liver: Some effects of salt and specific divalent cations. Proc. Nat. Acad. Sci. U.S.A. **57**, 743—750.

PONG, R. S., WOGAN, G. N., 1970: Time course and dose-response characteristics of aflatoxin B_1 effect on rat liver RNA polymerase and ultrastructure. Cancer Res. **30**, 294—304.

POTASHKIN, J. A., SCHLEGEL, R. A., 1980: A possible mechanism by which SV 40 T-antigen stimulates rRNA synthesis. Cell Biol. Int. Rep. **4**, 399—406.

POUSADA, C. R., MARCAUD, L., PORTIER, M. M., HAYES, D. D., 1975: Rapidly labeled RNA in *Tetrahymena pyriformis*. Eur. J. Biochem. **56**, 117—122.

PRESCOTT, D. M., 1964: Cellular sites of RNA synthesis. Progr. Nucl. Acids Res. Mol. Biol. **3**, 33—57.

Prescott, D. M., Murti, K. G., Bostock, C. J., 1973: Genetic apparatus of *Stylonychia* sp. Nature (London) **242**, 597—600.

Prestayko, A. W., Lewis, B. C., Busch, H., 1972: Endoribonuclease activity associated with nucleolar ribonucleoprotein particles from Novikoff hepatoma. Biochim. Biophys. Acta **269**, 90—103.

— — — 1973: Purification and properties of a nucleolar endoribonuclease from Novikoff hepatoma. Biochim. Biophys. Acta **319**, 323—335.

— Klomp, G. R., Schmoll, D. J., Busch, H., 1974: Comparison of proteins of ribosomal subunits and nucleolar preribosomal particles from Novikoff hepatoma ascites cells by two-dimensional polyacrylamide gel electrophoresis. Biochemistry **13**, 1945—1951.

Procunier, J. D., Dunn, R. J., 1978: Genetic and Molecular Organization of 5 S Locus and Mutants in *D. melanogaster*. Cell **15**, 1087—1093.

— Tartof, K. D., 1975: Genetic analysis of the 5 S RNA genes in *Drosophila melanogaster*. Genetics **81**, 515—523.

— — 1976: Restriction maps of 5 S rRNA genes of *Drosophila melanogaster*. Nature **263**, 255—257.

Pruitt, S. C., Grainger, R. M., 1981: A mosaicism in the higher order structure of *Xenopus* oocyte nucleolar chromatin prior to and during ribosomal gene transcription. Cell **23**, 711—720.

Pukkila, P. J., 1975: Identification of the lampbrush chromosome loops which transcribe 5 S rRNA in *Notophthalmus (Triturus) viridescens*. Chromosoma **53**, 71—89.

Purtell, M. J., Anthony, D. D., 1975: Changes in rRNA processing paths in resting and phytohaemagglutinin-stimulated guinea pig lymphocytes. Proc. Nat. Acad. Sci. U.S.A. **72**, 3315—3319.

Puvion, E., Moyne, G., Bernhard, W., 1976: Action of 3′-deoxyadenosine (cordycepin) on the nuclear ribonucleoproteins in isolated liver cells. J. Micr. Biol. Cell **25**, 17—32.

— Viron, A., Bernhard, W., 1977: Unusual accumulation of ribonucleoprotein constituents in the nucleus of cultured rat liver cells after hypothermal shock. Biol. Cell **29**, 81—88.

— Puvion-Dutilleul, F., Leduc, E. H., 1981: The formation of nucleolar perichromatin granules. J. Ultrastr. Res. **76**, 181—190.

Puvion-Dutilleul, F., Bachellerie, J.-P., 1979: Ribosomal transcriptional complex in subnuclear fractions of chinese hamster ovary cells after short-term actinomycin D treatment. J. Ultrastr. Res. **66**, 190—199.

— — Zalta, J.-P., Bernhard, W., 1977a: Morphology of ribosomal transcription units in isolated subnuclear fractions of mammalian cells. Biol. Cell **30**, 183—194.

— Bernadac, A., Puvion, E., Bernhard, W., 1977b: Visualization of two different types of nuclear transcriptional complexes in rat liver cells. J. Ultrastr. Res. **58**, 108—117.

Quincey, R. V., 1971: The number and location of genes for 5 S RNA within the genome of *Drosophila melanogaster*. Biochem. J. **123**, 227—233.

— Wilson, S. H., 1969: The utilization of genes for ribosomal RNA, 5 S RNA, and transfer RNA in liver cells of adult rats. Proc. Nat. Acad. Sci. U.S.A. **64**, 981—988.

RAE, P. M. M., 1981: Coding region deletions associated with the major form of rDNA interruption in *Drosophila melanogaster*. Nucl. Acids Res. **9**, 4997—5010.

— STEELE, R. E., 1979: Absence of cytosine methylation at C-C-G-G and G-C-G-C sites in the rDNA coding regions and intervening sequences of *Drosophila* and the rDNA of other higher insects. Nucl. Acids Res. **6**, 2987—2995.

— KOHORN, B. D., WADE, R. P., 1980: The 10 kb *Drosophila virilis* 28 S rDNA intervening sequence is flanked by a direct repeat of 14 base pairs of coding sequence. Nucl. Acids Res. **8**, 3491—3504.

RECH, J., CATHALA, G., JEANTEUR, P., 1980: Isolation and characterization of a ribonuclease activity specific for double-stranded RNA (RNase D) from Krebs II ascites cells. J. Biol. Chem. **255**, 6700—6705.

RECHER, L., WHITESCARVER, J., BRIGGS, L., 1969: The fine structure of a nucleolar component. J. Ultrastr. Res. **29**, 1—14.

— CHAN, H., BRIGGS, L., PARRY, N., 1972: Ultrastructural changes inducible with the plant alkaloid camptothecin. Cancer Res. **32**, 2495—2501.

— SYKES, J.-A., CHAN, H., 1976: Further studies on the mammalian cell nucleolus. J. Ultrastr. Res. **56**, 152—163.

REDDY, R., BUSCH, H., 1981: Small nuclear RNA's of nuclear SnRNP. In: The Cell Nucleus (BUSCH, H., ed.), Vol. VIII, pp. 261—306. New York: Academic Press.

— ROTHBLUM, L. I., SUBRAHMANYAM, C. S., LIU, M.-H., HENNING, D., CASSIDY, B., BUSCH, H., 1983: The nucleotide sequence of 8 S RNA bound to preribosomal RNA of Novikoff hepatoma. J. Biol. Chem. **258**, 584—589.

REEDER, R. H., 1974: Ribosomes from eucaryotes: genetics. In: Ribosomes (NOMURA, M., TISSIERES, A., LENGYEL, P., eds.), pp. 489—518. Cold Spring Harbor, N.Y.: Cold Spring Harbor Laboratory.

— HIGASHINAKAGAWA, T., MILLER, O. L., JR., 1976: The 5′-3′ polarity of the *Xenopus* rRNA precursor molecule. Cell **8**, 449—454.

— ROEDER, R. G., 1972: Ribosomal RNA synthesis in isolated nuclei. J. Mol. Biol. **67**, 433—439.

— SOLLNER-WEBB, G., WAHN, H., 1977: Sites of transcription initiation *in vivo* on *Xenopus laevis* rDNA. Proc. Nat. Acad. Sci. U.S.A. **74**, 5402—5406.

— MCKNIGHT, S. L., MILLER, O. L., 1978: Contraction ratio of the nontranscribed spacer of *Xenopus* rDNA chromatin. Cold Spring Harbor Symp. Quant. Biol. **42**, 1174—1177.

REEVES, R., 1978 a: Structure of *Xenopus* ribosomal gene chromatin during changes in genomic transcription rates. Cold Spring Harbor Symp. Quant. Biol. **42**, 709—722.

— 1978 b: Nucleosome structure of *Xenopus* oocyte amplified ribosomal genes. Biochemistry **17**, 4908—4916.

— JONES, A., 1976: Genomic transcriptional activity and the structure of chromatin. Nature **260**, 495—499.

REICHEL, R., BENECKE, B.-J., 1980: Reinitiation of synthesis of small cytoplasmic RNA species K and L in isolated nuclei *in vitro*. Nucl. Acids Res. **8**, 225—234.

REICHEL, R., MONSTEIN, H.-J., JANSEN, H.-W. PHILIPSON, L., BENECKE, B.-J., 1982: Small nuclear RNAs are encoded in the nontranscribed region of ribosomal spacer DNA. Proc. Nat. Acad. Sci. U.S.A. **79**, 3106—3110.

RENER, J., NARDONE, R. M., 1980: Nucleolar RNA synthesis in heterokaryons derived from senescent and pre-senescent human fibroblasts. Exp. Cell Res. **129**, 297—301.

RENKAWITZ, R., GERBI, S., GLÄTZER, K., 1979: Ribosomal DNA of the fly *Sciara coprophila* has a very small and homogeneous repeat unit. Mol. Gen. Genet. **173**, 1—13.

RENKAWITZ-POHL, R., GLÄTZER, K. H., KUNZ, W., 1980: Characterization of cloned ribosomal DNA from *Drosophila hydei*. Nucl. Acids Res. **8**, 4593—4611.

— — — 1981 a: Ribosomal RNA genes with an intervening sequence are clustered within the X chromosomal ribosomal DNA of *Drosophila hydei*. J. Mol. Biol. **148**, 95—101.

— MATSUMOTO, L., GERBI, S., 1981 b: Two distinct intervening sequences in different ribosomal DNA repeat units of *Sciara coprophila*. Nucl. Acids Res. **9**, 3747—3764.

RETEL, J., PLANTA, R. J., 1968: The investigation of the ribosomal RNA sites in yeast DNA by the hybridization technique. Biochim. Biophys. Acta **169**, 416—429.

REYNOLDS, R., MONTGOMERY, P. O., HUGHES, B., 1974: Nucleolar "caps" produced by actinomycin D. Cancer Res. **24**, 1269—1278.

RICHARDS, B., PARDON, J., LILLEY, D., COTTER, R., WOOLEY, J., WORCESTER, D., 1978: Nucleosome sub-structure during transcription and replication. Phil. Trans. R. Soc. (London) **B 283**, 287—289.

RIEGER, R., NICOLOFF, H., ANASTASSOVA-KRISTEVA, M., 1979: "Nucleolar dominance" in interspecific hybrids and translocation lines. Biol. Zbl. **98**, 385—398.

RINKE, J., STEITZ, J., 1982: Precursor molecules of both human 5 S rRNA and tRNA are bound by a cellular protein reactive with Anti-La Lupus antibodies. Cell **29**, 149—159.

RITOSSA, F., 1972: Procedure for magnification of lethal deletions of genes for rRNA. Nature (New Biol.) **240**, 109—111.

— 1976: The bobbed locus. In: The Genetics and Biology of *Drosophila* (ASHBRUNER, M., NOVITSKI, E., eds.), Vol. 1 a, pp. 801—846. New York: Academic Press.

— SPIEGELMAN, S., 1965: Localization of DNA complementary to rRNA in the nucleolus organizer region of *Drosophila melanogaster*. Proc. Nat. Acad. Sci. U.S.A. **53**, 737—745.

— ATWOOD, K., LINDSLEY, D., SPIEGELMAN, S., 1966: On the chromosomal distribution of DNA complementary to ribosomal and soluble RNA. Nat. Cancer Inst. Monogr. **23**, 449—472.

— MALVA, C., BONCINELLI, E., GRAZIANI, F., POLITO, L., 1971: On the first steps of rDNA magnification. Proc. Nat. Acad. Sci. U.S.A. **68**, 1580—1584.

RIZZO, A. J., WEBB, T. E., 1968: Concurrent changes in the concentration of monomeric ribosomes and the rate of ribosome synthesis in rat liver. Biochim. Biophys. Acta **169**, 163—174.

RIZZO, A. J., WEBB, T. E., 1972: Regulation of ribosome formation in regenerating rat liver. Eur. J. Biochem. **27**, 136—144.

ROCHAIX, J. D., BIRD, A., BAKKEN, A., 1974: Ribosomal RNA gene amplification by rolling circles. J. Mol. Biol. **87**, 473—487.

RODRIGUES-POUSADA, C., CYRNE, M. L., HAYES, D., 1979: Characterization of preribosomal ribonucleoprotein particles from *Tetrahymena pyriformis*. Eur. J. Biochem. **102**, 389—397.

ROEDER, R. G., 1974: Multiple forms of DNA-dependent RNA polymerase in *Xenopus laevis*. Levels of activity during oocyte and embryonic development. J. Biol. Chem. **249**, 249—256.

— 1976: Eucaryotic nuclear RNA polymerase. In: RNA Polymerase (LOSICK, R., CHAMBERLIN, M., eds.), pp. 285—329. Cold Spring Harbor, N.Y.: Cold Spring Harbor Lab.

— RUTTER, W. J., 1969: Multiple forms of DNA-dependent RNA polymerase in eucaryotic organisms. Nature **224**, 234—240.

— — 1970 a: Specific nucleolar and nucleoplasmic RNA polymerases. Proc. Nat. Acad. Sci. U.S.A. **65**, 675—683.

— — 1970 b: Multiple ribonucleic acid polymerases and ribonucleic acid synthesis during sea urchin development. Biochemistry **9**, 2543—2550.

ROGERS, M., 1968: Ribonucleoprotein particles in amphibian oocyte nucleus. Possible intermediates in ribosome synthesis. J. Cell Biol. **36**, 421—432.

ROIHA, H., GLOVER, D. M., 1980: Characterization of complete type II insertions in cloned segments of ribosomal DNA from *Drosophila melanogaster*. J. Mol. Biol. **140**, 341—355.

— — 1981: Duplicated rDNA sequences of variable length flanking the short type I insertions in the rDNA of *Drosophila melanogaster*. Nucl. Acids Res. **9**, 5521—5532.

— MILLER, J. R., WOODS, L. C., GLOVER, D. M., 1981: Arrangements and rearrangements of sequences flanking the two types of rDNA insertion in *D. melanogaster*. Nature **290**, 749—753.

ROMEN, W., ALTMANN, H. W., 1977: Die Struktur des funktionsgestörten Nucleolus. Klin. Wschr. **55**, 563—567.

— KNOBLOCH, U., ALTMANN, H. W., 1977: Vergleichende Untersuchungen der Kernveränderungen von Ratten-Hepatocyten nach Actinomycin D- und α-Amanitin-Vergiftung. Virchows Arch. Cell Pathol. **B 23**, 93—108.

ROSBASH, M., HARRIS, P. K. W., WOOLFORD, J., L., JR., TEEM, J. L., 1981: The effect of temperature-sensitive RNA mutants on the transcription products from cloned ribosomal protein genes of yeast. Cell **24**, 679—686.

ROSE, K. M., STETLER, D. A., JACOB, S. T., 1981: Protein activity of RNA polymerase I purified from a rat hepatoma: probable function of Mr 42,000 and 24,600 polypeptides. Proc. Nat. Acad. Sci. U.S.A. **78**, 2833—2837.

ROSKIN, G. I., 1949: The exchange of substances between the nucleolus and cytoplasm during mitosis (russ.). Dokl. Akad. Nauk (SSSR) **69**, 585—590.

ROTH, H., BOLLA, R., COX, G. S., REDFIELD, B., WEISSBACH, H., BROT, N., 1976: Uptake of ribosomal proteins by isolated HeLa nuclei. Biochem. Biophys. Res. Commun. **69**, 608—612.

ROTHBLUM, L. I., MAMRACK, P. M., KUNKLE, H. M., OLSON, M. O. J., BUSCH, H.,

1977: Fractionation of nucleoli. Enzymatic and two-dimensional polyacrylamide gel electrophoretic analysis. Biochemistry **16**, 4716—4721.

Rothblum, L. I., Reddy, R., Cassidy, B., 1982: Transcription initiation site of rat ribosomal DNA. Nucl. Acids Res. **10**, 7345—7362.

Royal, A., Simard, R., 1975: RNA synthesis in the ultrastructural and biochemical components of the nucleolus of chinese hamster ovary cells. J. Cell Biol. **66**, 577—585.

— — 1980: Nucleolar RNA metabolism in CHO-tsHl, a mutant cell line with a thermosensitive leucyl-t-RNA synthetase. Biol. Cellulaire **38**, 27—35.

Rubin, G. M., Hogness, D. S., 1975: Effect of heat shock on the synthesis of low molecular weight RNAs in *Drosophila*: accumulation of a novel form of 5 S RNA. Cell **6**, 207—213.

Rubtsov, P. M., Musakhanov, M. M., Zakharyev, V. M., Krayev, A. S., Skryabin, K. G., Bayev, A. A., 1980: The structure of the yeast ribosomal RNA genes. I. The complete nucleotide sequence of the 18 S rRNA gene in *Saccharomyces cerevisiae*. Nucl. Acids Res. **8**, 5779—5794.

Rudland, P. S., 1974: Control of translation in cultured cells: continued synthesis and accumulation of messenger RNA in nondividing cultures. Proc. Nat. Acad. Sci. U.S.A. **71**, 750—754.

— Weil, S., Hunter, A. R., 1975: Changes in RNA metabolism and accumulation of presumptive mRNA during transition from the growing to the quiescent state of cultured mouse fibroblasts. J. Mol. Biol. **96**, 745—760.

Rungger, D., Crippa, M., 1977: The primary ribosomal DNA transcript in eucaryotes. Prog. Biophys. Mol. Biol. **31**, 247—269.

— — Trendelenburg, M., Scheer, U., Franke, W. W., 1978: Visualization of rDNA spacer transcription in *Xenopus* oocytes treated with fluorouridine. Exp. Cell Res. **116**, 481—486.

— Achermann, H., Crippa, M., 1979: Transcription of spacer sequences in genes coding for ribosomal RNA in *Xenopus laevis*. Proc. Nat. Acad. Sci. U.S.A. **76**, 3957—3961.

Sacristian-Garate, A., Navarrete, M. H., Torre, C. de la, 1974: Nucleolar development in the interphase of the cell cycle. J. Cell Sci. **16**, 333—347.

Saha, B. K., Schlessinger, D., 1978: Separation and characterization of two activities from HeLa cell nuclei that degrade double-stranded RNA. J. Biol. Chem. **253**, 4537—4543.

Saiga, H., Higashinakagawa, T., 1979: Properties of *in vitro* transcription by isolated *Xenopus* oocyte nucleoli. Nucl. Acids Res. **6**, 1929—1940.

— Mizumoto, K., Matsui, T., Higashinakagawa, T., 1982: Determination of the transcription initiation site of *Tetrahymena pyriformis* rDNA using *in vitro* capping of 35 S pre-rRNA. Nucl. Acids Res. **10**, 4223—4236.

Sakharov, V. N., Voronkova, L. N., Chentsov, U. S., 1972: Ultrastructure of intranuclear bodies formed during cell division of cells irradiated with an ultraviolet microbeam (russ.). Reports Moscow Univ., Biol. Sci. N **5**, 56—59.

Sakonju, S., Brown, D. D., 1981: The binding of a transcription factor to deletion mutants of a 5 S ribosomal RNA gene. Cell **23**, 665—669.

SAKONJU, S., BROWN, D. D., 1982: Contact points between a positive transcription factor and the *Xenopus* 5 S RNA gene. Cell **31**, 395—405.

— BOGENHAGEN, D. F., BROWN, D. D., 1980: A control region in the center of the 5 S RNA gene directs specific initiation of transcription. I. The 5′ border of the region. Cell **19**, 13—25.

SALIM, M., MADEN, B. E. H., 1980: Nucleotide sequence encoding the 5′end of *Xenopus laevis* 18 S rRNA. Nucl. Acids Res. **8**, 2871—2884.

— — 1981: Nucleotide sequence of *Xenopus laevis* 18 S ribosomal RNA inferred from gene sequence. Nature **291**, 205—208.

SAMAL, B., BALLAL, N. R., CHOI, Y. C., BUSCH, H., 1978: Effect of sarkosyl on the fidelity of preribosomal RNA synthesis in isolated nucleoli. Biochem. Biophys. Res. Commun. **84**, 328—334.

SAMESHIMA, M., IZAWA, M., 1975: Properties and synthesis of multiple components in native small ribosomal subunits of mouse ascites tumor cells. Biochim. Biophys. Acta **378**, 405—415.

SAMOLS, D. R., HAGENBÜCHLE, O., GAGE, L. P., 1979: Homology of the 3′ terminal sequences of the 18 S rRNA of *Bombyx mori* and the 16 S rRNA of *E. coli*. Nucl. Acids Res. **7**, 1109—1119.

SANKOFF, D., CEDERGREN, R. J., LAPALME, G., 1976: Frequency of insertion-deletion, transversion, and transition in the evolution of 5 S rRNA. J. Mol. Evol. **7**, 133—149.

SARASIN, A., MOULÉ, Y., 1975: Translational step inhibited *in vivo* by aflatoxin B_1 in rat liver polysomes. Eur. J. Biochem. **54**, 329—340.

SARIN, P. S., GALLO, R. C. (eds.), 1980: Inhibitors of DNA and RNA polymerases, p. 252. Oxford: Pergamon Press.

SAWADOGO, M., SENTENAC, A., FROMAGEOT, P., 1981: *In vitro* transcription of cloned yeast ribosomal DNA by yeast RNA polymerase A. Biochem. Biophys. Res. Commun. **101**, 250—257.

SCHÄFER, U., KUNZ, W., 1975: Two separated nucleolus organizers on the *Drosophila hydei* Y chromosome. Mol. Gen. Genet. **137**, 365—368.

— — 1976: Ribosomal DNA content and bobbed phenotype in *Drosophila hydei*. Heredity **37**, 351—355.

SCHÄFER, M., WYMAN, A. R., WHITE, R., 1981: Length variation in the non-transcribed spacer of *Calliphora erythrocephala* ribosomal DNA is due to a 350 base-pair repeat. J. Mol. Biol. **146**, 179—199.

SCHEER, U., 1973: Nuclear pore flow rate of ribosomal RNA and chain growth rate of its precursor during oogenesis of *Xenopus laevis*. Develop. Biol. **30**, 13—28.

— 1978: Changes in nucleosome frequency in nucleolar and non-nucleolar chromatin as a function of transcription: An electron microscopic study. Cell **10**, 525—549.

— 1982: A novel type of chromatin organization in lampbrush chromosomes of *Plourodeles waltlii*: visualization of clusters of tandemly repeated, very short transcriptional units. Biol. Cell **44**, 213—220.

— ZENTGRAF, H., 1978: Nucleosomal and supranucleosomal organization of transcriptionally inactive rDNA circles in *Dytiscus* oocytes. Chromosoma **69**, 243—254.

— TRENDELENBURG, M. F., FRANKE, W. W., 1973: Transcription of rRNA cistrons.

Correlation of morphological and biochemical data. Exp. Cell Res. **80**, 175—190.

Scheer, U., Trendelenburg, M. F., Franke, W. W., 1975: Effects of actinomycin D on the association of newly formed ribonucleoproteins with the cistrons of rRNA in *Triturus* oocytes. J. Cell Biol. **65**, 163—179.

— — — 1976: Regulation of transcription of genes of ribosomal RNA during amphibian oogenesis. J. Cell Biol. **69**, 465—489.

— — Krohne, G., Franke, W. W., 1977: Lengths and patterns of transcriptional units in the amplified nucleoli of oocytes of *Xenopus laevis*. Chromosoma **60**, 147—167.

— Lanfranchi, G., Rose, K. M., Franke, W. W., Ringertz, N. R., 1983: Migration of rat RNA polymerase I into chick erythrocyte nuclei undergoing reactivation in chick-rat heterokaryons. J. Cell Biol. **97**, 1641—1643.

Scherrer, K., Darnell, J. E., 1962: Sedimentation chracteristics of rapidly labeled RNA from HeLa cells. Biochem. Biophys. Res. Commun. **7**, 486—490.

Schibler, U., Wyler, T., Hagenbuchle, O., 1975: Changes in size and secondary structure of the ribosomal transcription unit during vertebrate evolution. J. Mol. Biol. **94**, 505—517.

Schmid, W., Sekeris, C., 1973: Possible involvement of nuclear DNA-like RNA in the control of ribosomal RNA synthesis. Biochim. Biophys. Acta **312**, 549—554.

Schneider, E. L., Shorr, S. S., 1975: Alteration in cellular RNAs during the *in vitro* lifespan of cultured human diploid fibroblasts. Cell **6**, 179—184.

Scholla, C. A., Tedeschi, M. V., Fausto, N., 1980: Gene expression and the diversity of polysomal mRNA sequences in regenerating liver. J. Biol. Chem. **255**, 2855—2860.

Schwarzacher, H. G., Mikelsaar, A. V., Schnedl, W., 1978: The nature of Ag-staining of nucleolus organizer regions. Electron and lightmicroscopic studies on human cells in interphase, mitosis, and meiosis. Cytogenet. Cell Genet. **20**, 24—39.

Schweizer, E., MacKechnie, C., Halvorson, H. O., 1969: The redundancy of ribosomal and transfer genes in *Saccharomyces cerevisiae*. J. Mol. Biol. **40**, 261—277.

Seifart, K., Schurrer, A., Krüger, C., 1978: Ribosomal RNA synthesis in rat liver cells. Z. Physiol. Chem. **359**, 1419—1425.

Sekeris, C. E., Schmid, W., 1972: Action of α-amanitin *in vivo* and *in vitro*. FEBS Lett. **27**, 41—45.

Selker, E. U., Yanofsky, C., Driftmier, K., Metzenberg, R. L., Alzner-De Weerd, B., RajBhandary, U., 1981 a: Dispersed 5 S RNA genes in *N. crassa*: Structures, expression, and evolution. Cell **24**, 819—828.

— Free, S. J., Metzenberg, R. L., Yanofsky, C., 1981 b: An isolated pseudogene related to the 5 S RNA genes in *Neurospora crassa*. Nature **294**, 576—578.

Serfling, E., Wobus, U., Panitz, R., 1972: Effect of α-amanitin on chromosomal and nucleolar RNA synthesis in *Chironomus thummi* polytene chromosomes. FEBS Lett. **20**, 148—152.

Shaaya, E., Clever, U., 1973: *In vivo* effects of α-amanitin on RNA synthesis in *Calliphora erythrocephala*. Biochim. Biophys. Acta **272**, 373—381.

SHANMUGAM, G., 1978: Partial purification and characterization of Ribonuclease III like enzyme activity from cultured mouse embryo cells. Biochemistry **17**, 5052—5057.

SHAPER, J. H., PARDOLL, D. M., KAUFMANN, S. H., BARRACK, E. R., VOGELSTEIN, B., COFFEY, D. S., 1979: The relationship of the nuclear matrix to cellular structure and function. Adv. Enzyme Regulation **17**, 213—248.

SHASTRY, B. S., NG, S.-Y., ROEDER, R. G., 1982: Multiple factors involved in the transcription of class III genes in *Xenopus laevis*. J. Biol. Chem. **257**, 12979—12986.

SHELDON, S., SPEERS, W. C., LEHMAN, J. M., 1981: Nucleolar persistence in embryonal carcinoma cells. Exp. Cell Res. **132**, 185—192.

SHERMOEN, A. W., KIEFER, B. I., 1975: Regulation in rRNA deficient *Drosophila melanogaster*. Cell **4**, 275—280.

SHI, X.-P., WINGENDER, E., BÖTTRICH, J., SEIFART, K. H., 1983: Faithful transcription of ribosomal 5 S RNA *in vitro* depends on the presence of several factors. Eur. J. Biochem. **131**, 189—194.

SHIELDS, D., TATA, J. R., 1976: Variable stabilities and recoveries of rat liver RNA polymerases A and B according to growth status of the tissue. Eur. J. Biochem. **64**, 471—480.

SHINOZUKA, H., GOLDBLATT, P. J., FARBER, E., 1968: The disorganization of hepatic cell nucleoli induced by ethionine and its reversal by adenine. J. Cell Biol. **36**, 313—328.

— MARTIN, J. F., FARBER, J. L., 1973: The induction of fibrillar nucleoli in rat liver cells by D-galactosamine and their subsequent reformation into normal nucleoli. J. Ultrastr. Res. **44**, 279—292.

SHULMAN, R. W., WARNER, J. R., 1978: Ribosomal RNA transcription in a mutant of *Saccharomyces cerevisiae* defective in ribosomal protein synthesis. Molec. Gen. Genet. **161**, 221—223.

— SRIPATI, C. E., WARNER, J. R., 1977: Noncoordinated transcription in the absence of protein synthesis in yeast. J. Biol. Chem. **252**, 1344—1349.

SIEGEL, A., KOLACZ, K., 1983: Heterogeneity of pumpkin ribosomal DNA. Plant Physiol. **72**, 166—171.

SIEV, M., WEINBERG, R., PENMAN, S., 1969: The selective interruption of nucleolar RNA synthesis in HeLa cells by cordycepin. J. Cell Biol. **41**, 510—520.

SIGMUND, J., SCHWARZACHER, H. G., MIKELSAAR, A. V., 1979: Satellite association frequency and number of nucleoli depend on cell cycle duration and NOR activity. Hum. Genet. **50**, 81—91.

SILLEVIS-SMITT, W. W., VERMEULEN, C. A., VLAK, J. M., ROZIJN, T. H., MOLENAAR, I., 1972: Electron microscopic autoradiographic study of RNA synthesis in yeast nucleus. Exp. Cell Res. **70**, 140—144.

SIMARD, R., 1970: The nucleus: Action of chemical and physical agents. Intern. Rev. Cytol. **28**, 169—211.

— BERNHARD, W., 1966: Le phénomène de la ségrégation nucléolaire: spécificité d'action de certains antimétabolites. Int. J. Cancer **1**, 463—479.

— LANGELIER, Y., ROSEMONDE, M., MAESTRACCI, N., ROYAL, A., 1974: Inhibitors as tools in elucidating the structure and function of the nucleus. In: The Cell Nucleus (BUSCH, H., ed.), Vol. 3, pp. 447—487. New York: Academic Press.

SIMEONE, A., FALCO, A. DE, MACINO, G., BONCINELLI, E., 1982: Sequence organization of the ribosomal spacer of *D. melanogaster*. Nucl. Acids Res. **10**, 8263—8272.

SINCLAIR, G. D., BRASCH, K., 1978: The reversible action of α-amanitin on nuclear structure and molecular composition. Exp. Cell Res. **111**, 1—14.

SITZ, T. O., BANERJEE, N., NAZAR, R. N., 1981: Effect of point mutations on 5.8 S rRNA secondary structure and the 5.8 S–28 S rRNA junction. Biochemistry **20**, 4029—4033.

SKRYABIN, K. G., KRAYEV, A. S., RUBTSOV, P. M., BAYEV, A. A., 1979: Complete nucleotide sequence of the spacer region between 18 S and 5.8 S rRNA of yeast. Doklady Acad. Nauk SSSR **247**, 761—765.

SLACK, J. M., LOENING, U. E., 1974: 5′-ends of ribosomal and ribosomal precursor RNAs from *Xenopus laevis*. Eur. J. Biochem. **43**, 59—67.

SMETANA, K., BUSCH, H., 1974: The nucleolus and nucleolar DNA. In: The Cell Nucleus (BUSCH, H., ed.), Vol. 1, pp. 75—147. New York: Academic Press.

— GYORKEY, F., GYORKEY, P., BUSCH, H., 1970: Studies on the ultrastructure of nucleoli in human smooth muscle cells. Exp. Cell Res. **60**, 175—184.

— RAŠKA, I., SEBESTA, K., 1974: The effect of the *Bacillus thuringiensis* exotoxin on the fine nucleolar morphology and ultrastructure. Exp. Cell Res. **87**, 351—358.

SMITH, G. P., 1973: Unequal crossover and the evolution of multigene families. Cold Spring Harbor Symp. Quant. Biol. **38**, 507—513.

— 1976: Evolution of repeated DNA sequences by unequal crossover. DNA whose sequence is not maintained by selection will develop periodicities as a result of random crossover. Science **191**, 528—535.

SMUCKLER, E. A., HADJIOLOV, A. A., 1972: Inhibition of hepatic DNA-dependent RNA polymerases by the exotoxin of *Bacillus thuringiensis* in comparison with the effects of α-amanitin and cordycepin. Biochem. J. **129**, 153—166.

SOEIRO, R., BASILE, C., 1973: Non-ribosomal nucleolar proteins in HeLa cells. J. Mol. Biol. **79**, 507—519.

— VAUGHAN, M. H., DARNELL, J. E., 1968: The effect of puromycin on intranuclear steps in ribosome biogenesis. J. Cell Biol. **36**, 91—101.

SOLLNER-WEBB, B., MCKNIGHT, S., 1982: Accurate transcription of cloned *Xenopus* rRNA genes by RNA polymerase I: demonstration by S 1 nuclease mapping. Nucl. Acids Res. **10**, 3391—3405.

— REEDER, R. H., 1979: The nucleotide sequence of the initiation and termination sites for ribosomal RNA transcription in *X. laevis*. Cell **18**, 485—499.

SPEAR, B. B., 1974: Differential replication of rRNA genes in eucaryotes. In: Ribosomes (NOMURA, M., TISSIERES, A., LENGYEL, P., eds.), pp. 841—853. Cold Spring Harbor, N.Y.: Cold Spring Harbor Laboratory.

SPEIRS, J., BIRNSTIEL, M. L., 1974: Arrangement of the 5.8 S cistrons in the genome of *Xenopus laevis*. J. Mol. Biol. **87**, 237—256.

SPRING, H., TRENDELENBURG, M. F., SCHEER, U., FRANKE, W. W., HERTH, W., 1974: Structural and biochemical studies of the primary nucleus of two green algal species, *Acetabularia mediterranea* and *Acetabularia major*. Cytobiologie **10**, 1—65.

— KROHNE, G., FRANKE, W., SCHEER, U., TRENDELENBURG, M. F., 1976:

Homogeneity and heterogeneity of sizes of transcriptional units and spacer regions in nucleolar genes of *Acetabularia*. J. Micr. Biol. Cell **25**, 107—116.

SPRING, H., GRIERSON, D., HEMLEBEN, V., STÖHR, M., KROHNE, G., STADLER, J., FRANKE, W., 1978: DNA contents and members of nucleoli and pre-rRNA-genes in nuclei of gametes and vegetative cells of *Acetabularia mediterranea*. Exp. Cell Res. **114**, 203—215.

STAHL, A., 1982: The nucleolus and nucleolar chromosomes. In: The Nucleolus (JORDAN, E. G., CULLIS, C. A., eds.), pp. 1—24. Cambridge: Cambridge Univ. Press.

STAHL, D. A., LUEHRSEN, K. R., WOESE, C. R., PACE, N. R., 1981: An unusual 5 S rRNA, from *Sulfolobus acidocaldarius*, and its implications for a general 5 S rRNA structure. Nucl. Acids Res. **9**, 6129—6137.

STALDER, J., SEEBECK, TH., BRAUN, R., 1978: Degradation of the ribosomal genes by DNAse I in *Physarum polycephalum*. Eur. J. Biochem. **90**, 391—395.

STAMBROOK, P. J., 1976: Organization of the genes coding for 5 S RNA in the Chinese hamster. Nature **259**, 639—641.

STANNERS, C. P., BECKER, H., 1971: Control of macromolecular synthesis in proliferating and resting Syrian hamster cells in monolayer culture I. Ribosome function. J. Cell Physiol. **77**, 31—42.

— ADAMS, M. E., HARKINS, J. L., POLLARD, J. W., 1979: Transformed cells have lost control of ribosome number through their growth cycle. J. Cell Physiol. **100**, 127—138.

STEER, W. M., MOLGAARD, H. V., BRADBURY, E. M., MATTHEWS, H. R., 1978: Ribosomal genes in *Physarum polycephalum*: transcribed and non-transcribed sequences have similar base compositions. Eur. J. Biochem. **88**, 599—605.

STEFFENSEN, D. M., DUFFEY, P., PRENSKY, W., 1974: Localization of 5 S rRNA genes on human chromosome I. Nature (London) **252**, 741—743.

STELLETSKAYA, N., SYROVA, T., PANCHENKO, L., SCHUPPE, N., 1973: Peculiarity of RNA turnover in the presence of inhibitors of energy metabolism. Biokhimya **38**, 1267—1274.

STENRAM, U., 1972: Relationship between nucleolar size and the synthesis and processing of pre-ribosomal RNA in the liver of rat. FEBS Symposia **24**, 131—141.

STIEGLER, P., CARBON, PH., EBEL, J.-P., EHRESMANN, CH., 1981 a: A general secondary-structure model for prokaryotic and eukaryotic RNAs of the small ribosomal subunits. Eur. J. Biochem. **120**, 487—495.

— — ZUKER, M., EBEL, J.-P., EHRESMANN, CH., 1981 b: Structural organization of the 16 S ribosomal RNA from *E. coli*. Topography and secondary structure. Nucl. Acids Res. **9**, 2153—2172.

STIRPE, F., FIUME, L., 1967: Studies on the pathogenesis of liver necrosis by α-amanitin. Effect of α-amanitin on RNA synthesis and on RNA polymerase in mouse liver nuclei. Biochem. J. **105**, 779—782.

STOWELL, R. E., 1949: Alterations in nucleic acids during hepatoma formation in rats fed p-dimethylaminoazobenzene. Cancer **2**, 121—131.

STOYANOVA, B. B., DABEVA, M. D., 1980: Pre-ribosomal RNA transcription in rat liver is not dependent on continuous synthesis of proteins. Biochim. Biophys. Acta **608**, 358—367.

Stoyanova, B. B., Hadjiolov, A. A., 1979: Alterations in the processing of rat liver rRNA caused by cycloheximide inhibition of protein synthesis. Eur. J. Biochem. **96**, 349—356.

— Petrov, P. T., Dabeva, M. D., 1980: Morphological and biochemical changes in rat liver nucleus induced by cycloheximide. Biol. Zbl. **99**, 171—182.

Stoykova, A. S., Dudov, K. P., Dabeva, M. D., Hadjiolov, A. A., 1983: Different rates of synthesis and turnover of rRNA in rat brain and liver. J. Neurochem. **41**, 942—949.

Strehler, B. L., Chang, M., Johnson, L. K., 1979: Loss of hybridizable ribosomal DNA from human post-mitotic tissues during aging I. Age-dependent loss in human myocardium. Mech. Ageing Developm. **11**, 371—378.

Strelkov, L. A., Kaffiani, K. A., 1978: Molecular biology of animal ribosomal RNA genes (russ.). Uspekhi Biol. Khimii **19**, 32—60.

Stuart, W. D., Bishop, J. G., Carson, H. L., Frank, M. B., 1981: Location of the 18/28 S ribosomal RNA genes in two Hawaiian *Drosophila* species by monoclonal immunological identification of RNA. DNA hybrids *in situ*. Proc. Nat. Acad. Sci. U.S.A. **78**, 3751—3754.

Stüber, D., Bujard, H., 1977: Electron microscopy of DNA: Determination of absolute molecular weights and linear density. Mol. Gen. Genet. **154**, 299—303.

Stumph, W. E., Wu, J.-R., Bonner, J., 1979: Determination of the size of rat ribosomal DNA repeating units by electron microscopy. Biochemistry **18**, 2864—2871.

Sturani, E., Sacco, G., 1982: Regulation of synthesis of ribosomal protein in *Neurospora crassa*. Exp. Cell Res. **142**, 357—364.

Subrahmanyam, N. C., Azad, A. A., 1978: Nucleoli and ribosomal RNA cistron numbers in *Hordeum* species and interspecific hybrids exhibiting suppression of secondary constriction. Chromosoma **69**, 265—273.

Subrahmanyam, Ch. S., Cassidy, B., Busch, H., Rothblum, L. I., 1982: Nucleotide sequence of the region between the 18 S rRNA sequence and the 28 S rRNA sequence of rat ribosomal DNA. Nucl. Acids Res. **10**, 3667—3680.

Sugden, B., Keller, W., 1973: Mammalian deoxyribonucleic acid-dependent ribonucleic acid polymerases. J. Biol. Chem. **248**, 3777—3785.

Suhadolnik, R. J., 1970: Nucleoside Antibiotics, pp. 1—420. New York: Wiley/Interscience.

— 1979: Naturally occurring nucleoside and nucleotide antibiotics. Progr. Nucl. Acids Res. Mol. Biol. **22**, 193—291.

Sun, I. Y.-C., Johnson, E. M., Allfrey, V. G., 1979: Initiation of transcription of rDNA sequences in isolated nuclei of *Physarum polycephalum*: studies using nucleoside-5′-[δ-S]triphosphates and labeled precursors. Biochemistry **18**, 4572—4580.

Sutton, C. A., Sylvom, P., Hallberg, R. L., 1979: Ribosome biosynthesis in *Tetrahymena thermophila*. IV. Regulation of rRNA synthesis in growing and growth arrested cells. J. Cell Physiol. **101**, 503—514.

Szostak, J. W., Wu, R., 1980: Unequal crossing over in the ribosomal DNA of *S. cerevisiae*. Nature **284**, 426—430.

Tabata, S., 1980: Structure of the 5 S ribosomal RNA gene and its adjacent regions in *Torulopsis utilis*. Eur. J. Biochem. **110**, 107—114.

— 1981: Nucleotide sequences of the 5 S ribosomal RNA genes and their adjacent regions in *Schizosaccharomyces pombe*. Nucl. Acids Res. **9**, 6429—6437.

Takatsuka, Y., Kohno, M., Higashi, K., Hirano, H., Sakamoto, Y., 1976: Redistribution of chromatin containing ribosomal cistrons during liver regeneration. Exp. Cell Res. **103**, 191—199.

Takami, H., Busch, H., 1979: Two-dimensional gel electrophoretic comparison of proteins of nuclear fractions of normal liver and Novikoff hepatoma. Cancer Res. **39**, 507—518.

Tamaoki, T., Lane, B. G., 1968: Methylation of sugars and bases in ribosomal and rapidly labeled ribonucleates from normal and puromycin-treated L cells. Biochemistry **7**, 3431—3440.

Tan, E. M., 1978: Autoimmunity to nuclear antigens. In: The Cell Nucleus (Busch, H., ed.), Vol. VII, pp. 457—478. New York: Academic Press.

Tantravahi, U., Guntaka, R. V., Erlanger, B. F., Miller, O. J., 1981: Amplified ribosomal RNA genes in a rat hepatoma cell line are enriched in 5-methylcytosine. Proc. Nat. Acad. Sci. U.S.A. **78**, 489—493.

Tartof, K. D. 1973: Unequal mitotic sister chromosomal exchange and disproportionate replication as mechanisms regulating ribosomal RNA gene redundancy. Cold Spring Harbor Symp. Quant. Biol. **38**, 491—500.

— 1974: Unequal mitotic sister chromatid exchange as the mechanism of ribosomal RNA gene magnification. Proc. Nat. Acad. Sci. U.S.A. **71**, 1272—1276.

— 1975: Redundant genes. Ann. Rev. Genet. **9**, 355—385.

— 1979: Evolution of transcribed and spacer sequences in the ribosomal RNA genes of *Drosophila*. Cell **17**, 607—614.

— Dawid, I. B., 1976: Similarities and differences in the structure of X and Y chromosome rRNA genes of *Drosophila*. Nature **263**, 27—30.

— Perry, R. P., 1970: 5 S RNA genes of *Drosophila melanogaster*. J. Mol. Biol. **51**, 171—183.

Tata, J. R., 1970: Regulation of protein synthesis by growth and developmental hormones. In: Biochemical Actions of Hormones (Litwack, G., ed.), pp. 89—133. New York: Academic Press.

— Baker, B., 1974: Sub-nuclear fractionation. II. Intranuclear compartmentation of transcription *in vivo* and *in vitro*. Exp. Cell Res. **83**, 125—138.

— Hamilton, M. J., Shields, D., 1972: Effects of α-amanitin *in vivo* on RNA polymerase and nuclear RNA synthesis. Nature New Biol. **238**, 161—164.

Tavitian, A., Uretsky, S., Acs, G., 1968: Selective inhibition of rRNA synthesis in mammalian cells. Biochim. Biophys. Acta **157**, 33—42.

— — — 1969: The effect of toyocamycin on cellular RNA synthesis. Biochim. Biophys. Acta **179**, 50—57.

Tekamp, P. A., Valenzuela, P., Maynard, T., Bell, G. I., Rutter, W. J., 1979: Specific gene transcription in yeast nuclei and chromatin by added homologous RNA polymerases I and III. J. Biol. Chem. **254**, 955—963.

Thomas, G., Gordon, J., 1979: Regulation of protein synthesis during the shift of quiescent animal cells into the proliferative state. Cell Biol. Int. Repts. **3**, 307—320.

THOMAS, G., PODESTA, E., GORDON, J. (eds.), 1979a: Protein Phosphorylation and Bioregulation, pp. 1—232. Basel: S. Karger.

— SIEGMANN, M., GORDON, J., 1979b: Multiple phosphorylation of ribosomal protein S 6 during transition of quiescent 3 T 3 cells into early G_1 and cellular compartmentation of the phosphate donor. Proc. Nat. Acad. Sci. U.S.A. **76**, 3952—3956.

TIOLLAIS, P., GALIBERT, F., BOIRON, M., 1971: Evidence for the existence of several molecular species in the "45 S fraction" of mammalian ribosomal precursor RNA. Proc. Nat. Acad. Sci. U.S.A. **68**, 1117—1120.

TOBLER, H., 1975: Occurrence and developmental significance of gene amplification. In: Biochemistry of Animal Development (WEBER, R., ed.), Vol. 3, pp. 91—143. New York: Academic Press.

TODARO, G. J., GREEN, H., 1963: Quantitative studies of the growth of mouse embryo cells in culture and their development into established cell lines. J. Cell Biol. **17**, 299—313.

TODOKORO, K., ULBRICH, N., CHAN, Y.-L., WOOL, I. G., 1981: Characterization of the binding of rat liver ribosomal proteins L 6, L 8, L 19, S 9, and S 13 to 5.8 S rRNA. J. Biol. Chem. **256**, 7207—7212.

TODOROV, I., HADJIOLOV, A. A., 1979: A comparison of nuclear and nucleolar matrix proteins. Cell Biol. Int. Rep. **3**, 753—757.

— — 1981: A precursor to the small ribosome in nucleoli of Friend erythroleukemia cells. Cell Biol. Int. Rep. **5**, 711—716.

— NOLL, F., HADJIOLOV, A. A., 1983: The sequential addition of ribosomal proteins during the formation of the small ribosomal subunit in Friend erythroleukemia cells. Eur. J. Biochem. **131**, 271—275.

TOLSTOSHEV, P., BERG, R. A., RENNARD, S. I., BRADLEY, K. H., TRAPNELL, B. C., CRYSTAL, R. G., 1981: Procollagen production and procollagen messenger RNA levels and activity in human lung fibroblasts during periods of rapid and stationary growth. J. Biol. Chem. **256**, 3135—3140.

TONIOLO, D., BASILICO, C., 1976: Processing of rRNA in a temperature sensitive mutant of BHK cells. Biochim. Biophys. Acta **425**, 409—418.

— MEISS, H. K., BASILICO, C., 1973: A temperature sensitive mutation affecting 28 S ribosomal RNA production in mammalian cells. Proc. Nat. Acad. Sci. U.S.A. **70**, 1273—1277.

TØNNESEN, T., ENGBERG, J., LEICK, V., 1976: Studies on the amount and location of the tRNA and 5 S RNA Genes in *Tetrahymena pyriformis* GL. Eur. J. Biochem. **63**, 399—407.

TORELLI, U., FERRARI, S., TORELLI, G., CADOSSI, R., FERRARI, S., MONTAGNANI, G., NARNI, F., 1977: *In vitro* cleavage of 45 S pre-rRNA and of giant heterogenous RNA extracted from human leukaemic cells. Mol. Biol. Rep. **3**, 403—411.

TOWBIN, H., RAMJONE, H.-P., KUSTER, H., LIVERANI, D., GORDON, J., 1982: Monoclonal antibodies against eucaryotic ribosomes. J. Biol. Chem. **257**, 12709—12715.

TRAPMAN, J., PLANTA, R. J., 1975: Detailed analysis of the ribosomal RNA synthesis in yeast. Biochim. Biophys. Acta **414**, 115—125.

Trapman, J., Planta, R. J., 1976: Maturation of ribosomal in yeast. I. Kinetic analysis by labeling of high molecular weight rRNA species. Biochim. Biophys. Acta **442**, 265—274.

— De Jonge, P., Planta, R. J., 1975a: On the biosynthesis of 5.8 S ribosomal RNA in yeast. FEBS Lett. **57**, 26—30.

— Retel, J., Planta, R. J., 1975b: Ribosomal precursor particles from yeast. Exp. Cell Res. **90**, 95—104.

— Planta, R. J., Raué, H. A., 1976: Maturation of ribosomes in yeast. II. Position of the low molecular weight rRNA species in the maturation process. Biochim. Biophys. Acta **442**, 275—284.

Trendelenburg, M. F., 1974: Morphology of rRNA cistrons in oocytes of the water beetle, *Dytiscus marginalis* L. Chromosoma **48**, 119—135.

— 1981: Initiations of transcription at distinct promotor sites in spacer regions between pre-rRNA genes in oocytes of *Xenopus laevis*; an electron microscopic analysis. Biol. Cell **42**, 1—12.

— 1982: Chromatin structure of *Xenopus* rDNA transcription termination sites. Chromosoma **86**, 703—715.

— 1983: Progress in visualization of eukaryotic gene transcription. Human Genet. **63**, 197—215.

— Gurdon, J. B., 1978: Transcription of cloned *Xenopus* ribosomal genes visualized after injection into oocyte nuclei. Nature (London) **276**, 292—294.

— McKinnell, R. G., 1979: Transcriptionally active and inactive regions of nucleolar chromatin of amplified nucleoli of fully grown oocytes of hibernating frogs, *Rana pipiens* (*Amphibia, Anura*). Differentiation **15**, 73—95.

— Scheer, U., Franke, W. W., 1973: Structural organization of the transcription of rDNA in oocytes of the house cricket. Nature (London), New Biol. **245**, 167—170.

— Spring, H., Scheer, U., Franke, W. W., 1974: Morphology of nucleolar cistrons in a plant cell, *Acetabularia mediterranea*. Proc. Nat. Acad. Sci. U.S.A. **71**, 3626—3630.

— Scheer, U., Zentgraf, H., Franke, W. W., 1976: Heterogeneity of spacer lengths in circles of amplified rDNA of two insect species, *Dytiscus marginalis* and *Achaeta domesticus*. J. Mol. Biol. **108**, 453—470.

— Franke, W. W., Scheer, U., 1977: Frequencies of circular units of nucleolar DNA in oocytes of two insects, *Achaeta domesticus* and *Dytiscus marginalis*, and changes of nucleolar morphology during oogenesis. Differentiation **7**, 133—158.

— Zentgraf, H., Franke, W. W., Gurdon, J. B., 1978: Transcription patterns of amplified *Dytiscus* genes coding for rRNA after injection into *Xenopus* oocyte nuclei. Proc. Nat. Acad. Sci. U.S.A. **75**, 3791—3795.

Truett, M. A., Gall, J. G., 1977: The replication of ribosomal DNA in the macronucleus of *Tetrahymena*. Chromosoma **64**, 295—303.

Tsanev, R. G., 1965: Direct spectrophotometric analysis of RNA fractionation by agar-gel electrophoresis. Biochim. Biophys. Acta **103**, 374—382.

Tschudi, Ch., Pirrotta, V., 1980: Sequence and heterogeneity in the 5 S RNA gene cluster of *Drosophila melanogaster*. Nucl. Acids Res. **8**. 441—451.

Tsurugi, K., Ogata, K., 1977: Preferential degradation of newly synthesized ribosomal proteins in rat liver treated with a low dose of actinomycin D. Biochem. Biophys. Res. Commun. **75**, 525—531.

Tsurugi, K., Ogata, K., 1979: Degradation of newly synthesized ribosomal proteins and histones in regenerating rat liver with and without treatment with a low dose of actinomycin D. Eur. J. Biochem. **101**, 205—213.

— Morita, T., Ogata, K., 1972: Effects of the inhibition of ribosomal RNA synthesis on the synthesis of ribosomal structural proteins in regenerating rat liver. Eur. J. Biochem. **29**, 585—592.

— — — 1973: Identification and metabolic relationships between proteins of nucleolar 60 S particles and of ribosomal large subunits of rat liver by means of two-dimensional disc electrophoresis. Eur. J. Biochem. **32**, 555—562.

— — — 1974: Mode of degradation of ribosomes in regenerating rat liver *in vivo*. Eur. J. Biochem. **45**, 119—126.

— Oyanagi, M., Ogata, K., 1983: The role of nuclear serine proteases in the degradation of newly synthesized histones and ribosomal proteins. J. Biochem. (Tokyo) **93**, 1231—1237.

Tushinski, R. J., Warner, J. R., 1982: Ribosomal proteins are synthesized preferentially in cells commencing growth. J. Cell Physiol. **112**, 128—135.

Udem, S. A., Warner, J. R., 1972: Ribosomal RNA synthesis in *Saccharomyces cerevisiae*. J. Mol. Biol. **65**, 227—242.

— — 1973: The cytoplasmic maturation of a ribosomal precursor RNA in yeast. J. Biol. Chem. **248**, 1412—1416.

Udvardy, A., Seifart, K. H., 1976: Transcription of specific genes in isolated nuclei from HeLa cells *in vivo*. Eur. J. Biochem. **62**, 353—363.

Ulbrich, N., Lin, A., Wool, I. G., 1979: Identification by affinity chromatography of the eucaryotic ribosomal proteins that bind to 5.8 S rRNA. J. Biol. Chem. **254**, 8641—8645.

— Todokoro, K., Ackerman, E. J., Wool, I. G., 1980: Characterization of the binding of rat liver ribosomal proteins L 6, L 7, and L 19 to 5 S rRNA. J. Biol. Chem. **255**, 7712—7715.

— Chan, Y.-L., Huber, P. W., Wool, I. G., 1982: Separate binding sites on rat liver ribosomal protein L6 for 5 S and 5.8 S rRNA and for tRNA. J. Biol. Chem. **257**, 11353—11357.

Unuma, T., Busch, H., 1967: Formation of microspherules in nucleoli of tumor cells treated with high doses of actinomycin D. Cancer Res. **27**, 1232—1242.

— Senda, R., Muramatsu, M., 1973: Mechanism of nucleolar segregation: differences in effect of actinomycin D and cycloheximide on nucleoli of rat liver cells. J. Electron Microsc. (Tokyo) **22**, 205—216.

Urano, Y., Kominami, R., Mishima, Y., Muramatsu, M., 1980: The nucleotide sequence of the putative transcription initiation site of a cloned ribosomal RNA gene of the mouse. Nucl. Acids Res. **8**, 6043—6058.

Ursi, D., Vandenberghe, A., De Wachter, R., 1982: The sequence of the 5.8 S ribosomal RNA of the crustacean *Artemia salina* with a proposal for a general secondary structure model for 5.8 S ribosomal RNA. Nucl. Acids Res. **10**, 3517—3530.

VAGNER-CAPODANO, A. M., STAHL, A., 1980: Relationship of chromatin to nucleolar fibrillar center revealed by action of actinomycin D. Biol. Cell. **37**, 293—296.

VALENZUELA, P., BELL, G., VENEGAS, A., SEWELL, E., MASIARZ, F., DE GENNARO, L., WEINBERG, F., RUTTER, W. J., 1977: Ribosomal RNA genes of *Saccharomyces cerevisiae*. II. Physical map and nucleotide sequence of the 5 S rRNA gene and adjacent intergenic regions. J. Biol. Chem. **252**, 8126—8135.

VAN KEULEN, H., RETEL, J., 1977: Transcription specificity of yeast RNA polimerase A. Highly specific transcription *in vitro* of the homologous ribosomal transcription units. Eur. J. Biochem. **79**, 579—588.

VAN VENROOIJ, W. J., JANSSEN, A. P. M., 1976: Heterogeneity of native ribosomal 60 S subunits in Ehrlich ascites tumor cells cultured *in vitro*. Eur. J. Biochem. **69**, 55—60.

— — HOEYMAKERS, J. H., DE MAN, B. M., 1976: On the heterogeneity of native ribosomal subunits in Ehrlich ascites tumor cells cultured *in vitro*. Eur. J. Biochem. **64**, 429—435.

VARSANYI-BREINER, A., GUSELLA, J., KEYS, C., HOUSEMAN, D., SULLIVAN, D., BRISSON, N., VERMA, D., 1979: The organization of a nuclear DNA sequence from a higher plant: molecular cloning and characterization of soybean rDNA. Gene **7**, 317—334.

VASLET, CH., O'CONNELL, P., IZQUIERDO, M., ROSBASH, M., 1980: Isolation and mapping of a cloned ribosomal protein gene of *Drosophila melanogaster*. Nature **285**, 674—676.

VAUGHAN, M. H., JR., 1972: Comparison of regulation of synthesis and utilization of 45 S pre-rRNA in diploid and heteroploid human cells in response to valine deprivation. Exp. Cell Res. **75**, 23—30.

— SOEIRO, R., WARNER, J. R., DARNELL, J. E., 1967: The effects of methionine deprivation on ribosome synthesis in HeLa cells. Proc. Nat. Acad. Sci. U.S.A. **58**, 1527—1534.

VELDMAN, G. M., BRAND, R. C., KLOOTWIJK, J., PLANTA, R. J., 1980 a: Some characteristics of processing sites in ribosomal precursor RNA of yeast. Nucl. Acids Res. **8**, 2907—2920.

— KLOOTWIJK, J., DE JONGE, P., LEER, R. J., PLANTA, R. J., 1980 b: The transcription termination site of the rRNA operon in yeast. Nucl. Acids Res. **8**, 5179—5192.

— — HEERIKHUIZEN, H. VAN, PLANTA, R. J., 1981 a: The nucleotide sequence of the intergenic region between the 5.8 S and 26 S rRNA genes of the yeast rRNA operon. Possible implications for the interaction between 5.8 S and 26 S rRNA and the processing of the primary transcript. Nucl. Acids Res. **9**, 4847—4857.

— — DE REGT, V. C. H. F., PLANTA, R. J., BRANLANT, CH., KROL, A., EBEL, J.-P., 1981 b: The primary and secondary structure of yeast 26 S rRNA. Nucl. Acids Res. **9**, 6935—6952.

VENKOV, P. V., VASILEVA, A. P., 1979: *Saccharomyces cerevisiae* mutants defective in the maturation of ribosomal ribonucleic acid. Mol. Gen. Genet. **173**, 203—210.

— MILCHEV, G. I., HADJIOLOV, A. A., 1975: Rifampin susceptibility of RNA synthesis in a fragile *Saccharomyces cerevisiae* mutant. Antimicr. Agents Chemother. **8**, 627—632.

VENKOV, P. V., STATEVA, L. I., HADJIOLOV, A. A., 1977: Toyocamycin inhibition of rRNA processing in an osmotic-sensitive adenosine-utilizing *Saccharomyces cerevisiae* mutant. Biochim. Biophys. Acta **474**, 245—253.

VOGT, V., BRAUN, R., 1976: Structure of rDNA in *Physarum polycephalum*. J. Mol. Biol. **106**, 567—587.

— — 1977: The replication of ribosomal DNA in *Physarum polycephalum*. Eur. J. Biochem. **80**, 557—566.

WACHTLER, F., ELLINGER, A., SCHWARZACHER, H. G., 1980: Nucleolar changes in human phytohaemagglutinin-stimulated lymphocytes. Cell Tissue Res. **213**, 351—360.

— SCHWARZACHER, H. G., ELLINGER, A., 1982: The influence of the cell cycle on structure and number of nucleoli in cultured human lymphocytes. Cell Tissue Res. **225**, 155—163.

WALDRON, C., LACROUTE, F., 1975: Effect of growth rate on the amounts of ribosomal and transfer RNA in yeast. J. Bacteriol. **122**, 855—865.

WALKER, T. A., PACE, N. R., 1977: Transcriptional organization of the 5.8 S rRNA cistron in *Xenopus laevis*. Nucl. Acids Res. **4**, 595—601.

— JOHNSON, K., OLSEN, G., PETERS, M., PACE, N. R., 1982: Enzymatic and chemical structure mapping of mouse 28 S rRNA contacts in 5.8 S rRNA. Biochemistry **21**, 2320—2330.

WALKER, W. F., 1981: Proposed sequence homology between the 5′-end regions of prokaryotic 23 S rRNA and eukaryotic 28 S rRNA. FEBS Lett. **126**, 150—151.

WALTSCHEWA, L. V., VENKOV, P. V., STOYANOVA, B. B., HADJIOLOV, A. A., 1976: Degradation of ribosomal precursor and poly(A)-containing RNA in *Saccharomyces cerevisiae* caused by actinomycin D. Arch. Biochem. Biophys. **176**, 630—637.

— GEORGIEV, O., VENKOV, P. V., 1983: Relaxed mutant of *Saccharomyces cerevisiae*: proper maturation of rRNA in absence of protein synthesis. Cell **33**, 221—230.

WANG, T. Y., KOSTRABA, N. C., 1978: Proteins involved in positive and negative control of chromatin function. In: The Cell Nucleus (BUSCH, H., ed.), Vol. 4, Part A, pp. 289—317. New York: Academic Press.

WARBURTON, D., HENDERSON, A. S., ATWOOD, K. C., 1975: Localization of rDNA and giemsa-banded chromosome complement of white-handed gibbon *Hylobates bar*. Chromosoma **51**, 35—40.

WARNER, J. R., 1974: The assembly of ribosomes in eucaryotes. In: Ribosomes (NOMURA, M., TISSIERES, A., LENGYEL, P., eds.), pp. 451—488. Cold Spring Harbor, N.Y.: Cold Spring Harbor Laboratory.

— 1977: In the absence of rRNA synthesis the ribosomal proteins of HeLa cells are synthesized normally and degraded rapidly. J. Mol. Biol. **115**, 315—333.

— 1979: Distribution of newly formed ribosomal proteins in HeLa cell fractions. J. Cell Biol. **80**, 767—772.

— 1982: The yeast ribosome: structure, function, and synthesis. In: The Molecular

Biology of the Yeast Saccharomyces (STRATHERN, J. N., JONES, E. W., BROACH, J. R., eds.). New York: Cold Spring Harbor Laboratory.

WARNER, J. R., SOEIRO, R., 1967: Nascent ribosomes from HeLa cells. Proc. Nat. Acad. Sci. U.S.A. **58**, 1981—1990.

— UDEM, S. A., 1972: Temperature sensitive mutations affecting ribosome synthesis in *S. cerevisiae*. J. Mol. Biol. **65**, 243—257.

— GIRARD, M., LATHAM, H., DARNELL, J. E., 1966: Ribosome formation in HeLa cells in the absence of protein synthesis. J. Mol. Biol. **19**, 373—386.

— KUMAR, A., UDEM, S. A., WU, R. S., 1973: Ribosomal proteins and the assembly of ribosomes in eukaryotes. Biochem. Soc. Symp. **37**, 3—18.

— TUSHINSKI, R. J., WEIJKSNORA, P. J., 1979: Coordination of RNA and proteins in eukaryotic ribosome production. In: Ribosome, Structure, Function, and Genetics (CHAMBLISS, G., *et al.*, eds.), pp. 889—902. Baltimore: Univ. Park Press.

WASYLYK, B., CHAMBON, P., 1979: Transcription by eukaryotic RNA polymerases A and B of chromatin assembled *in vitro*. Eur. J. Biochem. **98**, 317—327.

— THEVENIN, G., OUDET, P., CHAMBON, P., 1979: Transcription of *in vitro* assembled chromatin by *Escherichia coli* RNA polymerase. J. Mol. Biol. **128**, 411—440.

WEAVER, R. F., BLATTI, S. P., RUTTER, W. J., 1971: Molecular structures of DNA-dependent RNA polymerase II from calf thymus and rat liver. Proc. Nat. Acad. Sci. U.S.A. **68**, 2994—2999.

WEBER, M. J., 1972: Ribosomal RNA turnover in contact inhibited cells. Nature (London) **235**, 58—61.

WEGNEZ, M., MONIER, R., DENIS, H., 1972: Sequence heterogeneity in the 5 S RNA in *Xenopus laevis*. FEBS Lett. **25**, 13—20.

WEIL, P., LUSE, D., SEGALL, J., ROEDER, R., 1979: Selective and accurate initiation of transcription at the Ad 2 major late promoter in a soluble system dependent on purified RNA polymerase II and DNA. Cell **18**, 469—484.

WEINBERG, R. A., 1982: Oncogenes of spontaneous and chemically induced tumors. Adv. Cancer Res. **36**, 149—163.

— PENMAN, S., 1968: Small molecular weight monodisperse nuclear RNA. J. Mol. Biol. **38**, 289—304.

— — 1970: Processing of 45 S nucleolar RNA. J. Mol. Biol. **47**, 169—178.

WEINMANN, R., 1972: Regulation of rRNA and 5 S RNA synthesis in *Drosophila melanogaster*. I. Bobbed mutants. Genetics **72**, 267—276.

— ROEDER, R., 1974: Role of DNA-dependent RNA polymerase III in the transcription of the tRNA and 5 S RNA genes. Proc. Nat. Acad. Sci. U.S.A. **71**, 1790—1794.

WEINTRAUB, H., WORCEL, A., ALBERTS, B., 1976: A model for chromatin based upon two symmetrically paired half-nucleosomes. Cell **9**, 409—417.

WEISBROD, S., 1982: Active chromatin. Nature **297**, 289—295.

WEISS, J. W., PITOT, H. C., 1974: Inhibition of rRNA maturation in Novikoff hepatoma cells by toyocamycin, tubercidin, and 6-thioguanosine. Cancer Res. **34**, 581—587.

WEISS, S. B., 1960: Enzymatic incorporation of ribonucleoside triphosphates into the interpolynucleotide linkages of RNA. Proc. Nat. Acad. Sci. U.S.A. **46**, 1020—1030.

Wellauer, P. K., Dawid, I. B., 1973: Secondary structure maps of RNA. Processing of HeLa rRNA. Proc. Nat. Acad. Sci. U.S.A. **70**, 2827—2831.

— — 1974: Secondary structure maps of rRNA and rDNA. I. Processing of *Xenopus laevis* rRNA and structure of single-stranded rDNA. J. Mol. Biol. **89**, 379—395.

— — 1975: Structure and processing of rRNA: A comparative electron microscopic study in three animals. Brookhaven Symp. Biol. **26**, 214—223.

— — 1977: The structural organization of rDNA in *Drosophila melanogaster*. Cell **10**, 193—212.

— — 1978: Ribosomal DNA in *Drosophila melanogaster*. II. Heteroduplex mapping of cloned and uncloned rDNA. J. Mol. Biol. **126**, 769—782.

— — 1979: Isolation and sequence of human ribosomal DNA. J. Mol. Biol. **128**, 289—303.

— Reeder, R. H., 1975: A comparison of the structural organization of amplified ribosomal DNA from *Xenopus mulleri* and *Xenopus laevis*. J. Mol. Biol. **94**, 151—160.

— Dawid, I. B., Kelley, D. E., Perry, R. P., 1974 a: Secondary structure maps of rRNA. II. Processing of mouse L-cell rRNA and variations in the processing pathway. J. Mol. Biol. **89**, 397—407.

— Reeder, R. H., Carroll, D., Brown, D., Deutch, A., Higashinakagawa, T., Dawid, I., 1974 b: Amplified rDNA from *Xenopus laevis* has heterogenous spacer lengths. Proc. Nat. Acad. Sci. U.S.A. **71**, 2823—2827.

— Dawid, I. B., Reeder, R. H., 1976 a: The molecular basis for length heterogeneity in rDNA from *Xenopus laevis*. J. Mol. Biol. **105**, 461—486.

— Reeder, R. H., Dawid, I. B., Brown, D. D., 1976 b: The arrangement of length heterogeneity in repeating units of amplified and chromosomal rDNA from *Xenopus laevis*. J. Mol. Biol. **105**, 487—505.

— Dawid, I. B., Tartof, K. D., 1978: X and Y chromosomal rDNA of *Drosophila*: Comparison of spacers and insertions. Cell **14**, 269—278.

Wells, D. J., Stoddard, L. S., Getz, M. J., Moses, H. L., 1979: α-Amanitin and 5-fluorouridine inhibition of serum stimulated DNA synthesis in quiescent AKR-2 B mouse embryo cells. J. Cell Physiol. **100**, 199—214.

Wensink, P. C., Brown, D. D., 1971: Denaturation map of the ribosomal DNA of *Xenopus laevis*. J. Mol. Biol. **60**, 235—247.

Whelly, S., Ide, T., Baserga, R., 1978: Stimulation of RNA synthesis in isolated nuclei by preparations of SV 40 T-antigen. Virology **88**, 82—91.

White, R. L., Hogness, D. S., 1977: R-loop mapping of the 18 S and 28 S sequences in the long and short repeating units in *Drosophila melanogaster* DNA. Cell **10**, 177—192.

Widnell, C. C., Tata, J. R., 1964: Evidence for two DNA-dependent RNA polymerase activities in isolated rat liver nuclei. Biochim. Biophys. Acta **87**, 531—533.

— — 1966: Studies on the stimulation by ammonium sulfate of the DNA-dependent RNA polymerase of isolated rat liver nuclei. Biochim. Biophys. Acta **123**, 478—492.

Wiegers, U., Kramer, G., Klapproth, K., Hilz, H., 1976: Separate pyrimidine nucleotide pools for messenger RNA and rRNA synthesis in HeLa S 3 cells. Eur. J. Biochem. **64**, 535—540.

WIESLANDER, L., LAMBERT, B., EGYHÁZI, E., 1975: Localization of 5 S genes in *Chironomus tentans*. Chromosoma **51**, 49—56.

WILD, M. A., GALL, J. G., 1979: An intervening sequence in the gene coding for 25 S ribosomal RNA of *Tetrahymena pigmentosa*. Cell **16**, 565—573.

— SOMMER, R., 1980: Sequence of a ribosomal RNA gene intron from *Tetrahymena*. Nature **283**, 693—694.

WILKES, P. R., BIRNIE, G. D., PAUL, J., 1979: Changes in nuclear and polysomal polyadenylated RNA sequences during rat liver regeneration. Nucl. Acids Res. **6**, 2193—2208.

WILKINSON, D., CIHAK, A., PITOT, H. C., 1971: Inhibition of rRNA maturation in rat liver by 5-fluoroorotic acid resulting in the selective labeling of cytoplasmic messenger RNA. J. Biol. Chem. **246**, 6418—6427.

— TLSTY, T. D., HANAS, R. J., 1975: The inhibition of rRNA synthesis and maturation in Novikoff hepatoma cells by 5-fluorouridine. Cancer Res. **35**, 3014—3020.

WILKINSON, J. K., SOLLNER-WEBB, B., 1982: Transcription of *Xenopus* rRNA genes by RNA polymerase I *in vitro*. J. Biol. Chem. **257**, 14375—14383.

WILLEMS, M., PENMAN, P., PENMAN, S., 1969: The regulation of RNA synthesis and processing in the nucleolus during inhibition of protein synthesis. J. Cell Biol. **41**, 177—187.

WILLIAMS, M. A., KLEINSCHMIDT, J. A., KROHNE, G., FRANKE, W., 1982: Argyrophilic nuclear and nucleolar proteins of *Xenopus laevis* oocytes identified by gel electrophoresis. Exp. Cell Res. **137**, 341—351.

WILLIAMSON, J. H., PROCUNIER, J. D., 1975: Disproportionately replicated, nonfunctional rDNA in compound chromosomes of *Drosophila melanogaster*. Mol. Gen. Genet. **139**, 33—37.

WILLIAMSON, P., FELSENFELD, G., 1978: Transcription of histone-covered T_7DNA by *E. coli* RNA polymerase. Biochemistry **17**, 5695—5705.

WIMBER, D. E., STEFFENSEN, D. M., 1970: Localization of 5 S RNA genes on *Drosophila* chromosomes by RNA-DNA hybridization. Science **170**, 639—641.

— WIMBER, D. R., 1977: Sites of the 5 S ribosomal genes in *Drosophila*. I. The multiple clusters in the *virilis* group. Genetics **86**, 133—148.

WINICOV, I., PERRY, R. P., 1974: Characterization of a nucleolar endonuclease possibly involved in ribosomal RNA maturation. Biochemistry **13**, 2908—2614.

— — 1975: Enzymological aspects of processing of mammalian rRNA. Brookhaven Symp. Biol. **26**, 201—213.

— 1976: Alternate temporal order in rRNA maturation. J. Mol. Biol. **100**, 141—155.

WOESE, C. R., MAGRUM, L. J., GUPTA, R., SIEGEL, R. B., STAHL, D. A., 1980: Secondary structure model for bacterial 16 S rRNA: phylogenetic, enzymatic, and chemical evidence. Nucl. Acids Res. **8**, 2275—2293.

WOLF, S., SCHLESSINGER, D., 1977: Nuclear metabolism of rRNA in growing, methionine-limited and ethionine-treated HeLa cells. Biochemistry **16**, 2783—2791.

WOLF, S., SAMESHIMA, M., LIEBHABER, S. A., SCHLESSINGER, D., 1980: Regulation of rRNA levels in growing, ^{3}H-arrested and crisis-phase WI-38 human diploid fibroblasts. Biochemistry **19**, 3484—3490.

WONG, Y.-CH., O'CONNELL, P., ROSBASH, M., ELGIN, S. C. R., 1981: DNase I hypersensitive sites of the chromatin for *Drosophila melanogaster* ribosomal protein 49 gene. Nucl. Acids Res. **9**, 6749—6762.

WOODCOCK, C. L. F., STANCHFIELD, J. E., GOULD, R. R., 1975: Morphology and size of ribosomal cistrons in two plant species: *Acetabularia mediterranea* and *Chlamydomonas reinhardii.* Plant Sci. Lett. **4**, 17—23.

WOOL, I. G., 1979: The structure and function of eucaryotic ribosomes. Ann. Rev. Biochem. **48**, 719—754.

— 1980: The structure and function of eucaryotic ribosomes. In: Ribosomes (CHAMBLISS, G., *et al.*, eds.), pp. 797—824. Baltimore: Univ. Park Press.

— STÖFFLER, G., 1976: Determination of the size of the pool of free ribosomal proteins in rat liver cytoplasm. J. Mol. Biol. **108**, 201—218.

WOOLFORD, J. L., ROSBASH, M., 1981: Ribosomal protein genes rp 39 (1078), rp 39 (11–40), and rp 52 are not contiguous to other ribosomal protein genes in the *Saccharomyces cerevisiae* genome. Nucl. Acids Res. **9**, 5021—5036.

— HEREFORD, L. M., ROSBASH, M., 1979: Isolation of cloned DNA sequences containing ribosomal protein genes from *Saccharomyces cerevisiae.* Cell **18**, 1247—1259.

WORMINGTON, W. M., BOGENHAGEN, D. F., JORDAN, E., BROWN, D. D., 1981: A quantitative assay for *Xenopus* 5 S RNA gene transcription *in vitro.* Cell **24**, 809—817.

WRAY, W., STUBBLEFIELD, E., 1970: A new method for the rapid isolation of nuclei from mammalian fibroblasts at near neutral pH. Exp. Cell Res. **59**, 469—478.

WU, B. C., SPOHN, W. H., BUSCH, H., 1981: Comparison of nuclear proteins of several human tumors and normal cells by two-dimensional gel electrophoresis. Cancer Res. **41**, 336—342.

WU, G. R., 1978: Adenovirus DNA directed transcription of 5.5 S RNA *in vitro.* Proc. Nat. Acad. Sci. U.S.A. **75**, 2175—2179.

WU, R. S., WARNER, J. R., 1971: Cytoplasmic synthesis of nuclear proteins. J. Cell Biol. **41**, 177.

— KUMAR, A., WARNER, J. R., 1971: Ribosome formation is blocked by camptothecin, a reversible inhibitor of RNA synthesis. Proc. Nat. Acad. Sci. U.S.A. **68**, 3009—3014.

WUNDERLICH, F., 1981: Nucleocytoplasmic transport of ribosomal subparticles: interplay with the nuclear envelope. In: The Cell Nucleus (BUSCH, H., ed.), Vol. IX, pp. 249—288. New York: Academic Press.

WYDRO, R., BROT, N., WEISSBACH, H., 1980: RNA synthesis in permeable HeLa cells. Arch. Biochem. Biophys. **201**, 73—80.

YAGURA, T., YAGURA, M., MURAMATSU, M., 1979: *Drosophila melanogaster* has different rRNA sequences on X and Y chromosomes. J. Mol. Biol. **133**, 533—547.

YAMAMOTO, M., SEIFART, K. H., 1977: Synthesis of ribosomal 5 S RNA by isolated nuclei from HeLa cells *in vitro.* Biochemistry **16**, 3201—3209.

— — 1978: Heterogeneity in the 3′-terminal sequence of 5 S rRNA synthesized by isolated HeLa cell nuclei *in vitro.* Biochemistry **17**, 457—461.

YAMAMOTO, M., JONAS, D., SEIFART, K., 1977: Transcription of ribosomal 5S RNA by RNA polymerase C in isolated chromatin from HeLa cells. Eur. J. Biochem. **80**, 243—253.

YAO, M.-CH., 1981: Ribosomal RNA gene amplification in *Tetrahymena* may be associated with chromosome breakage and DNA elimination. Cell **24**, 765—774.

— GALL, J. G., 1977: A single integrated gene for rRNA in a eucaryote, *Tetrahymena pyriformis*. Cell **12**, 121—132.

— BLACKBURN, E. H., GALL, J. G., 1978: Amplification of the rRNA genes in *Tetrahymena*. Cold Spring Harbor Symp. Quant. Biol. **43**, 1293—1296.

YOUNGER, L. R., GELBOIN, H. V., 1970: The electrophoretic distribution of RNA synthesized *in vitro* by isolated rat liver nuclei. Biochim. Biophys. Acta **204**, 168—174.

YU, F.-L., 1974: Two functional states of the RNA polymerases in the rat hepatic nuclear and nucleolar fractions. Nature (London) **251**, 344—346.

— 1975: An improved method for the quantitative isolation of rat liver nuclear RNA polymerases. Biochim. Biophys. Acta **395**, 329—336.

— 1976: Nuclear RNA polymerase—a positive gene control factor. In: Ribosomes and RNA Metabolism (ZELINKA, J., BALAN, J., eds.), pp. 75—87. Bratislava: Slovak Academy of Science.

— 1977: Mechanism of aflatoxin B_1 inhibition of rat hepatic nuclear RNA synthesis. J. Biol. Chem. **252**, 3245—3251.

— 1980: High concentration of RNA polymerase I is responsible for the high rate of nucleolar transcription. Biochem. J. **188**, 381—385.

— 1981: Studies on the mechanism of aflatoxin B_1 inhibition of rat liver nucleolar RNA synthesis. J. Biol. Chem. **256**, 3292—3297.

— FIEGELSON, P., 1972: The rapid turnover of RNA polymerase of rat liver nucleolus and its messenger RNA. Proc. Nat. Acad. Sci. U.S.A. **69**, 2833—2837.

ZAMB, T. J., PETES, TH. D., 1982: Analysis of the junction between rRNA genes and single copy chromosomal sequences in the yeast *S. cerevisiae*. Cell **28**, 355—364.

ZARDI, L., BASERGA, R., 1974: rRNA synthesis in WI-38 cells stimulated to proliferate. Exp. Mol. Pathol. **20**, 69—77.

ZAYETZ, V. W., BAVYKIN, S. G., KARPOV, V. L., MIRZABEKOV, A. D., 1981: Stability of the primary organization of nucleosome core particles upon some conformational transitions. Nucl. Acids Res. **9**, 1053—1068.

ZENTGRAF, H., SCHEER, U., FRANKE, W. W., 1975: Characterization and localization of the RNA synthesized in mature avian erythrocytes. Exp. Cell Res. **96**, 81—95.

ZUCHOWSKI, CH., I., HARFORD, A. G., 1977: Chromosomal rearrangement which affect chromosomal integration of the ribosomal genes in *Drosophila melanogaster*. Cell **11**, 383—388.

ZWIEB, CH., GLOTZ, C., BRIMACOMBE, R., 1981: Secondary structure comparisons between small subunit ribosomal RNA molecules from six different species. Nucl. Acids Res. **9**, 3621—3639.

ZYLBER, E., PENMAN, S., 1971: Products of RNA polymerases in HeLa cell nuclei. Proc. Nat. Acad. Sci. U.S.A. **68**, 2861—2865.

Subject Index

Cell Biology Monographs

Volume 1: **The Lytic Compartment of Plant Cells**
By **Ph. Matile**
1975. 59 figures and 40 plates. XIII, 183 pages.
ISBN 3-211-81296-2

Volume 2: **The Golgi Apparatus**
By **W. G. Whaley**
1975. 97 figures. XII, 190 pages.
ISBN 3-211-81315-2

Volume 3: **Lysosomes: A Survey**
By **E. Holtzman**
1976. 56 figures. XI, 298 pages.
ISBN 3-211-81316-0

Volume 4: **Mitochondria: Structure, Biogenesis and Transducing Functions**
By **H. Tedeschi**
1976. 18 figures. X, 164 pages.
ISBN 3-211-81317-9

Volume 5: **Microbodies/Peroxisomen pflanzlicher Zellen**
Morphologie, Biochemie, Funktion und Entwicklung eines Zellorganells
Von **B. Gerhardt**
1978. 59 Abbildungen. IX, 283 Seiten.
ISBN 3-211-81436-1

Volume 6: **Patterns of Chloroplast Reproduction**
A Developmental Approach to Protoplasmic Plant Anatomy
By **Th. Butterfass**
1979. 28 figures. XIV, 205 pages.
ISBN 3-211-81541-4

Cell Biology Monographs

Volume 7: **Peroxisomes and Related Particles in Animal Tissues**
By **P. Böck, R. Kramar,** and **M. Pavelka**
1980. 60 figures. XIII, 239 pages.
ISBN 3-211-81582-1

Volume 8: **Cytomorphogenesis in Plants**
Edited by **O. Kiermayer**
1981. 202 figures. X, 439 pages.
ISBN 3-211-81613-5

Volume 9: **The Protozoan Nucleus**
Morphology and Evolution
By **I. B. Raikov**
Translated from the 1978 Russian Edition
by N. Bobrov and M. Verkhovtseva
1982. 1 portrait and 116 figures. XV, 474 pages.
ISBN 3-211-81678-X

Volume 10: **Sialic Acids**
Chemistry, Metabolism and Function
Edited by **R. Schauer**
1982. 66 figures. XIX, 344 pages.
ISBN 3-211-81707-7

Volume 11: **DNA Methylation and Cellular Differentiation**
By **J. H. Taylor**
1984. 17 figures. VIII, 135 pages.
ISBN 3-211-81761-1

Springer-Verlag Wien NewYork

Plant Gene Research

Basic Knowledge and Application

Edited by
E. S. Dennis, B. Hohn, Th. Hohn (Managing Editor), P. J. King, J. Schell, D. P. S. Verma

The first volume

Genes Involved in Microbe-Plant Interactions

Edited by **D. P. S. Verma,** Department of Biology, McGill University, Montreal, Canada, and **Th. Hohn,** Friedrich Miescher-Institut, Basel, Switzerland.

1984. 54 figures. XIV, 393 pages. ISBN 3-211-81789-1

For the practical use of new techniques of gene manipulation in plant breeding it is very important to understand the molecular and physiological aspects of the close interactions which occur between higher plants and microorganisms. The nature of these interactions ranges from harmless through symbiotic to parasitic; they also include direct gene transfer from bacteria to plants, which has become a subject of research thanks to the new techniques applied in molecular biology. Knowledge of this gene transfer occurring in nature opens new perspectives for its future utilization in plant breeding. Possibly the strongest influence in making plant gene research a growing field of scientific interest and activity is the expectation that this research will open new and very effective ways for the breeding of agriculturally and biotechnologically important plants. The first volume in the series "Plant Gene Research" provides an overview of the important aspects of plant-microbe interactions and of various research methods.